The Mammalian Fetus *in vitro*

The Mammalian Fetus
in vitro

Edited by
C. R. AUSTIN

Charles Darwin Professor of Animal Embryology
University of Cambridge

Springer-Science+Business Media, B.V.

ISBN 978-0-412-11030-6   ISBN 978-1-4899-7212-5 (eBook)
DOI 10.1007/978-1-4899-7212-5

Library of Congress Catalog Card Number 73–6352

Contents

Contributors

C. R. AUSTIN — Physiological Laboratory, Cambridge, U.K.

B. J. N. Z. DANESH — Western General Hospital, Edinburgh, U.K.

E. N. HEY — The Children's Clinic, The Royal Infirmary, Newcastle-upon-Tyne, U.K.

T. KOLOBOW — Laboratory of Technical Development, National Heart and Lung Institute, National Institutes of Health, Bethesda, Maryland, U.S.A.

R. A. McCANCE — Sidney Sussex College, Cambridge, U.K.

M. C. MACNAUGHTON — Department of Obstetrics and Gynaecology, The Royal Maternity Hospital, Glasgow, U.K.

D. A. T. NEW — Physiological Laboratory, Cambridge, U.K.

D. A. NIXON — Department of Physiology, St. Mary's Hospital Medical School, London, U.K.

G. B. SHARMAN — School of Biological Sciences, Macquarie University, New South Wales, Australia.

C. H. M. WALKER   Department of Child Health, University of Dundee, Dundee, U.K.

W. M. ZAPOL   Anesthesia Laboratories of the Harvard Medical School Massachusetts General Hospital, Boston, Massachusetts, U.S.A.

Preface

'Fetus' is an adopted Latin word, which had a rather different meaning to that of the familiar English term. According to Smith's Latin Dictionary it signified firstly a 'bringing forth, breeding, dropping or hatching', and secondly by metonymy it came to mean 'the young or progeny'. The closest modern equivalents would seem to be parturition and neonate, respectively. With adoption the meaning has been pushed back in ontogeny—the third edition of the Shorter Oxford Dictionary gives: 'the young of viviparous animals in the womb, and of oviparous animals in the egg, when fully developed.' Various medical texts imply progressively earlier beginnings for the human fetus: 'from the end of the third month to birth', '. . . of the second month . . .' and even '. . . of the fifth week . . .'.

Chapters in the present book deal with stages covered by both Latin and English meanings of 'fetus', and indeed the treatment encompasses developmental stages from implantation onwards. This is not done with any wish to extend the meaning still further, but has the intention of providing continuity and breadth to the study of what is commonly understood by the familiar term.

Comments in the public media often convey the impression that culture of the human conceptus through full development *in vitro*, from fertilization to 'decantation', represents a technical achievement to be anticipated in the near future. Such a view, of course, is quite unwarranted, and very many difficulties must first be overcome before any mammal can be cultured through development. Nevertheless progress in this field of endeavour is continuous, and this book describes recent advances and current status for the information and interest of basic scientists as well as clinical investigators.

Chapters are devoted to the technical procedures involved in sustaining, in particular, mid-term sheep, dog and human fetuses, the preterm human fetus and the premature human infant. The nature of physiological and other investigations that can be made on these subjects is described in some detail. Account is given of research on rat, mouse and opossum post-implantation embryos and fetuses maintained *in vitro*, and of anatomical and physiological observations on marsupial pouch young, which show the adaptations needed by a 'fetus' to survive in an external environment. Special preparations such as the exteriorized fetus, placenta and feto-placental unit are described and discussed, and attention is given to work on organ and tissue culture, and the contributions that all these studies make towards solving the problems encountered with the isolated fetus. Some facts and reflections are also offered on the pros and cons of viviparity, on the kinds of observation that can be made with the fetus *in vivo*, and on the social, ethical and legal aspects, and future potentialities, of research on the human fetus. No doubt there are sins of omission as well as commission in the present text, but a conscientious attempt has been made to deal informatively and comprehensively with the challenge presented by the mammalian fetus *in vitro*.

C. R. A.

Physiological Laboratory,
University of Cambridge
September 1972.

It is with deep regret that I have to record the death on the 9th December 1972 of Dr. D. A. Nixon, author of Chapter 4 and internationally known for his work on fetal physiology.

C. R. A.

1 Implications of Viviparity

C. R. AUSTIN

The many problems we face in our efforts to maintain the mammalian fetus *in vitro* justify a brief scrutiny of the special niche normally occupied by the mammalian fetus, and some consideration of its principal features. Such a survey reasonably begins in the realm of phylogeny, with the evolutionary aspects of viviparity, and the developmental problems faced by primitive oviparous forms. We may infer that the progeny of our primordial ancestors, developing externally in a marine environment, rapidly reached a free-swimming stage, like that of the modern sea urchin blastula, at which they hatched from the egg investments and became fully independent organisms. They had thus, from almost the start of development, to attend to their own needs in nutrition, respiration and excretion, and the avoidance of predators and inclement living conditions. In their small and simple form such embryos, or more correctly larvae, were very much at the mercy of the environment, and wastage must have been immense.

Representative stages in evolution

Subsequent evolutionary history probably entailed a succession of stages similar to those for which representatives can be found among present-day animals (Table 1.1) (see Harrison Matthews, 1955; Amoroso, 1968). A major first step would have been the adoption of the cleidoic egg—essentially an extension of the intra-ovular period made possible by the provision of a protective shell or casing and, in the megalecithal egg, of a more generous parental contribution of food material or yolk. These devices immediately confer several important advantages, but at the same time they introduce difficulties—they tend

1

to impede gaseous exchange and the removal of waste products, effects that have required compensatory changes in the embryo and its associated structures. By virtue of the advantages inherent in the cleidoic egg, an animal becomes better equipped before being cast adrift in the world, achieving some of the capabilities that enable the adult form to cope with the problems of life. This stage of evolution corresponds broadly to that of present-day oviparous fish, ranging from teleosts that spawn large numbers of relatively small jelly-covered eggs, to elasmobranchs whose large yolky products enjoy the protection of tough chitinous envelopes.

Table 1.1. *Representative stages in the evolution of viviparity*

	Microlecithal egg	*Megalecithal egg*	*Embryonic development* in vitro	*Embryonic development* in vivo	*Provision of embryotrophe*	*Yolk-sac placenta*	*Chorio-allantoic placenta*
Sea urchin	+	−	+	−	−	−	−
Fish, bird, monotreme	−	+	+	−	−	−	−
Fish	−	+	−	+	+	−	−
Fish, amphibian	−	+	−	+	+	+	+
Reptile	−	+	−	+	+	+	+
Marsupial	+	−	−	±*	+	+	(+)†
Placental mammal	+	−	−	+	+	+	+

* Development partly in marsupium
† In a few members, e.g. *Perameles*

The next step could have been one that is again represented in fish and also in reptilian species, namely reproduction by ovoviviparity. It takes many forms; a simple condition is seen in several elasmobranchs in which the megalecithal eggs are not encased and shed but, surrounded only by their primary investments, remain in the oviduct where hatching eventually occurs. The added degree of enclosure increases the hindrance to gaseous exchange and excretion, and accordingly vascularity in adjacent maternal and fetal tissues is extensively developed, promoting transfer between fetal and maternal circulatory systems. Thus the maternal organism began to carry more of the burden of meeting embryonic needs, and a type of placentation found its origin: the vascularity about the yolk mass constituted the primitive yolk-sac placenta.

Further advance (still among elasmobranchs) is marked by the persistence of internal development until after the yolk mass is exhausted;

the later part of fetal development then depends upon nutritive material provided by the mother in a new way, mostly in the form of yolk-like 'embryotrophe' consisting of oviduct secretion supplemented by breakdown products from unfertilized eggs, dead fetuses and desquamated tissue. But ovoviviparity clearly has limited potential, and even among mammals the yolk-sac placenta seems incapable of satisfying the requirements of prolonged gestation—as witness the early birth of marsupials.

The new feature in evolution that seems pre-eminently responsible for advanced viviparity entails extension in the function of the allantois. Initially achieving significance as the homologue of the amphibian urinary bladder—in birds the breakdown products of protein metabolism are concentrated in this diverticulum, which is jettisoned at hatching—the allantois is evidently fit for better things; uniting with the chorion and becoming richly vascularized, it has a second function as the chorio-allantoic respiratory membrane of bird and oviparous reptile eggs, and the chorio-allantoic (or more commonly 'allantoic') placenta of viviparous reptiles and 'placental' mammals.

Why the allantoic placenta should come to have powers superior to those of the yolk-sac placenta presents something of a mystery. The advantage cannot be held to spring from a closer intimacy between maternal and fetal systems—the epitheliochorial placenta has as many intervening layers as the yolk-sac placenta, yet (in Equidae) supports far more advanced fetal development than do the haemochorial placentae of mouse and man. Close intimacy, therefore, is not uniquely important for prolonged gestation. Not only that, but it introduces further difficulties, and these relate especially to the immunological implications. In some way the advantage offered by the allantoic placenta seems to derive from a capacity to produce secretions (possibly originally excretory products) which have come to play an essential role in the hormonal control of pregnancy. Specifically the function has to do with maintenance of progesterone levels required for continued gestation. Accordingly we find that viviparity has both immunological and endocrine aspects of great importance.

Immunological problems in viviparity

The rise of the allantoic placenta, with its potential for close relations with the maternal system—in syndesmochorial and endotheliochorial placentae there is actual interruption of the uterine epithelium and invasion of deeper layers by fetal cells—has exacerbated another problem of viviparity, namely the maintenance of the fetus as an allograft on the maternal tissues. Indeed, the maternal organism sometimes does produce destructive antibodies against fetal tissues, as in erythroblastosis, but generally antibody production is small and innocuous. Numerous attempts have been made to explain the persistence of the developing fetus, to identify the fetal and maternal adaptations that make mammalian viviparity possible. According to the present consensus of opinion, the most important factors are the following:

(1) The low antigenicity of trophoblast cells, which can be introduced into ectopic sites of immunized animals and not suffer rejection, unless first treated with appropriate lytic enzymes (see Billington, 1971). Electron microscopic studies have provided some support for the existence of a protective surface layer of mucoprotein on trophoblast cells, and they thus could constitute a barrier between fetal and maternal tissues. The action is thought both to impede the passage of fetal antigens into the mother as well as protect fetal cells from immunological rejection. While 'deportation' of trophoblast is regarded as a normal process, the severe nature of haemolytic disease following damage to the feto-maternal barrier, usually during parturition, testifies to the seriousness of admission of large numbers of unprotected fetal cells into the maternal circulation (see Walker, 1971).

Actual migration of maternal cells into the fetal system, as reported in certain mice, would denote major breaches in the barrier, but appears to be a highly exceptional occurrence. Nevertheless the possibility that it could happen raises the question whether the fetus might sometimes produce antibodies against maternal antigens. Immunological competence, however, is evidently a fairly late acquisition. The newborn rat or mouse readily responds to foreign antigens, and homografts are rejected, but the machinery is not fully established: immunological tolerance can easily be induced with small doses of antigen. By contrast, the neonatal guinea-pig, more advanced in development at birth, resists the induction of tolerance. The neonatal opossum, which makes its appearance at a much earlier developmental stage than the rat and mouse, seems to be immunologically quite inert, for it lacks the ability to reject homografts and even xenografts are accepted (points that are discussed further in Chapter 3).

4

(2) The protective action of the uterine decidual tissue, at least in the early stages of pregnancy, before the feto-maternal barrier is established. Rejection of grafted tissues from the uterus seems clearly attributable to lack of a decidual reaction. On the other hand the decidua has an undoubted restraining effect on tropho-blast invasion, which can be most destructive in locations other than the uterus (see Finn, 1971). The role of the decidua must, however, have limited significance because fetal development can proceed to term in ectopic sites.

(3) The possibilities that the maternal immune response is incom-pletely aroused, or involves the production of protective anti-bodies, or both. The former effect is referred to as low-dose tolerance and is attributed to the passage into the maternal system of too few fetal cells to provoke a full response (except, of course, for the invasions that lead to haemolytic disease). These ideas found recent support in the work of Anderson (1970). Protective or 'enhancing' antibodies, on the other hand, seem to coat the fetal cells with which they come into contact, and prevent their cellular rejection. The theory was set up by Breyere & Barrett (1960), who were concerned also with the apparently privileged status of tumours, and supported by Currie (1969); shortly afterwards, however, Currie (1970) found it inadequate and preferred the explanation of a feto-maternal barrier.

Probably all these mechanisms play some role in the maintenance of normal pregnancy, the emphasis on one mechanism or another reflecting essentially species differences.

Endocrine functions in viviparity

Internal maintenance of young necessarily requires co-ordination be-tween the maternal and fetal systems, and this has devolved mainly upon activities of certain endocrine organs. Evidence of such a co-ordinating mechanism exists already in lower vertebrates, and corpora lutea for example are found in certain cyclostomes, teleosts and amphibians; not always, however, are these structures associated with viviparity. They occur (though in a 'regressing' state from the beginning) in the hagfish *Myxine* which is oviparous, and so their primitive function is not clear. Among reptiles and birds corpora lutea are much more common features. In some viviparous snakes their duration coincides with the length of gestation, which they may therefore play a part in determining; consistently, progesterone has been extracted from such structures, and ovariectomy early in pregnancy interrupts gestation.

A much more detailed and coherent picture of endocrine function emerges with the higher mammals, though even here many problems remain and striking intergeneric, and also interspecific, differences make a unifying concept difficult to propound (see Heap, 1972). In the marsupials, the luteal phase of the oestrous cycle is about the same length as the very brief pregnancy, so that uterine preparation for gestation does not exceed that achieved in a normal cycle. Adaptation for the more advanced viviparity of placental mammals has required great lengthening of the luteal phase, brought about through prolonged secretion of progesterone. The overriding feature in all placental species investigated is the evidence that the fetus and its membranes (the allantoic placenta) are essentially in command of the maternal system, appropriately influencing behaviour, energy and protein metabolism, cardiovascular function, glomerular filtration rate and hormone secretion rate, suppressing ovulatory and cyclic activity, and contributing to ligament relaxation and mammary development. This fetal influence is mediated mainly through the production of progesterone. Uniformly among placental mammals this hormone appears responsible for preserving gestation by promoting the secretory activity of the endometrium and the proliferation of the myometrium, and suppressing the contractility of the myometrium. Progesterone for these functions is primarily a product of the corpora lutea, which may be maintained by the luteotrophic complex from the pituitary; in several species, however, the placenta sooner or later assumes control of, or even takes over, progesterone synthesis. The change in role is signified by the fact that in woman, monkey and guinea-pig ovariectomy as early as in the first half of pregnancy is compatible with the continuance of that state; in the horse, sheep, cow, ferret and rat ovariectomy can be tolerated in the second half; in the goat, sow and rabbit the ovaries are needed throughout (Table 1.2).

During pregnancy the fetus is largely protected from maternal hormones by the permeability characteristics and enzymic activity of the placenta; in addition the fetus itself is capable of rapidly metabolizing steroids. One aspect of these functions that is of special clinical importance lies in the conversion of maternal (or placental) progesterone into dehydroepiandrosterone in the fetal adrenal cortex, and this compound to 16-OH-dehydroepiandrosterone in the fetal liver, and this in its turn to oestriol in the placenta. Maternal excretion levels of oestriol so serve as an indication of normal function in the human fetus.

Fetal dominance is again well shown at parturition, the moment of which seems to be determined by the fetal hypothalamus acting through the pituitary, though the hormonal picture differs between species, as it

Table 1.2. *Stage of pregnancy at which ovariectomy can be tolerated.* + = *Some or all fetuses survive;* — = *abortion or resorption*

Animal	First half	Second half
Man	+	+
Monkey	+	+
Guinea-pig	+	+
Horse	—	+
Sheep	—	+
Cat	—	+
Cow	—	+
Ferret	—	+
Rat	—	+
Goat	—	—
Sow	—	—
Rabbit	—	—

does in pregnancy. The position is perhaps clearest in the sheep (Fig. 1.1) in which species release of ACTH has been shown to favour the output of adrenal corticosteroids; these then provoke (*a*) reduction in placental release of progesterone and apparently also increase in that of oestrogen, (*b*) secretion of surfactant which is vital for fetal lung function after birth, (*c*) glycogen storage in fetal heart muscle and liver, important to the fetus for resisting parturiant hypoxia and inanition, and (*d*), with other hormones, growth of mammary glands and initiation of milk secretion. These points are well supported by the results of appropriate ablation and injection experiments (see Heap, 1972; Liggins, 1972; Liggins *et al.*, 1972). There is also an increase in placental output of prostaglandins PGF_2a and PGE_2 which powerfully stimulate myometrial contractions in the sheep; in the goat, a corpus-luteum-dependent species, the prostaglandins exert a luteolytic action. The shift in the progesterone: oestrogen ratio evidently increases uterine reactivity to oxytocin, which is important in the later stages of labour.

The endocrine co-ordination of pregnancy and parturition, though not the exclusive controlling mechanism, typifies the close integration that exists between fetal and maternal systems and is mediated by the chorio-allantoic placenta. The existence of such a delicate balance between two metabolic systems poses a major technical problem for *in-vitro* experimentation, for one may suppose that success could depend not only on providing means for the supply of appropriate materials and removal of wastes (and the solution of the numerous technical problems of organ perfusion), but also on the maintenance of a state of dynamic equilibrium in which changes in one system are met by

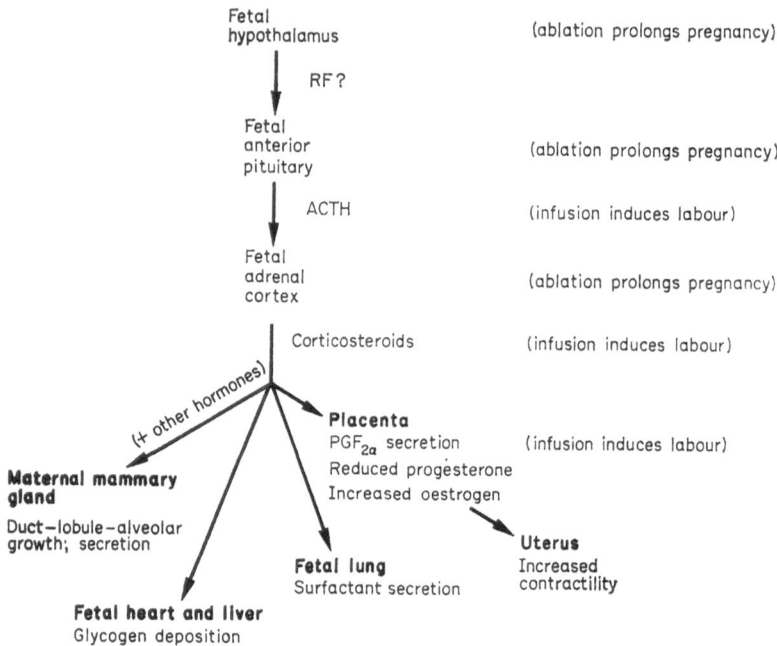

Fig. 1.1. The presiding role of the fetal hypothalamus in controlling events at parturition

equivalent responses in the other, conferring an essential resilience on the union. Most of the chapters in this book are concerned with aspects of this challenging problem; others deal with the natural circumstances wherein a fetus completes the later part of its development *in vitro* as shown in marsupial reproduction, and with the problems attendant on precocious termination of intrauterine life, as in the premature birth of human children. But before leaving the subject of fetal development *in vivo*, some thought should be given to the techniques that have been developed for the study of this system in its intact condition—techniques that have proved of special value for monitoring human fetal development. The applications are chiefly of clinical interest but nevertheless yield basic information which can fruitfully supplement data from investigations performed *in vitro*.

Techniques applicable to the fetus *in vivo*

The most productive monitoring techniques have been the use of radiations (X-rays and ultrasonic vibrations) as long-range probes, the sampling of fluid and cellular components from the amnion (amniocentesis), and the direct viewing of the fetus and amnion (amnioscopy). Other techniques that are much in use, but which are not really relevant in the present context, are those applied to the infant during parturition —scalp blood sampling, recording of ECG and EEG, etc.—and these will not be considered here (but see Waisman & Kerr, 1970).

X-radiography, though limited in its application by the need to keep radiation hazard at a minimum during pregnancy, can be usefully employed by the sixteenth week to provide information on a variety of morphological and physiological features of the fetus. Fetal size and state of bone development (and thus gestational age) can readily be determined. Twin pregnancies can be detected and a variety of anomalous conditions affecting both bony as well as soft tissues, such as hydrops fetalis, can be diagnosed. The procedure may be combined with the use of contrast media, commonly introduced into the amnion, and entail several successive exposures of short duration. Observations may then include such things as rate of fetal swallowing and efficiency of gastro-intestinal transport. Another application involves injection of contrast medium into the femoral artery, which with radiography yields information on placental size, position and vasculature, including some indication of flow rates in the placental circulation (see Ramsey, 1970a, b).

The use of *ultrasonic vibrations* or 'sonar' (ultrasonography), employed on the echo-sounding principle for obtaining images from various surfaces or interfaces, is much safer in pregnancy (indeed there appears to be no likelihood of deleterious effect with competent application). Although the resolving power is very much lower than with X-rays, sonar is said to be capable of permitting measurements such as that of the distance between the parietal bones of the fetal head to an accuracy of about 0·5 mm. Pregnancy can often be diagnosed as early as six weeks, by detection of the gestation sac, and its position—whether intra-uterine or ectopic—can be determined. Twins can be identified from about seven-and-a-half weeks, and at about the same time the persistence of residues from incomplete abortion, or the presence of a fibroma or a hydatidiform mole can readily be seen. Used on the Doppler principle ultrasonography yields estimates of fetal heart beat at the thirteenth week of gestation (the earliest possible by auscultation is about 24 weeks); from variations in this parameter in late pregnancy inferences on the operation of abnormal pressures on the fetus can be drawn. Reflex responses are also observable—the normal fetus shows a transient tachycardia if surprised by a loud sound (90 dB at the abdominal surface). Also, in late pregnancy, the placenta can be accurately located and its limits defined. A detailed account of ultrasonics in obstetrics is given by Donald (1968).

Amniocentesis involves the tapping of the amnion with the aid of a simple lumbar-puncture needle to obtain samples of amniotic fluid. The procedure is feasible from about the fourteenth week of pregnancy but is not advisable before the sixteenth week, at which time 15–20 ml of fluid can safely be removed.

The turnover in amniotic fluid is normally rapid, up to half the total volume being renewed hourly; most of the water and electrolytes are exchanged between fetus and mother. The cycling of amniotic fluid is due mainly to fetal swallowing, and replacement by fetal micturition or secretion through the amniotic epithelium. Amniocentesis should always be preceded by prior survey, preferably with ultrasonography, to eliminate the possibility of twins, which would add materially to the risks of the operation, and to locate the placenta. The operation can be done under local anaesthesia. Samples recovered by amniocentesis are commonly centrifuged to separate the cellular components, which can be examined directly for blood-group determination and diagnosis of sex by the presence or absence of sex chromatin, or may be cultured *in vitro* for chromosome or enzyme studies. The fluid is subjected to biochemical analyses for steroids, proteins, amino acids and enzymes. With the material obtained by amniocentesis it is now possible to diagnose the existence in the fetus of any one of about 40 single-gene defects (Table 1.3) and a variety of chromosome anomalies (Table 1.4) (see Emery, 1970; Ferguson-Smith *et al.*, 1971; Castelazo-Ayala, Karchmer & Shor-Pinsker, 1972; Milunsky *et al.*, 1972). A detraction from the usefulness of procedures that can provide evidence of inborn errors of metabolism is that 6–8 weeks are needed to grow enough cells for the tests; with amniocentesis at 16 weeks, the answer would not be available until well after the minimum age of viability (now put at 20 weeks). The information is thus useless for selective abortion.

Table 1.3. *Single gene defects that can be diagnosed from analysis of material recovered by amniocentesis*

Acatalasaemia	Hypervalinaemia
Adrenogenital syndrome	I-Cell disease
Argininosuccinic aciduria	Juvenile GM$_1$ gangliosidosis
Citrullinaemia	Ketotic hyperglycinaemia
Congenital erythropoietic porphyria	Lysosomal acid phosphate
Cystinosis	Mannosidosis
Fabry's disease	Maple-syrup urine disease
Fucosidosis	Metachromatic leukodystrophy
Galactosaemia	Methylmalonic aciduria
Gaucher's disease	Moriquio syndrome
Generalized gangliosidosis	Niemann-Pick disease
Glucose-6-PO$_4$ deficiency	Ornithine-alpha-ketoacid
Glycogen-storage disease–II	transaminase deficiency
Glycogen-storage disease–III	Orotic aciduria
Glycogen-storage disease–IV	Pompe's disease
Histidinaemia	Pyruvate decarboxylase
Homocystinuria	Refsum's disease
Hurler syndrome	San Filippo syndrome
Hyperammonaemia	Scheie syndrome
Hyperlysinaemia	Tay-Sachs disease
	Xeroderma pigmentosum

Table 1.4. *The more common chromosomal anomalies detectable at amniocentesis*

Nature of anomaly	Common name	Approximate frequency at birth
47 XXX	'Super' female or triplo-X female	1:700
45 XO	Turner's syndrome	1:700
47 XXY	Klinefelter's syndrome	1:400
47 XYY	–	very low
Trisomy-21	Mongolism, Down's syndrome	1:700 (rising to 1:50 at 45 years)
15/21 translocation	Mongolism, Down's syndrome	very low
13/15 translocation	–	very low
Trisomy-6	–	very low ⎤
Trisomy-17	–	very low ⎟
Trisomy-(13-15)	–	very low ⎟ *
Triploidy	–	very low ⎦
Mosaicism	–	very low

* Trisomies and triploidies characterize about 25 % of spontaneous human abortuses.

Amnioscopy is done by passing a telescopic device equipped with fibre-optics illumination through the cervix and into the near vicinity of the fetus. The method is especially valuable for the diagnosis of numerous structural anomalies, including spina bifida, and for the identification of meconium in the amniotic fluid, a strong indication of fetal pathology.

Clearly *in vivo* techniques have their limitations, though for certain purposes they are invaluable, and there is no doubt that this field of inquiry will generate more and better methods in the course of time. Equally obvious is that the constraints operating in the clinical field inevitably render progress extremely slow, and basic research plainly demands the manipulative freedom that comes with the mammalian fetus studied *in vitro*.

References

AMOROSO, E. C. (1968) The evolution of viviparity. *Proceedings of the Royal Society of Medicine* (Symp. No. 10), **61**, 1188–1198.

ANDERSON, J. M. (1970) The transplantation immunology of certain mammalian mothers and progeny. *Proceedings of the Royal Society*, Series B, **176**, 115–129.

BILLINGTON, W. D. (1971) Biology of the trophoblast. *Advances in Reproductive Physiology*, **5**, 27–66.

BREYERE, E. J. & BARRETT, M. K. (1960) Prolonged survival of skin homografts in parous female mice. *Journal of the National Cancer Institute*, **23**, 1405–1410.

CASTELAZO-AYALA, L., KARCHMER, S. & SHOR-PINSKER, V. (1972) The biochemistry of amniotic fluid during normal pregnancy correlation with maternal and fetal blood. In *Physiological Biochemistry of the Fetus*, ed. Hodari, A. A. & Mariona, F., Ch. 2. Springfield: Thomas.

CURRIE, G. A. (1969) The foetus as an allograft: the role of maternal unresponsiveness to paternally derived foetal antigens. In *Foetal Autonomy*, ed. Wolstenholme, G. E. W. & O'Connor, M., Ciba Foundation Symposium, pp. 32–58. London: Churchill.

CURRIE, G. A. (1970) Effects of interstrain pregnancy on the immune status of female mice sensitized to paternal antigens. *Journal of Reproduction and Fertility*, **23**, 501–503.

DONALD, I. (1968) Ultrasonics in obstetrics. *British Medical Bulletin*, **24**, 71–75.

EMERY, A. E. H. (1970) Antenatal diagnosis of genetic disease. In *Modern Trends in Human Genetics*, ed. Emery, A. E. H., Vol. 1, Ch. 9. London: Butterworth.

FERGUSON-SMITH, M. E., FERGUSON-SMITH, M. A., NEVIN, C. &

STONE, M. (1971) Chromosome analysis before birth and its value in genetic counselling. *British Medical Journal*, **4**, 69–74.

FINN, C. A. (1971) The biology of decidual cells. In *Advances in Reproductive Physiology*, ed. Bishop, M. W. H., pp. 1–26. London: Logos Press.

HARRISON MATTHEWS, L. (1955) The evolution of viviparity in vertebrates. In *Comparative Physiology of Reproduction* (Memoirs of the Society for Endocrinology, No. 4), ed. Chester Jones, I. & Eckstein, P., pp. 129–148. Cambridge University Press.

HEAP, R. B. (1972) Role of hormones in pregnancy. In *Reproduction in Mammals*, ed. Austin, C. R. & Short, R. V., Vol. 3, pp. 74–105. Cambridge University Press.

LIGGINS, G. C. (1972) The fetus and birth. In *Reproduction in Mammals*, ed. Austin, C. R. & Short, R. V., Vol. 2, pp. 7–109. Cambridge University Press.

LIGGINS, G. C., GRIEVES, S. A., KENDALL, J. Z. & KNOX, B. S. (1972) The physiological roles of progesterone, oestradiol-17β and prostaglandin $F_{2\alpha}$ in the control of ovine parturition. *Journal of Reproduction and Fertility*, Supplement 16, 85–103.

MILUNSKY, A., LITTLEFIELD, J. W., KANFER, J. N., KOLODNY, E. H., SHIH, V. E. & ATKINS, L. (1972) Prenatal genetic diagnosis. *New England Journal of Medicine*, **283**, 1370–1381, 1441–1446, 1448–1504.

RAMSEY, E. M. (1970a) Current techniques in placentography. In *Fetal Growth and Development*, ed. Waisman, H. A. & Kerr, G., pp. 67–78. New York: McGraw-Hill.

RAMSEY, E. M. (1970b) Methods for the demonstration of placental circulation. In *Methods in Mammalian Embryology*, ed. Daniel, J. C. Jr., pp. 419–430. San Francisco: W. H. Freeman.

WAISMAN, H. A. & KERR, G. (ed.) (1970) *Fetal Growth and Development*. New York: McGraw-Hill.

WALKER, W. (1971) Haemolytic disease of the newborn. In *Recent Advances in Paediatrics*, ed. Gairdner, D. & Hall, D., pp. 119–170. London: Churchill.

2 Studies on Mammalian Fetuses *in vitro* During the Period of Organogenesis

D. A. T. NEW

General problems of organ and fetal culture

The long evolution of mammalian viviparity has resulted in such a close integration between mother and fetus that the fetus can in many ways be regarded as an extra organ temporarily acquired by the mother. The analogy should not be pressed too far. Unlike most organs, the fetus receives most of the benefits of the association and pays very little, at least in physiological currency, for its keep. Nevertheless, in removing the fetus from the maternal environment and attempting to maintain it *in vitro*, we immediately encounter many of the problems of culturing explanted organs, and should first consider how far organ culture methods are applicable. The methods and results of different types of organ and tissue culture have been fully described in several excellent books in recent years (e.g. Paul, 1970; Willmer, 1965), and no attempt will be made here even to summarize them. But a few points can be noted of particular relevance to the culture of fetuses during the period of organogenesis.

Size of fetus

One of the main limiting factors in organ culture is the size of the explant (see discussion by Moscona, Trowell & Willmer, 1965). If the explant exceeds a certain size, and lacks any internal circulation system, the central cells die, either from lack of oxygen or glucose, or from poisoning by accumulated carbon dioxide, lactic acid or other waste products. In mature organs the limiting factor has been shown both in theory and experiment to be the diffusion rate of oxygen (Trowell, 1961). In practice, explants of no greater thickness than 0·5–1·0 mm can be maintained in air, but the limit increases to about 2 mm in oxygen (or sometimes more in oxygen under pressure). Thus various whole organs from adult mice or young rats can be maintained in culture, including thyroid, spinal ganglia, ureter, ovary, oviduct or uterus. In embryonic organs, which often obtain more of their energy from anaerobic respiration, the diffusion rate of one of the other substances may be more critical than that of oxygen, but here also the limiting thickness is, at most, a few millimetres.

Organogenesis in mammals mostly occurs while the fetus is still very small. It begins when the conceptus consists of little more than a few cell layers, and is largely completed before any part of the developing fetus exceeds 5 mm in thickness. Differentiation at the tissue and cellular level continues throughout—and beyond—fetal life, but the main organ systems are all laid down while the fetus is still measurable in millimetres. The fetus at this time is therefore very similar in size to the organs that can be successfully maintained in culture, and some of the *in-vitro* techniques used for the explanted fetus are adaptations of organ culture techniques. But special problems arise from the extremely rapid growth of the fetus during organogenesis—in rats, for example, there is a fivefold increase or more in weight every 24 h (Fig. 2.7). This growth rate is far higher than those with which the organ culturist is usually concerned, and a culture method that may be adequate for a fetus at the time of explantation may, within a few hours, be incapable of meeting its rocketing demands. As a result, cultures of *growing* fetuses are still usually measured only in hours or days, in contrast to the weeks or months of many organ culture studies. (Accounts do appear from time to time claiming longer maintenance of fetal development in culture, but it is usually clear that the author has spent most of the time nursing a dead or dying fetus, presumably in the hope of resurrection.)

Static and circulating culture systems
The material needs of the fetus *in vivo* are supplied by its own blood circulation and by the blood circulation of the mother, with a special organ, the placenta, to facilitate exchanges between the two. In a static culture system, which fails to maintain or provide a substitute for either of these circulations, the fetus is dependent both internally and externally on diffusion gradients for its respiration, nutrition and excretion, and under these conditions only the earliest stages of organogenesis can be maintained. Fortunately, in rodents, the yolk-sac placenta becomes functional during organogenesis and continues to grow when explanted with the fetus into culture. Rats and mice, therefore, have proved

particularly valuable for studies of fetal development at this stage. The fetal blood circulates through the capillaries of the yolk-sac and provides an effective system for rapid transfer of materials between the culture medium and all parts of the fetus. If the culture medium itself is then kept circulating over the yolk-sac, the system approximates to that *in vivo* and can support fetal organogenesis to advanced stages.

The capacity of the yolk-sac to provide this supporting role gives fetal culture an advantage over most forms of organ culture. The persistence of the fetus-yolk-sac blood circulation in culture in effect provides the fetus with its own 'perfusion' system. As a result, rodent fetuses can be grown in culture both more rapidly and to a larger size than is usually possible with isolated organs. But *in vivo* the placental function of the yolk-sac is soon supplemented, and eventually largely taken over, by the allantoic placenta. (In many other mammals the allantoic placenta is from the start the main, or only, placenta.) Unlike the yolk-sac, the allantoic placenta is a complex structure comprised of both fetal and maternal tissues and has not yet been successfully grown in culture. In later chapters methods will be described for replacing the allantoic placenta by perfusion of the fetus through the umbilical blood vessels, or for maintaining the allantoic placenta together with the fetus as a 'feto-placental' unit. These techniques are valuable for physiological studies on the larger fetuses obtained at later stages of gestation. But they have not yet been successfully applied to very small fetuses during the period of organogenesis. The technical problems of perfusion through the tiny umbilical vessels of a fetus only a few millimetres in length are formidable, and although a fetus of this size can be explanted with its allantoic placenta intact and 'maintained' for some time in culture as a feto-placental unit, the placenta fails to grow.

In the absence of a functional allantoic placenta, present culture techniques aim at maximum exploitation of the yolk-sac to support fetal organogenesis as long as possible. Culture systems maintaining a continuous flow of medium over the yolk-sac, or involving an increased proportion and pressure of oxygen in the gas phase, have proved very effective and will be described in more detail later in this chapter.

Why grow fetuses in culture?

The inaccessibility of the mammalian fetus *in vivo* has always been a problem for embryologists, and the experimental advantages of a system that would support development of an isolated fetus in a controlled environment are too obvious to need labouring. But there are further reasons for wishing to culture fetuses during organogenesis. The very small size of the fetus at this period makes it extremely difficult to observe in any other way. It is too small to be examined by most of the techniques—e.g. X-radiography, ultrasonography, amniocentesis, amnioscopy—used for monitoring development of older fetuses (see Chapter 1). Excision, grafting and other micro-operations that have been used so successfully in the analysis of embryonic development in other vertebrates have been impossible to apply to the small fetus *in utero,* and our knowledge of the mechanisms of mammalian organogenesis still lags far behind that available for the amphibian or chick. Yet the early stages of mammalian development when the main organ systems are being formed are particularly in need of study, especially as they represent the time of onset of some of the most serious congenital malformations.

Culture methods may also have important applications in the study of teratogenic agents, and in testing new drugs for teratogenicity. During organogenesis the fetus is particularly susceptible to teratogenic agents, but the sundry effects produced by such agents vary very widely from one mammalian species to another, and drugs that are harmless when injected into pregnant laboratory animals may be teratogenic in women. Many factors may influence this variation including differences in placental permeability and maternal metabolism. In culture many of these variables can be avoided and the drug applied directly to the fetus.

Fetuses grown in culture

For the purpose of this account 'fetal' development will be regarded as beginning with the appearance of the first organ primordia. Fetal age will be given as days after the time of fertilization. Where mating occurs during the night the embryos will be regarded as $\frac{1}{2}$-d old at noon on the following day. This day (recognized in rodents by the presence of a vaginal plug and spermatozoa in the vaginal smear) will be designated the first day or Day 1. Thus a $10\frac{1}{2}$-d-old fetus = a fetus on the eleventh day or Day 11.

Rabbit

One of the earliest attempts to obtain fetal development *in vitro* was made by Waddington & Waterman (1933). They explanted rabbit blastodiscs of primitive-streak to three-somite stages onto plasma clots,

18

and in the most successful cultures a six- to nine-somite fetus was obtained, with neural tube and beating heart. The technique involved separating the blastodisc from the rest of the blastocyst, which may have partly accounted for the rather limited development in culture.

Glenister (1967, 1971, and earlier papers) has developed methods for culturing whole rabbit blastocysts together with pieces of explanted endometrium. The blastocyst is placed above the endometrium which is supported by a piece of rayon resting on a plasma clot or steel grid in the culture dish. The method has been used primarily for studying implantation and placental development in culture, but many of the explanted blastocysts have begun to form fetal structures. Rabbit blastocysts are large (3–4 mm at explantation) and, in this type of culture, lack sufficient support and tend to flatten. This may account for the rather irregular organogenesis that often occurs. Glenister (1971) has recently devised a method for suspending the explant from an arch of polyurethane foam; in this preparation the blastocyst hangs from the endometrium and is subjected to less mechanical distortion.

In the culture system of Daniel (1971) the blastocysts are grown in a flask of medium which is kept oxygenated by a mixture of 5% carbon dioxide, 10% oxygen and 85% nitrogen continuously bubbled through it. This solves the problem of mechanical support for the blastocysts (since they are totally immersed), and the flow of bubbles maintains a steady circulation of the medium over the developing embryo. Explanted at $6\frac{1}{2}$ d of gestation, the blastocysts expand in culture at approximately the normal rate for about 48 h, developing an amnion, blood islands and a fetus with closed neural tube, optic vesicles, 8–11 somites and a beating heart. Daniel (1970) has also grown somewhat more advanced rabbit fetuses in culture by adapting the 'circulator' system described on pp. 25–26.

A particularly interesting feature of this work with rabbit blastocysts is the maintenance of growth and development during the very early stages of fetal development, around and immediately following the time of implantation (in the rabbit implantation occurs at about $7\frac{1}{2}$ d of gestation). But although many of the explanted embryos form a beating heart, development of a functioning blood circulation in culture has apparently not yet proved possible, and the most advanced stage of fetal development that can be achieved is less than that obtainable from rodents.

Hamster

Fetuses of the Syrian hamster (*Cricetus auratus*) have been successfully grown in culture by Givelber & di Paolo (1968). Pregnant hamsters were killed on the eighth day of gestation and the embryos explanted with yolk-sac, amnion and ectoplacental cone intact, but with the Reichert membrane opened. At the time of explantation the stage of development varied from head-fold to about seven somites, with a protein content per embryo of $84 \pm 22 \mu g$. The explants were cultured in circulating (diluted) serum and, under the best conditions, more than half the fetuses continued growing for 48 h with a final protein content of $325 \pm 94 \mu g$. In most of the survivors a blood circulation was established and many developed limb buds. Givelber & di Paolo assessed the amount of growth obtained as equivalent to 24–30 h development *in vivo*. However, they reported much greater variation between the embryos grown in culture than is found *in vivo* and the frequent development of a malformation characterized by a greatly dilated and thinned thorax. Also the degree of success obtained was restricted to fetuses explanted on the eighth day; little development was shown by explants on the seventh or ninth days.

Opossum

Although live embryos of the opossum (*Didelphis marsupialis*) are much more difficult to obtain than those of the standard laboratory animals, the short gestation period and simple placental system of this marsupial make it an attractive animal for studying the possibilities of fetal development *in vitro*. A preliminary study has been made by New & Mizell (1972). The best results were obtained from fetuses explanted with their fetal membranes at 10–11 d gestation. At 10 d the opossum fetus has about 28 somites, the heart has just begun to beat, and the forelimbs are flattening to the paddle shape. The large yolk-sac (diameter about 17 mm) has a vascular area covering a third to a half of its surface. At this stage the yolk-sac lightly adheres to the endometrium, but can be pulled away easily and transferred undamaged to the culture vessel. By 11 d the vascular area of the yolk-sac adheres more firmly to the endometrium, and the avascular areas of adjacent yolk-sacs are fused. The yolk-sacs at this stage can be pulled away in a mass from the endometrium, but are damaged and collapse when separated one from another. However, the vascular area can be kept intact and transferred to the culture vessel with the fetus.

The explanted fetuses were cultured in roller tubes (p. 24) in a medium of 20% opossum serum and 80% tissue-culture medium 199 or Ham's F10. The rate of differentiation was about the same as *in vivo*. Among the externally visible developments in culture were differentiation of the limbs (e.g. formation of digits on the forefoot) and head, and enlargement of the allantois. The most successful fetuses appeared well developed at the end of the 20-h culture period (Plate 2.1, p. 36). But the increase in weight of the fetus during the culture period was rather small (up to 50%) and, although a good blood circulation was maintained for a few hours in many of the explants, only two out of 33 still had a functioning circulation at 18 h.

Rat, mouse and guinea pig

The early rodent conceptus is an oval to spherical structure, made up of the yolk-sac and other embryonic membranes, with the fetus at one end (Fig. 2.1). In several studies of the earliest stages of fetal development of rats and mice, the rudimentary 'fetus' has been separated from the yolk-sac, or the yolk-sac has been opened to give a flat layer of tissue in culture. Grobstein (1950) examined the behaviour of the isolated mouse embryonic shield explanted between the sixth and eighth days of gestation (prestreak to head-fold stages), and cultured on plasma clots. The younger explants soon died, but the older showed spreading and outgrowth of the tissue with the formation of rhythmically beating areas. However, there was no clear development of a heart or other organs. N. Skreb (personal communication, 1972) has recently studied the development of rat embryos explanted at 7–9 d (primitive streak) and cultured in rat or calf serum. The extra-embryonic part was cut off at the level of the amnion and discarded. By a longitudinal cut through the primitive streak, the cup-shaped embryonic part was transformed into a flat sheet and supported on lens paper on a steel grid at the surface of the serum. The cultures were incubated for two weeks, and cartilage, gland, muscle and other tissues differentiated, but were not organized as in a normal fetus. Smith (1964) obtained more regular development from mouse embryos explanted at four to eight somites. The embryos were cultured on plasma clots, after the yolk-sac and amnion had been torn open so that the explants flattened out. The cultures were incubated for 18–20 h, and development continued up to the 16-somite stage, but without a blood circulation.

Better fetal development at these stages can be obtained when the fetus is explanted with the visceral yolk-sac intact, as in the early experiments of Jolly & Lieure (1938). These workers cultured rat and guinea-

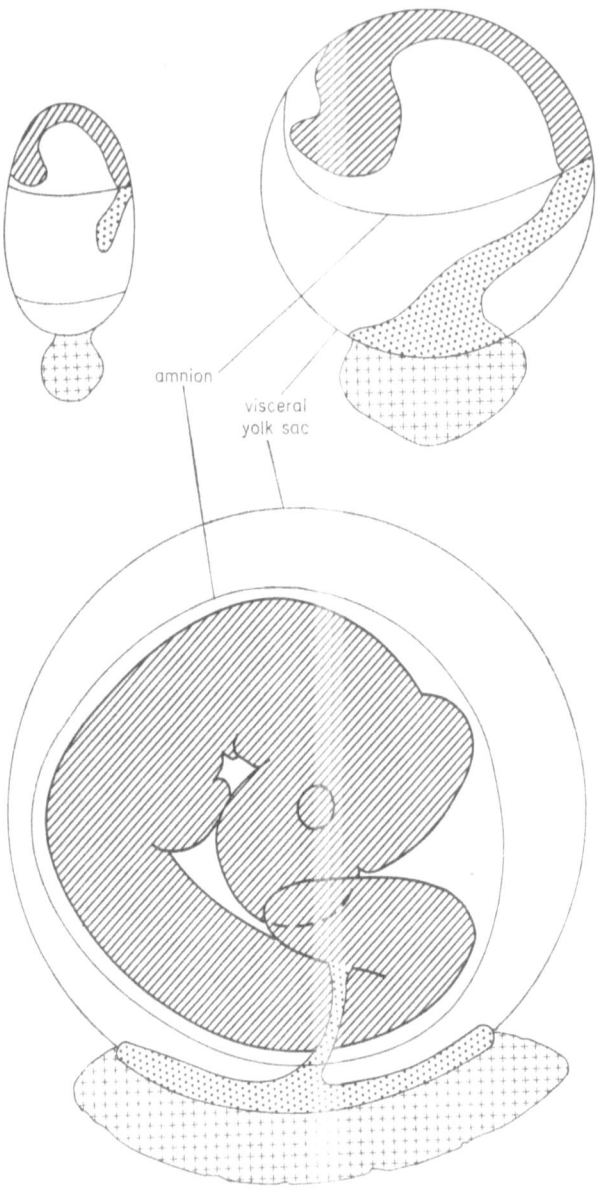

amnion

visceral
yolk sac

Fig. 2.1. The fetus and membranes of the rat at $9\frac{1}{2}$, $10\frac{1}{2}$ and $11\frac{1}{2}$ d gestation, drawn to same scale, $\times$ 20. (The Reichert membrane omitted.) Hatched area, fetus. Stipple, allantois. Crosses, ectoplacental cone. Between $10\frac{1}{2}$ and $11\frac{1}{2}$ d the fetus twists from a dorsally-concave to a ventrally-concave posture

pig embryos explanted at stages between primitive streak and a few somites. The explants grew well in homologous serum, and 37% of the rat fetuses formed a rhythmically beating heart, and 9% a functioning blood circulation. Nicholas & Rudnick (1934, 1938) appear to have had a similar degree of success with rat embryos cultured in heparinized rat plasma and embryo extract. Nicholas (1938) also claimed good results with a culture method involving circulating medium, but gave no indication of the proportion of fetuses developing or stage of development attained.

In recent years methods have been improved and, with the currently available techniques, rat and mouse fetuses can be grown in culture during the period of organogenesis much better than those of any other mammal that has been studied. The techniques for rats and mice, and the results that can be obtained will therefore be described in more detail.

Culture methods for rat and mouse fetuses

Methods for obtaining and explanting rat and mouse fetuses are described elsewhere (New, 1966a, 1971). The uterus from each pregnant animal is transferred to a dish of saline solution and carefully torn open with forceps. This exposes the decidual masses from which the fetuses and their membranes can be dissected under low-power magnification. In rodents the outermost layer of membranes is Reichert's membrane with attached trophoblast and parietal endoderm. This is opened (but need not be removed) because it does not expand in culture and restricts expansion of the other membranes. The visceral yolk-sac, amnion and ectoplacental cone are left intact and explanted with the fetus (Fig. 2.1).

Watch-glass cultures
Watch-glass cultures or similar cultures with static (non-flowing) medium give good growth of rat fetuses explanted up to the eleventh day of gestation. In these cultures the explanted conceptus is placed with nutrient medium in a watch-glass inside a petri dish, as in the standard method for organ culture. Each petri dish is lined with wet gauze or cotton to provide a humid atmosphere and, if necessary, the petri dishes can be housed in a gas-tight chamber containing any desired mixture of oxygen, nitrogen and carbon dioxide, or hyperbaric oxygen. The younger fetuses (9-d rat) develop adequately in 5% carbon dioxide in air, but older fetuses require higher concentrations of oxygen.

The nutrient medium can be a plasma clot (New & Stein, 1964), homologous serum (New, 1966b), or mixtures of serum and chemically defined medium (Clarkson, Doering & Runner, 1969). For most purposes the liquid media are to be preferred; they are simpler to prepare and more convenient for chemical analysis. The rat conceptus of 10 d or older floats near the surface of whole serum, so that its respiration is unaffected by variations in depth of the serum in the watch-glass. Younger explants, or explants in mixtures of serum and synthetic media, may sink to the bottom of the watch-glass, and the oxygen level required in the gas phase then varies with the depth of serum.

The method has been adapted by Le Goascogne & Brun (1969) for the study of developing 9½-d rat fetuses by time-lapse cinemicrography.

Roller tubes

In this method the culture chambers are small, stoppered, cylindrical glass containers (e.g. specimen tubes 2.5×5 cm) in which about half

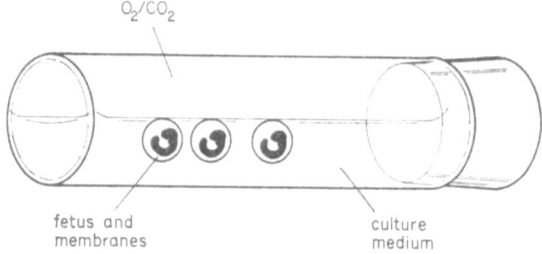

Fig. 2.2. Roller tube. The culture medium, with fetuses, half fills the tube and the space above is filled with the required gas mixture. The tube is rotated at 60 rev/min during incubation

the available volume is filled with the nutrient medium, and the other half with a suitable oxygen, nitrogen and carbon dioxide mixture (Fig. 2.2). The tubes are laid horizontally on rollers, and rotated at about 60 rev/min during incubation. This promotes oxygenation of the medium by continuously exposing a fresh layer to the gas phase, and assists respiration by keeping the explants gently swirling about in the medium. Oxygenation is uniform throughout the medium, and whether the explants sink or float is therefore irrelevant.

The improved oxygenation makes possible the growth in culture of more advanced fetuses (up to 40 somites) than can be obtained from watch-glass cultures. The method is particularly useful in experiments where it is desired to test the effect of an agent, e.g. a teratogen, on fetal development; the continual movement of the explant ensures maximum exposure to the agent (New & Brent, 1972). The main disadvantage of the method is that it is inconvenient for examining the fetuses during the course of the culture; the rotation has to be stopped, with possible effects on respiration, and only low-power observation is possible through the curved wall of the culture tubes.

Further details, and a comparison of roller tubes with other methods providing circulating medium, are given in New, Coppola & Terry (1973).

Circulators

The circulator shown in Fig. 2.3 was designed to allow continuous observation of the fetus, and to circulate the culture medium without

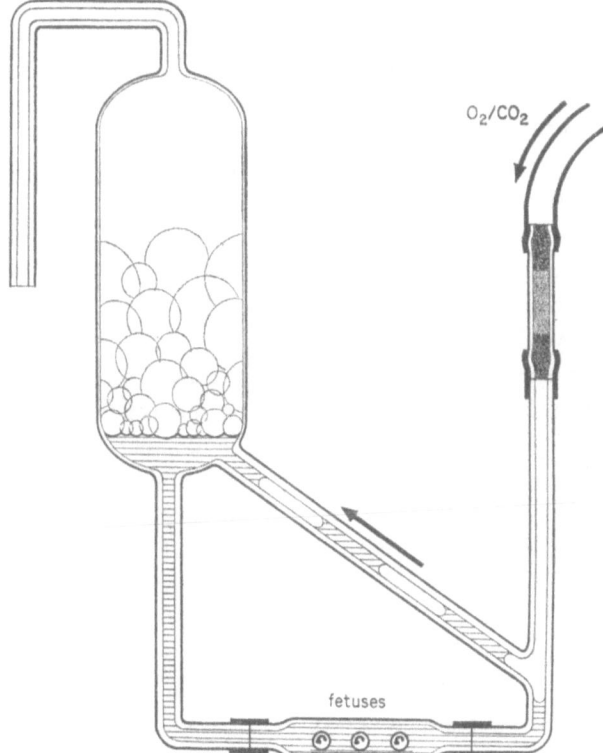

Fig. 2.3. Circulator. The triangle of glass tube is filled with the culture medium, and circulation is maintained by O_2/CO_2 entering through the filter on the right and flowing as a stream of bubbles up the sloping tube. The bubbles collapse in the chamber on the left. The fetuses are contained in the detachable chamber in the horizontal tube. Drawn to scale, $\times \frac{1}{2}$.

the use of mechanical pumps (New, 1967). It consists essentially of a triangle of glass tubing through which the medium is continuously re-circulated and oxygenated by a stream of bubbles flowing up the sloping tube. The bubbles are discharged into the large chamber, where the medium drains from them and they collapse. If necessary an anti-foamant can be added (Shepard, Tanimura & Robkin, 1970).

The gas is obtained from a cylinder of oxygen and carbon dioxide or oxygen, nitrogen and carbon dioxide, mixed in the required proportions, and equipped with a regulator valve to maintain the pressure at about 1 or 2 lbf/in^2. It is first humidified by being bubbled through water, then passed through a filter. The purpose of the filter is to provide a barrier against infection, and to control the flow of gas so that only a few millilitres per minute pass into the circulator, forming a steady flow of bubbles.

The fetuses are contained in a detachable chamber. To prevent them circulating with the medium, each explant is anchored by means of the Reichert membrane to a strip of gauze attached to a piece of glass cover-slip or stainless steel wire frame (for further details see New, 1971).

If the fetus chamber is made with flat sides, the explants can be observed without optical distortion and visibility is good (Plate 2.2). Continuous observation is possible throughout the culture period, with-out cooling the fetuses or interrupting the circulation of the culture medium. (D. L. Cockroft and the author have successfully filmed live rat fetuses by this method.)

Although the circulators are less simple in construction than roller tubes, they are sufficiently compact to permit numbers of them to be placed side by side on racks in a small incubator. They can therefore conveniently be used for comparative studies of fetal development under a variety of experimental conditions. They may also be housed in small pressure chambers for studies involving hyperbaric oxygen (Plate 2.3).

Other methods

Two other systems have recently been devised for growing rat and mouse embryos in circulating medium. Both employ more complex apparatus than the circulators, but have advantages for special applica-tions.

In the system of Tamarin & Jones (1968), flow is maintained by a peristaltic pump and the medium is exposed to the oxygenating gas mixture in a separate unit. The embryo chamber is made from a con-verted 10-ml syringe, held horizontally; a series of wells cut along the length of the syringe plunger contain the explants, and are connected to the pump by tubes passing through the plunger head. During incubation the plunger, with its growing fetuses, is housed inside the barrel of the syringe but can be withdrawn at any time that access to the fetuses is desired.

The 'Plasmom' apparatus of Robkin, Shepard & Tanimura (1972) also involves a peristaltic pump to provide circulation. For oxygenation the culture medium is passed through a bundle of fine silicone rubber tubes (combined length 15 m) immersed in a water bath. The required gas mixture is bubbled through the water in the bath, and the high permeability of the silicone rubber to oxygen and carbon dioxide ensures rapid equilibration with the medium flowing through the tubes. At no stage is gas bubbled through the culture medium itself and the problems of foaming, or of adding anti-foamants, are therefore avoided. As it passes through the silicone tubes in the water bath, the medium is warmed as well as oxygenated. The temperature of the water bath is controlled by a thermistor inserted into the circulating medium near the embryo chamber. No incubator is needed and all parts of the apparatus are readily accessible for observation and manipulation.

Development of rat and mouse fetuses *in vitro*

Up to the time of implantation the eggs and embryos of mice grow much more readily in culture than those of the rat. The culture techniques for mouse eggs are now well established (e.g. Whittingham, 1971), but the requirements for growing rat eggs are still largely unknown and reports of even limited success (e.g. Folstad, Bennett & Dorfman, 1969) are few. Surprisingly, this difference between the two species rapidly disappears after implantation, and throughout organogenesis both behave very similarly in culture. Rats have larger numbers of embryos per litter and have been used more extensively than mice for studies of fetal development *in vitro*, but when mouse fetuses of equivalent stages of development have been cultured the results have been much the same.

Embryos and fetuses explanted before somite formation (up to 9 d of gestation in the rat)
The youngest stages of development have proved particularly exacting in their culture requirements. Blastocysts in culture usually cease development at about the stage equivalent to that of implantation *in vivo*,

and there is some evidence that early fetal development can occur only if the blastocyst implants. Jenkinson & Wilson (1970) obtained improved development, including endoderm and pro-amnion formation in mouse blastocysts that were allowed to 'implant' in culture in pieces of bovine eye lens. The differentiation of these blastocysts resembled that of eggs transferred to ectopic sites. Hsu (1971) reported blood-island and heart differentiation in 5–20% of blastocysts that 'implanted' in culture on a collagen layer; development was, however, somewhat abnormal and the heart contractions appeared only after a prolonged period *in vitro* (10–14 d).

More extensive fetal development can be obtained if the embryos are explanted two or three days later. New & Daniel (1969) studied rat embryos explanted at $7\frac{1}{2}$–$8\frac{1}{2}$ d gestation. At this stage the embryos ('egg-cylinders') have implanted in the uterine wall (implantation in the rat occurs at about $6\frac{1}{2}$ d), and have developed an ectoplacental cone, but no organ rudiments have yet formed. The explants were cultured in circulators in homologous serum equilibrated with 5% carbon dioxide in air. After 2–3 d in culture, somite-stage embryos (fetuses) had formed, some with small anterior limb buds. The fetuses acquired beating hearts, but very few developed a functioning blood circulation, although a capillary network with conspicuous masses of erythrocytes was invariably present. This circulation failure seemed to be associated with abnormal heart development in most of the fetuses. The two lateral heart primordia, instead of fusing to form a single heart, developed into two more-or-less-separate hearts, often beating independently of each other.

Steele (1972) has recently made the important observation that heart formation in these fetuses appears to be strongly affected by the way the nutrient serum has been prepared. Fetuses grown in serum obtained from blood that had been left standing for a day after extraction from the donor rats usually developed double hearts. But fetuses grown in serum obtained by immediate centrifugation of fresh blood, or in heparinized plasma, formed normal single hearts. The fetuses with single hearts developed better than those with double hearts, and some of them formed a functioning blood circulation.

Present evidence suggests that prospects for growing the very young fetuses in culture are promising, but so far the results have not been as good as with later stages. Development is frequently abnormal; the somites are not well formed, and often the posterior part of the embryonic axis protrudes from the yolk-sac (Steele, 1972). Also, protein synthesis by these fetuses *in vitro* is much less than *in vivo*.

Fetuses explanted between head-fold and 20 somites (9–11 d gestation in the rat)

Fetuses of this age are the least exacting in their culture requirements. They can be grown successfully (up to 25–30 somites) on plasma clots and in static liquid media, although somewhat more advanced development (up to 30–35 somites) can be obtained in circulating medium (New, 1967). The oxygen concentration needs to be adjusted carefully for the type of culture and stage of development of the fetus (New & Coppola, 1970a). In general the older embryos benefit from raised oxygen concentrations, but excess oxygen at any stage can rapidly be fatal.

Differentiation of the fetus and yolk-sac in culture follows closely, in both form and rate, that *in vivo* (Plate 2.4). A blood circulation becomes established in the fetus, and in the network of blood capillaries formed in the rapidly expanding yolk-sac. Blood begins to circulate at about the 10-somite stage, and the yolk-sac then becomes a functional placenta. Soon after this the fetus 'turns', i.e. twists from being dorsally concave to the ventrally concave posture characteristic of the later stages of development (Fig. 2.1). The neural tube closes and the earlier stages of brain, eye and ear formation are initiated. The anterior limb-buds become prominent, and then in older fetuses the posterior limb-buds begin to form.

Although the ectoplacental cone often forms a circle of outgrowth over any surface of the culture chamber with which it is in contact, there is no organized differentiation of the allantoic placenta. An allantoic blood circulation is rarely established (*in vivo* it becomes functional at about the 20-somite stage), and a characteristic feature of the development of fetuses of this stage is the abnormal formation of a large blood vessel bypassing the umbilical vessels to the allantoic placenta (Fig. 2.9).

Fetuses explanted at 20–35 somites (11–12 d gestation in the rat)

Fetuses explanted after the 20-somite stage have an allantoic blood circulation, and fetal blood continues to circulate through the placenta in culture. But there is no maternal blood circulation in the explanted allantoic placenta, and to compensate for this the culture medium must be equilibrated with raised oxygen concentrations, so that more oxygen reaches the fetus via the yolk-sac. In circulating medium equilibrated with 5% carbon dioxide in air, the yolk-sac expands normally but there is little growth of the fetus; in 95% oxygen and 5% carbon dioxide both yolk-sac and fetus develop well, often attaining 40–45 somites after 40 h in culture, and with four or five times the initial protein content (New, 1967; New & Coppola, 1970a). More limited development can be obtained in static medium in hyperbaric oxygen (New & Coppola, 1970b).

Plate 2.5 shows the development usually obtained from fetuses explanted at about 25 somites and grown under optimum conditions. The most conspicuous advances externally are the extension of the limb- and tail-buds, and the further elaboration of the brain and other organs of the head. The blood circulation is well maintained *in vitro* for about 18 h, and the heart beat steadily increases in rate to about 160–200 beats per minute. By 24 h the heart rate has often begun to slow down, and areas of blood stasis appear in the yolk-sac capillary network; this is usually the first sign that the culture is deteriorating. By 40 h the heartbeat has failed and fetal development has ceased.

Fetuses explanted at more than 35 somites (12 d gestation in the rat)

Rat fetuses explanted at $12\frac{1}{2}$ d (about 40 somites) have been grown in circulating serum in hyperbaric oxygen to 50–55 somites (New & Coppola, 1970b). A vigorous heart beat and yolk-sac blood circulation was maintained for over 18 h, and both rate of development and final appearance of the fetuses appeared to be similar to that *in vivo* (Plate 2.6). These results were achieved in oxygen at two atmospheres pressure (gauge reading 15 lbf/in^2).

Attempts to improve development by the use of higher pressure were unsuccessful (Fig. 2.4), apparently because of an increasingly damaging effect on the yolk-sac. In oxygen at normal pressure the yolk-sacs expanded to 8–9 mm in diameter, but in oxygen at two atmospheres pressure they remained at about the same size (6–7 mm) as at explantation, and at three atmospheres they were found collapsed and partially disintegrated at the end of the culture period. As the yolk-sac is in direct contact with the culture medium, the fact that this part of the explant should be the first to show deterioration from excessive oxygen in the medium is not surprising. Evidently the optimum oxygen concentration for these older fetuses in culture is the result of a balance between the beneficial effects to the fetus of extra oxygen conveyed by the yolk-sac and the harmful effects of the extra oxygen on the survival and growth of the yolk-sac itself.

Cockroft (1973) has recently devised an ingenious method for increasing the area of respiratory surface instead of increasing the oxygen pressure. The yolk-sac and amnion are carefully opened without damage to the yolk-sac blood vessels, and are spread out on the floor of the culture chamber. This exposes the fetus to the circulating culture medium, and respiratory exchange can occur through the fetal surface as well as via the yolk-sac. With these arrangements the best results

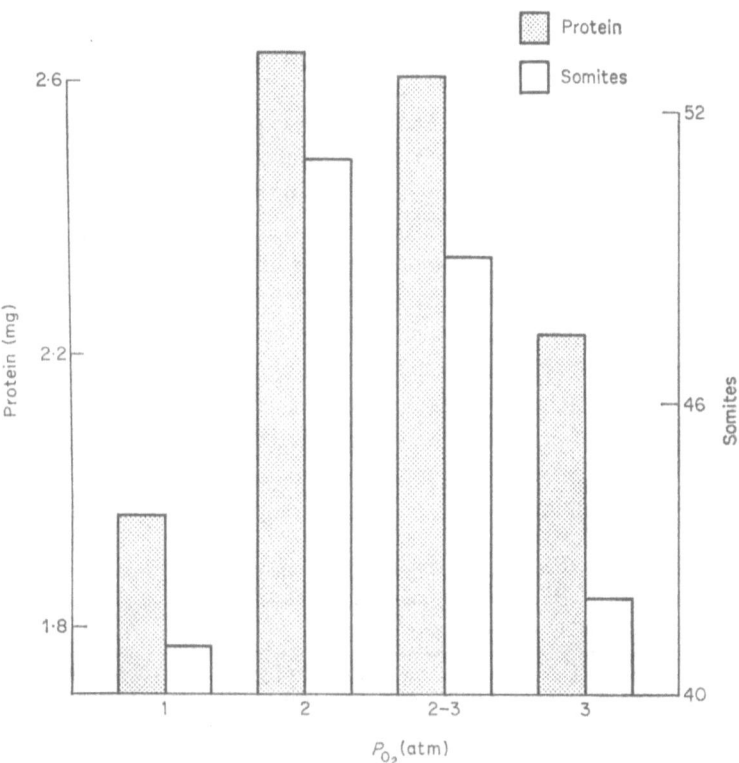

Fig. 2.4. Growth as determined by protein content, and differentiation as determined by somite number, of $12\frac{1}{2}$-d rat fetuses grown in circulating serum under different pressures of oxygen

have been obtained in culture medium equilibrated with 95% oxygen and 5% carbon dioxide at normal pressure. Development of $12\frac{1}{2}$-d fetuses was better than that of closed yolk-sac explants in hyperbaric oxygen, and for the first time growth of fetuses explanted at $13\frac{1}{2}$ d has also proved possible. Explanted at about 50–55 somites (3·5–4·0 mg protein), these fetuses have attained over 60 somites (6·5–7·5 mg protein) and are the most advanced fetuses that have yet been grown in culture (Plate 2.7).

Respiration and energy pathways

Oxygen consumption

The method of Warburg, Posener & Negelein (1924) has been used by several investigators to determine the oxygen consumption of the mid-term rat fetus, and De Plaen (1970) gives a useful summary of this work. The results from Warburg determinations (and other methods) are reasonably consistent. Thus Kleiber, Cole & Smith (1943), Boell (1958) and De Plaen (1970) all give values of oxygen consumption between 7 and 10 μl/h mg dry weight, when fetuses explanted at 13–14 d of gestation are maintained in oxygen. If the fetuses are maintained in air the oxygen consumption is only about 4 μl/h mg dry weight (Netzloff *et al.*, 1968a, b; De Plaen, 1970).

The condition of the explanted fetus in these experiments is not always clearly described. However, Netzloff *et al.* (1968a, b) are careful to explain that their measurements were made on the fetus explanted with the yolk-sac intact, but with the chorio-allantoic blood vessels ligated and the chorio-allantoic placenta removed. Their criteria for viability were (*a*) a beating heart, and circulation of fetal blood in the vitelline vessels, throughout the experiments, and (*b*) a constant rate of oxygen consumption for the 1-h measurement period. Table 2.1 sum-

Table 2.1. *Oxygen consumption per milligram dry weight of explanted rat embryos and yolk-sac, as determined by the Warburg method (gas phase: air) (Netzloff et al., 1968a, b)*

Age (d)	Number of determinations	Dry weight (mg) ±S.E.	μl O_2/h mg (Q_{O_2}) ±S.E.
11½	9	0·87 ± 0·07	8·21 ± 1·22
12½	9	4·01 ± 0·30	4·76 ± 0·34
13½	14	9·15 ± 0·34	3·90 ± 0·23
14½	23	17·53 ± 0·50	3·39 ± 0·10
15½	9	31·05 ± 1·10	2·79 ± 0·14

marizes the results for fetuses explanted between the twelfth and sixteenth day of gestation. The results show a fall of Q_{O_2} (oxygen consumed per hour per mg dry weight) with increasing gestational age.

Previous investigators had reported a decline in oxygen consumption per unit weight with increasing age for both chick (Romanoff, 1941) and rat embryos (Boell & Nicholas, 1939), and had interpreted it as a declining metabolic rate. Philips (1941) and others have criticized this because isolated embryos above a certain size contain a proportion of inner, non-respiring tissue remote from the external source of oxygen, and the proportion of such tissue rises with increased size of embryos. Netzloff *et al.* (1968b) comment that 'though such criticism is probably valid for previous studies with isolated embryos, our data from offspring in their yolk-sac with intact circulations appear to substantiate the belief that the metabolic rate decreases with advancing gestational age'. However, they also suggest (1968a) that the decline in oxygen uptake might reflect a declining importance of the yolk-sac as compared with the chorio-allantoic placenta.

Do these Warburg determinations indicate that the metabolic rate of the fetus *in vivo* declines with advancing gestational age? The answer clearly depends on whether the fetuses are respiring at the normal rate during the experiment. So far, the criteria of 'normality' used have been inadequate. To know that heart beat and blood circulation continue and that oxygen uptake is linear is not enough, because all these could be present at abnormally low values.

Table 2.2. *Growth (protein synthesis) of rat fetuses explanted, with visceral yolk-sac intact, at 9½–13½ d of gestation and cultured for 2 d in serum equilibrated with different oxygen concentrations. Results expressed as number of hours* in vivo *required for same amount of protein synthesis. (Based on data of New & Coppola, 1970a, b, and Fig. 2.7)*

Age	Oxygen concentration				
(d)	5 %	20 %	95 %	2 atm	3 atm
9½	28	37	<20		
10½		26	17		
11½		8	22	21	
12½			7	14	10
13½			<5		

Much better criteria would be normal rates of growth and differentiation of the fetus. These have not been established in Warburg experiments, but relevant to the problem is the work of New & Coppola (1970a, b) on the oxygen concentrations necessary for fetal development in culture. Rat fetuses were explanted with their embryonic membranes, and grown in circulators by the method described previously (p. 25). The circulating nutrient serum was equilibrated with different oxygen concentrations, and growth of fetuses was assessed at the end of the 2-d culture period by determining the amount of new protein synthesized. The results were compared with growth *in vivo*, and Table 2.2 shows the equivalent number of hours *in vivo* (as estimated from Fig. 2.7) required for the same increases in protein.

Of particular relevance to the Warburg determinations are the results of growing fetuses in 20% or 95% oxygen. Table 2.2 shows that 20% oxygen supports extensive growth of fetuses explanted at $9\frac{1}{2}$ and $10\frac{1}{2}$ d of gestation, but only very limited growth of those explanted at $11\frac{1}{2}$ d. This suggests that the Warburg determinations (in air) shown in Table 2.1 for fetuses of $12\frac{1}{2}$ d and older indicate levels of oxygen consumption lower than those of fetuses growing *in vivo*. Similarly Table 2.2 suggests that 95% oxygen supports growth only up to 12–13 d, and Warburg determinations on older fetuses, even in oxygen, have probably given values of oxygen consumption lower than those occurring *in vivo*. All that we can conclude at present about the rate of oxygen utilization by the rat fetus during organogenesis is that at 11–12 d of gestation (very early limb-bud stage) oxygen is probably consumed at about 8 µl/h mg dry weight; for later stages of organogenesis we lack valid measurements.

Why does the oxygen concentration required in culture increase with gestational age of the fetus, when the Warburg measurements appear to show a decline in oxygen consumption? The explanation almost certainly involves the lack of a normal allantoic placental transport system. *In vivo* the allantoic blood circulation becomes established at the end of the eleventh day (at about the 20-somite stage), and its oxygen-carrying capacity is then added to that of the yolk-sac circulation established earlier (at about 10 somites). The allantoic placenta fails to grow in culture, and it is usually removed before Warburg determinations are made. Fetuses explanted into culture after 11 d, therefore, require for prolonged growth a raised oxygen concentration, so that more oxygen is transported by the yolk-sac circulation; in the Warburg measurements of Table 2.1, the oxygen concentration remains the same, and the earlier suggestion of Netzloff *et al.* seems likely to be correct, namely, that the decline in Q_{O_2} reflects the increasing inadequacy of the yolk-sac to transport sufficient oxygen on its own.

The harmful effect of excessive oxygen, already noted for the older fetuses in hyperbaric oxygen (Fig. 2.4), is shown by Table 2.2 to apply to fetuses at all stages. The oxygen concentration required for maximum fetal growth in culture appears to have an optimum value for each stage of development. The increase in this optimum value as the fetus grows probably reflects the need to compensate for the lack of a functional allantoic placenta, and does not in itself imply any change in the metabolism of the fetal tissues. It can be explained most simply on the assumption that the fetal tissues require, at all stages of organogenesis, an oxygen pressure in the fetal blood similar to that of blood in equilibrium with air; to maintain this pressure at a constant level, blood returning to the fetus from the yolk-sac must carry more oxygen to allow for its dilution by the blood in the rest of the system.

Glycolysis, Krebs cycle and the electron-transport system
Recent studies by Shepard and his colleagues at the University of Washington (Aksu *et al.*, 1968; Shepard *et al.*, 1969, 1970; Tanimura

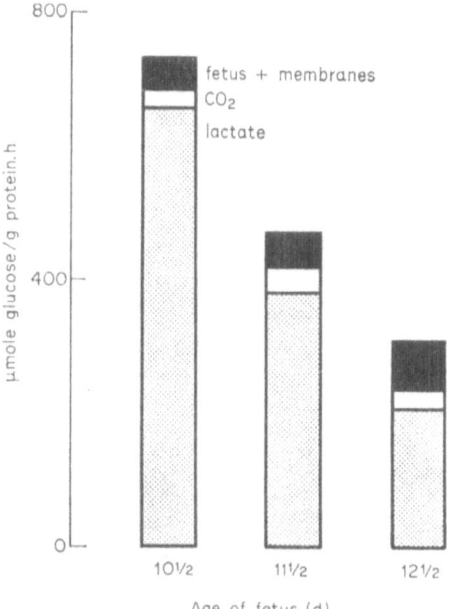

Fig. 2.5. Utilization of glucose by rat fetuses *in vitro*, as determined from studies with ^{14}C-glucose. (Courtesy of T. H. Shepard)

& Shepard, 1970) have demonstrated some of the pathways of energy metabolism in the early rat fetus. They have studied $10\frac{1}{2}$-,$11\frac{1}{2}$- and $12\frac{1}{2}$-d fetuses maintained in circulators, with labelled ^{14}C-glucose added to the nutrient serum. After 3 h of culture the metabolic fate of the glucose was measured by determining liberated $^{14}CO_2$, ^{14}C-lactate production in the medium, and incorporation of ^{14}C-glucose (or its products) into the tissues. The results (Fig. 2.5) showed that the rate of lactate production falls dramatically with advancing age of the fetus, while the rate of glucose incorporation into the tissues increases. The results refer specifically to products of glucose metabolism, and do not give information about possible contributions from fatty and amino acids. But they clearly indicate some major changes in energy pathways at this period of gestation. Furthermore, the fetuses were maintained in culture conditions that supported growth during the 3-h experimental period, and the conclusions therefore probably apply also to development *in vivo*.

The main energy-yielding pathways of glucose metabolism in the cell

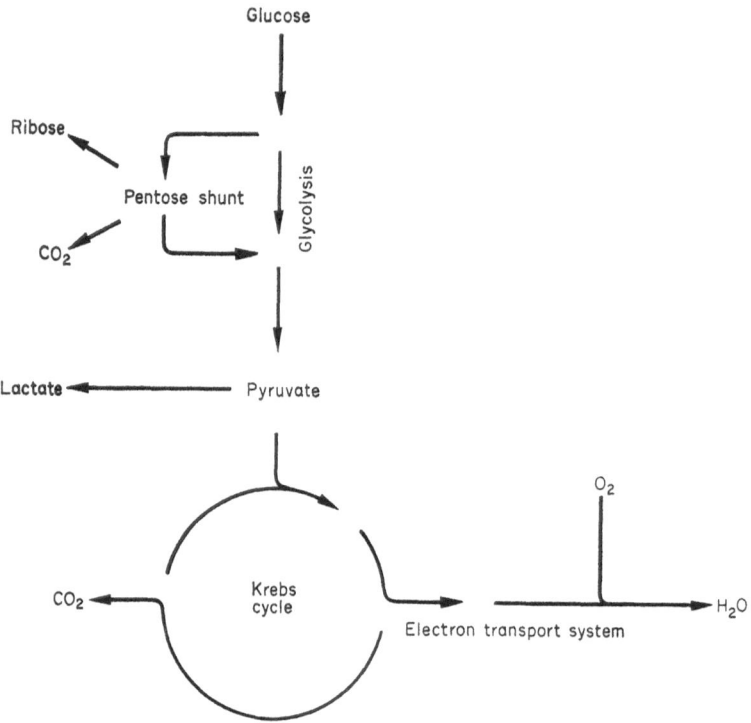

Fig. 2.6. The main energy-yielding pathways of glucose metabolism

are summarized in Fig. 2.6. The 6-carbon glucose molecule is first split in a series of enzymically catalysed steps into two 3-carbon molecules of pyruvic acid, a process known as glycolysis or the Embden–Meyerhof pathway. Glycolysis does not require oxygen and can proceed under aerobic or anaerobic conditions. However, in the absence of oxygen the pyruvic acid molecule cannot be further broken down, and is converted to lactic acid (3-carbon), which escapes from the cell into the blood stream or nutrient medium. The energy yield to the fetus from such anaerobic glycolysis is two molecules of ATP per molecule of glucose metabolized.

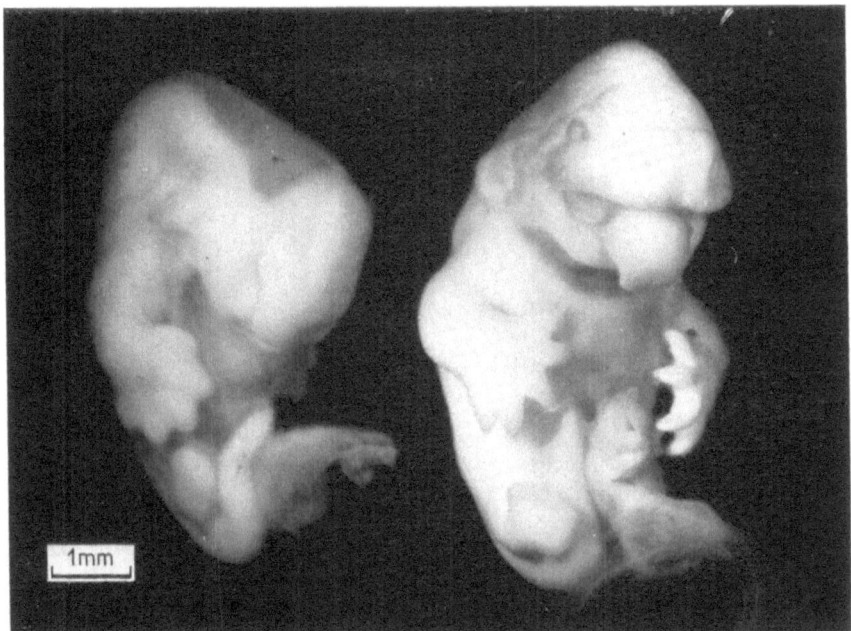

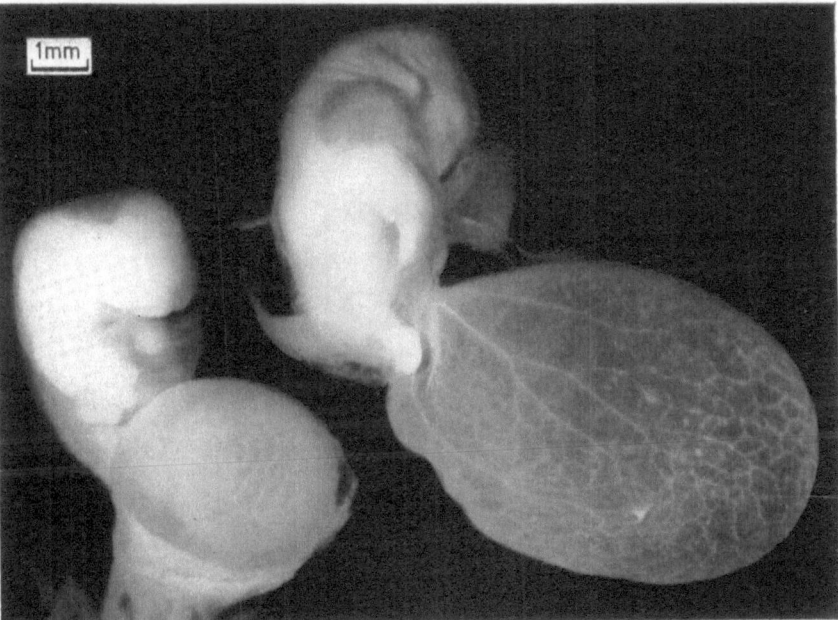

Plate 2.1. Opossum fetuses from the same litter, all explanted at 11 d gestation. Left, as explanted; right, after 20 h in culture. (*Top*) Dissected free of all membranes to show growth of the fetus. Among the externally visible developments in culture are the digits on the forefoot, the change of hind limb from bud to 'paddle' stage, and the formation of eyelid folds. Also, as *in vivo*, the head has become raised from the chest with mouth open and tongue protruding. (*Bottom*) Dissected free of the yolk-sac but photographed to show the growth of the allantois in culture. This membrane does not form a complex placenta in the opossum; towards the end of gestation it enlarges and fills with urine. (Copyright 1972 by the American Association for the Advancement of Science)

Plate 2.2. Rat fetuses, enclosed in the yolk-sac, growing in the fetus chamber of a circulator as shown in Fig. 2.3

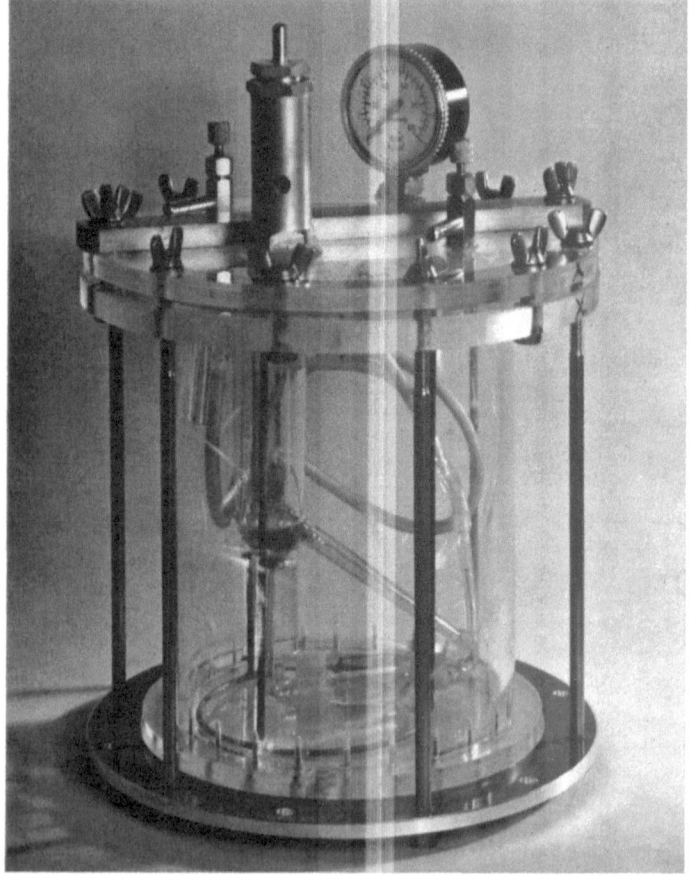

Plate 2.3. Apparatus for growing fetuses in circulating medium in hyperbaric oxygen. The perspex pressure chamber contains circulators similar to that shown in Fig. 2.3

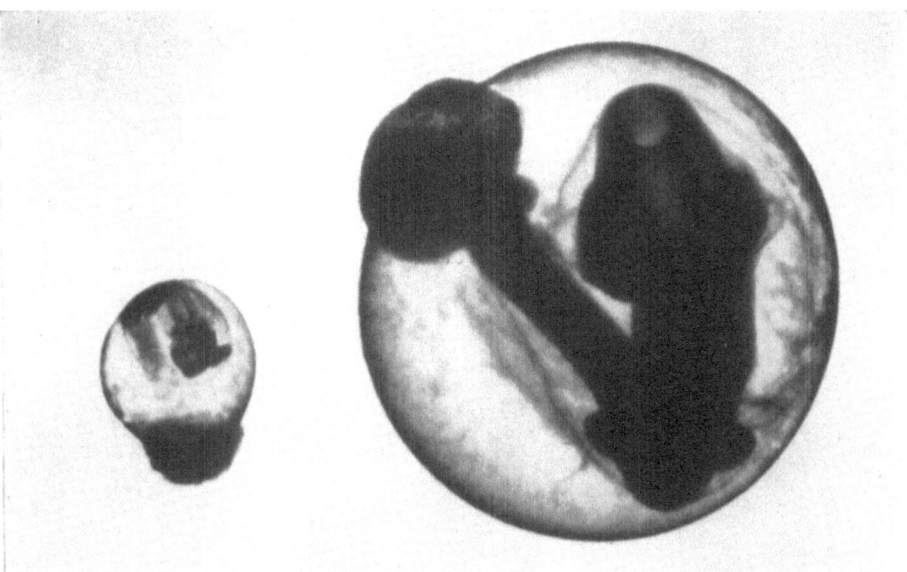

Plate 2.4. Rat fetus explanted at seven-somite stage (*left*), and a similar fetus after growth in culture (*right*). × 13

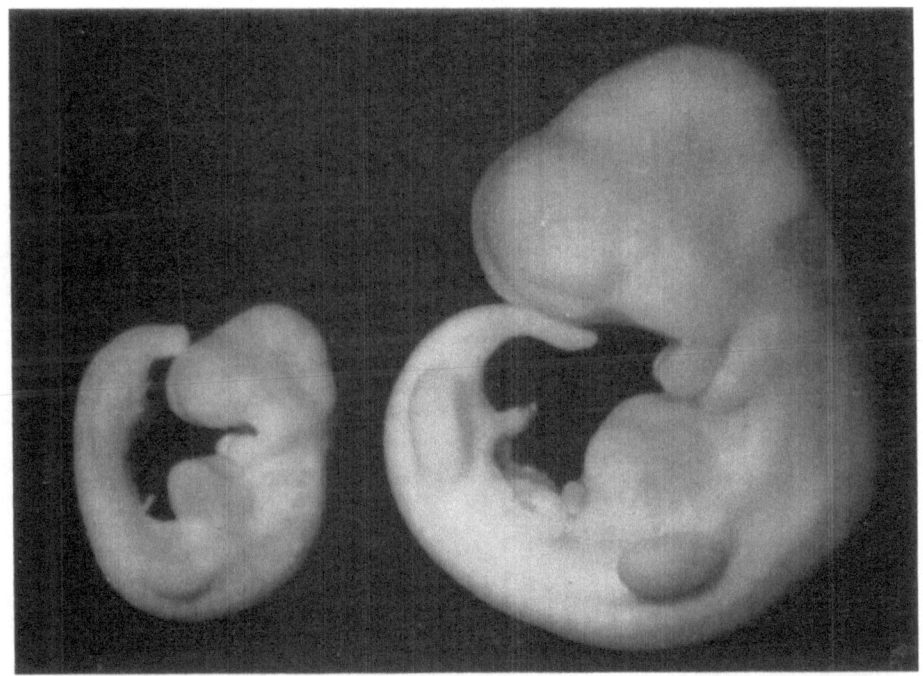

Plate 2.5. Rat fetus explanted at 11½ d gestation (*left*), and a similar fetus after growth in culture (*right*). The fetal membranes have been removed. × 16

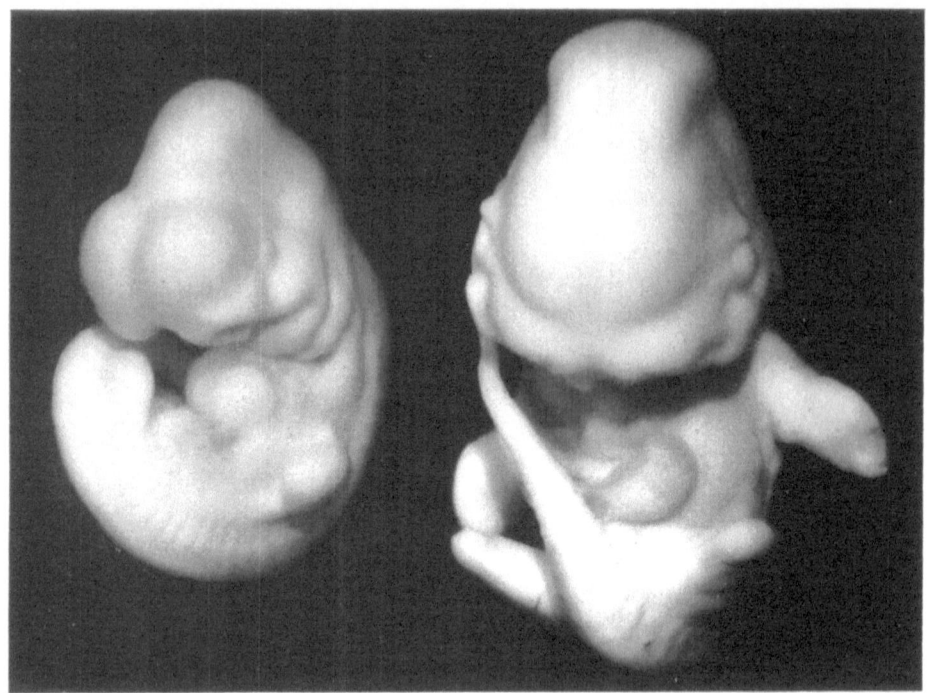

Plate 2.6. Rat fetus explanted at 12½ d gestation (*left*), and a similar fetus after growth in culture in hyperbaric oxygen (*right*). The fetal membranes have been removed. × 14

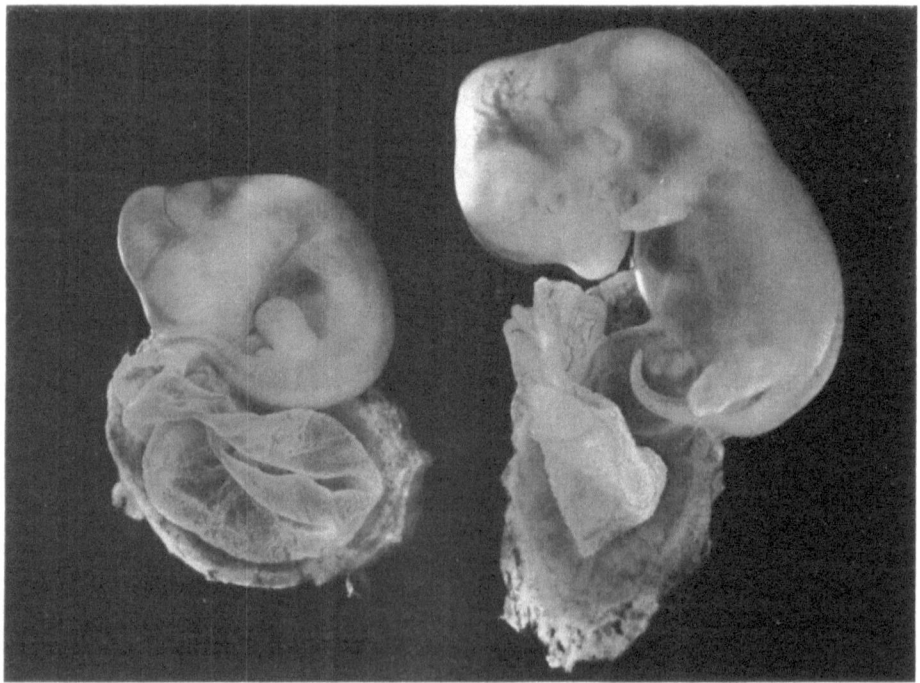

Plate 2.7. Rat fetus as explanted at 13½ d gestation with the yolk-sac opened (*left*), and a similar preparation after growth in culture (*right*). (Courtesy of D. L. Cockroft.) × 6

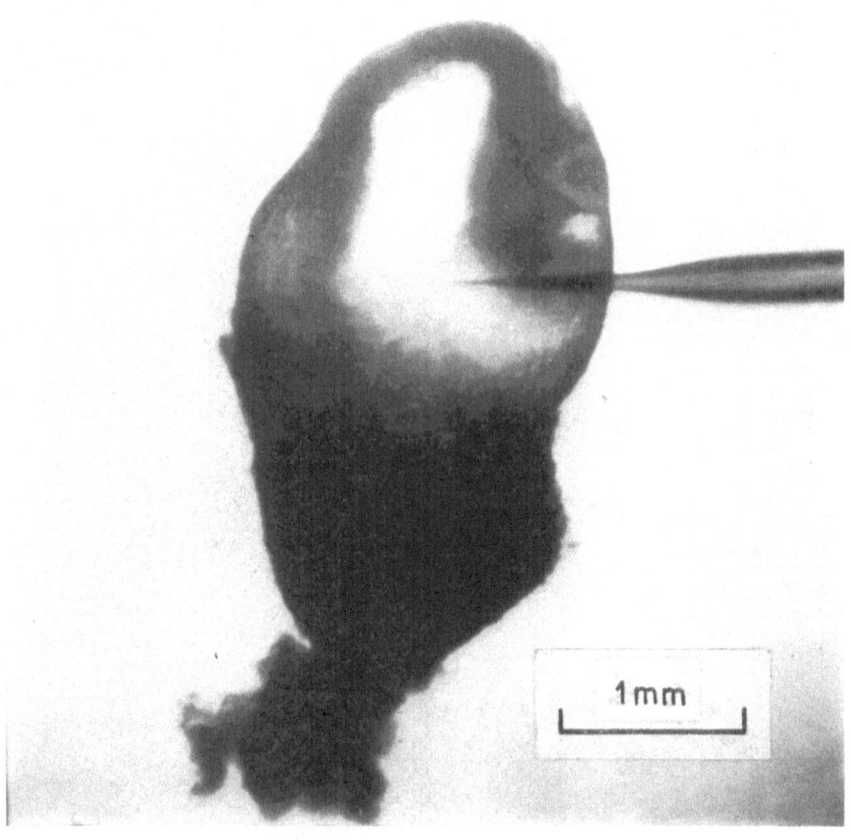

Plate 2.8. Rat fetus explanted at 10½ d gestation with glass injection pipette inserted into the extra-embryonic coelom

If oxygen and the appropriate (mitochondrial) enzymes are available, the pyruvic acid is completely oxidized by the linked processes of oxidative decarboxylation and oxidative phosophorylation to carbon dioxide and water. These two processes, each involving several steps, are usually known as the Krebs (or citric acid) cycle, and the electron- (or hydrogen-) transport system, respectively. Oxygen is consumed during their working and carbon dioxide is evolved. The energy yield from each molecule of glucose completely oxidized to carbon dioxide and water is 38 molecules of ATP, i.e. 19 times that derived from the conversion of glucose to lactic acid.

The data of Shepard et al. shown in Fig. 2.5 indicate a very high rate of glucose consumption by the $10\frac{1}{2}$-d fetus and its membranes. (The same authors give for comparison the following figures for glucose consumption in μmoles/g protein h by rat tissues: perfused heart 267 and 515; perfused liver 118; brain slices 107; diaphragm 66; lens 29.) About 90% of the glucose consumed by the $10\frac{1}{2}$-d fetus is converted to lactic acid. But by $12\frac{1}{2}$ d the rate of glucose consumption has fallen to less than half the $10\frac{1}{2}$-d value, and only about 68% of the glucose consumed is converted to lactic acid.

This strongly suggests that at $10\frac{1}{2}$ d fetal energy-yielding metabolism is characterized by high rates of glycolysis, with relatively little activity of the Krebs cycle and electron-transport system, and that during the following days the rate of glycolysis falls while the Krebs cycle-electron-transport activity increases. In support of this conclusion observations by Shepard's group have established two further lines of evidence: (i) when the fetuses are exposed to reduced oxygen concentrations the heart rate falls rapidly in the $11\frac{1}{2}$- and $12\frac{1}{2}$-d stages, but not in the $10\frac{1}{2}$-d stages, and (ii) studies of the electron-transport system in tissue fractions separated from homogenates of fetuses by differential centrifugation showed a threefold increase in activity between $10\frac{1}{2}$ and $12\frac{1}{2}$ d of gestation.

Carbon dioxide and the pentose shunt

If the Krebs cycle activity is increased in the older fetus, the reason why production of carbon dioxide stays fairly constant remains to be explained (Fig. 2.5). As the fetus becomes less dependent for its energy requirements on glycolysis (which does not produce carbon dioxide) and more dependent on the Krebs cycle and electron-transport system the rate at which carbon dioxide was evolved might be expected to rise. One explanation why it does not may be a possible incorporation of some carbon dioxide into other metabolic systems (there was increased incorporation of ^{14}C into the tissues of the $12\frac{1}{2}$-d fetuses), but a more readily demonstrable cause is the declining activity of the 'pentose shunt'. In the pentose-shunt pathway of glucose metabolism, glucose-6-phosphate is taken from the glycolytic sequence to yield ribulose-5-phosphate and carbon dioxide. Ribulose is a 5-carbon sugar readily convertible to ribose, and therefore of importance for nucleic acid synthesis, but alternatively, by a complex series of reactions, six molecules of ribulose-5-phosphate can be reconverted to five molecules of glucose-6-phosphate and recycled. The pentose-shunt gives a high yield of energy—36 molecules of ATP for each molecule of glucose expended.

The occurrence of the pentose-shunt can be detected by comparing $^{14}CO_2$ production from glucose labelled on the first carbon atom with that from glucose labelled on the sixth. The Krebs cycle (following glycolysis) yields $^{14}CO_2$ in similar amounts from C1- and C6-labelled glucose. With some exceptions this is characteristic of mature tissues. But Tanimura and Shepard (1970) found from determinations of $^{14}CO_2$ production by rat fetuses in culture that the C1:C6 ratio was 11·3 at $10\frac{1}{2}$ d, 3·6 at $11\frac{1}{2}$ d and 4·6 at $12\frac{1}{2}$ d. This indicates that the pentose-shunt is active at these stages, and that its activity is greatest in the $10\frac{1}{2}$-d fetuses. Carbon dioxide contributed by the pentose shunt is therefore highest in the youngest fetuses, and declines as the Krebs-cycle production of carbon dioxide increases.

Isolated hearts

Although this chapter is primarily concerned with studies on whole fetuses *in vitro*, the fact is worth noting that some of the conclusions of Shepard's group are supported by results from work on isolated fetal hearts. Clark (1971) and Wildenthal (1971) have demonstrated the importance of glucose metabolism for maintaining activity in isolated fetal rat and mouse hearts. Cox & Gunberg (1972a) have recently made a careful study of differences in metabolism of isolated hearts from 11-, 12- and 13-d rat fetuses. They have compared frequency of contraction of the hearts in the presence of various substrates. Glucose, and some other compounds involved in the Embden–Meyerhof glycolytic pathway, proved capable of maintaining cardiac contraction in the 11-d heart under aerobic or anaerobic conditions, strongly suggesting glycolysis as the main energy source. The 12- and 13-d hearts exhibited a shift in dependence towards other metabolic pathways; these hearts were able to use a wider variety of substrates, but under anaerobic conditions there was a significant decline in contraction frequency. Further studies (Cox & Gunberg, 1972b) have shown that, in the presence of an inhibitor of glycolysis, both 11- and 12-d hearts exhibit depressed contractile activity, but in the presence of inhibitors of the Krebs's cycle or oxidative phosphorylation, only the older hearts are affected.

Why do the energy pathways change?

The rapid change in energy metabolism between 10 and 12 d of gestation is paralleled by the activation during this period of the placental-blood circulation, first in the yolk-sac placenta ($10\frac{1}{2}$ d) and then in the allantoic placenta (11 d). The suggestion is sometimes made that during early gestation the embryo is in a relatively anaerobic environment favouring glycolysis as the main energy source, but that as the placental blood circulations become established they provide more oxygen and the fetus can obtain the higher energy yields of the Krebs cycle and oxidative phosphorylation. The close association of these events in time makes this an attractive hypothesis. But it supposes that the early fetus or embryo has difficulty in obtaining oxygen, and this is questionable. During its earliest stages of development, up to $10\frac{1}{2}$ d, the fetus is near the surface of the conceptus (Fig. 2.1), close to the maternal blood stream, and only later does it become surrounded by the yolk-sac. Moreover, the amount of tissue present in the $10\frac{1}{2}$-d fetus is very small and could respire aerobically with little oxygen consumption; but 2 d later the respiring tissue—and presumably the energy demand—has increased fiftyfold. Rather than glycolysis being forced on the early fetus by an anaerobic environment, the likelihood seems to be that the fetus adopts aerobic respiration only in order to satisfy its rapidly increasing energy needs.

As an alternative speculation the following may be considered. Glycolysis requires rather few enzymes and is therefore a valuable energy-yielding mechanism in a rapidly growing embryo, where there is competition for priorities in the synthesis of enzyme systems. It will therefore be to the advantage of the fetus to persist with this mechanism as long as the mother can supply the substrate (glucose) and remove the product (lactic acid). However, it is to the advantage of the mother that the fetus change to the Krebs cycle and electron-transport system, which requires much less glucose and yields the more easily disposable products, carbon dioxide and water. At the early stages of gestation the advantages of glycolysis to the fetus greatly outweigh the disadvantages to the mother, but the balance may well be expected to tip in the other direction during the period of exceptionally rapid fetal growth accompanying organogenesis.

Growth of the Fetus

Growth *in vivo*

A useful measure of growth in culture is the increase of fetal protein. As a standard of comparison, Fig. 2.7 shows rate of protein increase of the

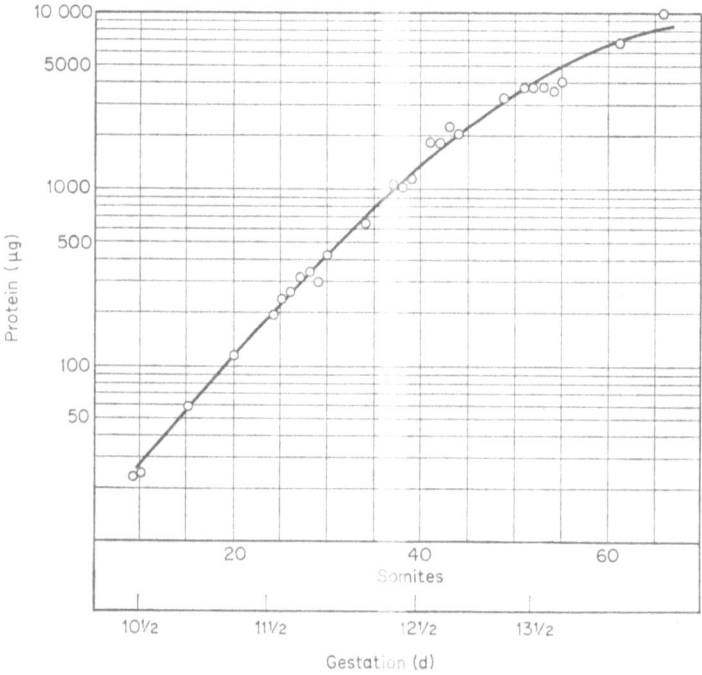

Fig. 2.7. Protein content and somite number of the rat fetus *in vivo* between 10 and 15 d gestation

rat fetus *in vivo*, as determined from 122 fetuses taken from pregnant rats of Wistar and Hooded strains on Days 11 to 15 of gestation (I am indebted to D. L. Cockroft for some of the data). The stage of development of each fetus was determined by counting the number of somites; in the older fetuses the somite immediately behind the anterior limb bud was regarded as somite 14 (Witschi, 1962), and the count was made posteriorly from this. Protein determinations were made on the fetuses alone (without membranes) by the colorimetric method of

Lowry *et al.* (1951). Each point on the graph gives the average of two to nine fetuses. Berry (1968) has published similar data for growth of the rat fetus between $10\frac{1}{2}$ and $12\frac{1}{2}$ d of gestation; the slope of the corresponding part of our curve agrees fairly well with his, but the protein values for each somite stage are somewhat lower.

To link stage of development with time of gestation, somite counts were made on 154 fetuses from 49 pregnant rats killed on days 11–14 (10–15 rats examined daily). The rats were killed between 9 a.m. and 3 p.m. each day, and the somite counts corrected to a 'noon value' by adding or subtracting at the rate of 1 somite/$1\frac{1}{2}$ h.

The rats had been mated overnight, and the days of gestation are therefore shown in Fig. 2.7 as $10\frac{1}{2}$, $11\frac{1}{2}$, $12\frac{1}{2}$, $13\frac{1}{2}$. The somite counts, expressed as mean and standard error, were: $10\frac{1}{2}$ d: 9.7 ± 0.3; $11\frac{1}{2}$ d: 23.5 ± 0.2; $12\frac{1}{2}$ d: 39.6 ± 0.8; $13\frac{1}{2}$ d: 51.7 ± 0.2. Somite formation evidently occurs at the rate of about 15 somites per day during most of organogenesis, probably slowing somewhat in the later stages. Protein increase per fetus accelerates rapidly over the whole period, but rate of increase *per unit weight* gradually declines.

Growth *in vitro*

In a study of several hundred rat fetuses explanted at $10\frac{1}{2}$ d gestation, Berry (1968) concluded that rate of somite formation in culture was within the range expected *in vivo*, but that growth, as indicated by increase in protein content, was retarded. The fetuses were examined after 6, 12 or 24 h in culture. The results suggest that during the first few hours protein increase may have continued at approximately the normal rate, but in the latter part of the culture period it had clearly fallen behind.

More data are now available, including results from several laboratories, and from rat fetuses explanted at different ages and grown in different culture systems. This is summarized in Table 2.3. All the fetuses were cultured by the methods described previously (p. 23). The $10\frac{1}{2}$- and $11\frac{1}{2}$-d fetuses were grown in culture for periods of 24 and 40–45 h, and comparison of the results from these two culture periods shows that growth continues for more than 24 h. But all the figures for final protein, and to a lesser extent those for final somites, indicate retarded development as compared with growth *in vivo* (Fig. 2.7). Similarly, Clarkson, Doering & Runner (1969) found that mouse fetuses, explanted at early somite stages and incubated for 22 h in culture, grew at less than the normal rate.

Table 2.3. *Somite number and protein content of rat fetuses grown in culture*

Age at explantation (d)	*Type of culture*	*Duration of culture* (h)	*Final somites*	*Final protein* (μg)	*Author*
9½	Circulator	48	21	90±	New & Coppola (1970a)
10½	Static	24	21–31	100–110	Berry (1968)
10½	Static	24	–	55±7	Payne & Deuchar (1972)
10½	Roller tubes	24	20–25	121±12	New & Brent (1972)
10½	Circulator	24	22	138±3	New *et al.* (1973)
10½	Static	43	27	227	New & Coppola (1970a)
10½	Circulator	43	32	334	New & Coppola (1970a)
11½	Circulator	22	32	about 500	Shepard *et al.* (1970)
11½	Plasmom	22	34	632	Robkin *et al.* (1972)
11½	Roller tube	24	34	576±25	New *et al.* (1973)
11½	Circulator	24	35	749±27	New *et al.* (1973)
11½	Circulator	40	43	1152±245	New (1967)
11½	Circulator	45	43	1306	New & Coppola (1970a)
11½	Circulator	42	41	1199±71	New & Coppola (1970b)
12½	Circulator	42	51	2640±90	New & Coppola (1970b)

How much is growth *in vitro* retarded? In Table 2.4 an attempt is made to assess this by estimating the equivalent times for similar growth (and differentiation) *in vivo*. In making the comparison, only growth under the best available culture conditions has been considered. Development in culture evidently becomes progressively more retarded with increasing age of the fetus at explantation, and protein increase is more affected than somite formation. Protein increase in the 9½-d

Table 2.4. *Equivalent durations* in vitro (*under best available culture conditions*) *and* in vivo *for somite formation and protein synthesis by rat fetuses* (*based on data of New & Coppola, 1970a, b, New* et al., *1973, and Fig. 2.7*)

Explantation age (d)	*Duration* in vitro (h)	*Equivalent duration* in vivo: *Somites* (h)	*Equivalent duration* in vivo: *Protein* (h)
10½	24	21	20
11½	24	16	15
9½	48	42	37
10½	43	32	26
11½	42	26	22
12½	42	22	14

fetuses is slowed down by only about half a day in 2 d in culture, while in the $11\frac{1}{2}$- and $12\frac{1}{2}$-d fetuses it is retarded by a day or more.

Why is protein increase retarded in culture?

Fetuses are normally explanted at laboratory temperatures and placed initially in cold culture medium. During this process the heart (if present) stops beating or beats only sluggishly. When cultures are placed in the incubator they must first warm up to the incubation temperature before fetal metabolism can return to its normal rate. The time required for warming varies with the type of culture and may be longer, for example, if the cultures are enclosed in gas chambers. But warming and restoration of the heart beat rarely require more than an hour of incubation and usually take much less. (Moreover, heart beat and blood circulation are resumed equally well whether the fetuses have been cooled for 10 min or 3–4 h, or stored for an hour in a refrigerator at 2°C.) Return to the normal metabolic rate possibly takes longer than restoration of the blood circulation, or cooling during explantation may even permanently impair fetal metabolism, but there is no evidence for this.

Table 2.5. *Final protein content and somite number of rat fetuses explanted at $11\frac{1}{2}$ d of gestation and grown in culture for 42 h at different temperatures and carbon dioxide concentrations*

	Number of fetuses	Protein (µg) (Mean ± S.E.)	Somites (Mean ± S.E.)
37·5°C (5 % CO_2)	8	1322 ± 104	40·9 ± 0·7
34·5°C (5 % CO_2)	8	1276 ± 94	41·5 ± 0·2
P for difference		>0·10	>0·10
37·5°C (5 % CO_2)	12	1261 ± 104	40·7 ± 0·8
32·0°C (5 % CO_2)	12	947 ± 78	38·9 ± 0·9
P for difference		$0·05 > P > 0·02$	>0·01
5 % CO_2 (37·5°C)	4	1102 ± 81	40·2 ± 1·0
2 % CO_2 (37·5°C)	4	1135 ± 61	39·5 ± 0·3
P for difference		>0·10	>0·10

If the incubation temperature (37°–38°C) were slightly lower than that which the fetuses normally experience in the uterus, it might be responsible for a reduced growth rate in culture. Robkin *et al.* (1972) have shown that the heart rate is closely affected by temperature, falling (in a 25-somite fetus) from about 150 beats/min at 38°C to about 70 at 30°C. Our observations agree with this, but give no evidence of reduced growth with small reductions of temperature. Table 2.5 shows final protein content and somite number of $11\frac{1}{2}$-d fetuses grown in culture for 42 h at 34·5°C and 32·0°C, compared with those grown at 37·5°C. Growth is significantly reduced at 32°C but not at 34·5°C.

The culture medium for the explanted rat fetuses is usually homologous serum. At the beginning of the culture period this medium can be assumed to resemble the plasma of the blood that normally bathes the conceptus *in vivo*. But Steele (1972) has shown differences in the development of very young fetuses grown in serum, as compared with those cultured in plasma. Also, serum probably deteriorates during the course of a culture, and may become depleted of essential nutrients or poisoned by fetal waste products. In an attempt to counteract such alterations in the medium, cultures have been made with much increased amounts of serum, or with renewal of the serum during the culture period (New, 1966, 1967). These procedures would be expected to compensate, at least to some extent, for most forms of 'deterioration' of the medium, but they have given, at best, only very small improvements in fetal growth.

There remains the possibility that some essential substances might be very unstable at incubation temperatures and therefore rapidly lost from the culture medium. Such (hypothetical) substances might not be restored sufficiently soon or frequently by the renewal procedures that have been tried. But before seeking an explanation along these lines, the more likely effects of placental failure in culture should be considered.

Functions of the allantoic and yolk-sac placentae

The relations of the rodent fetus and its placentae at mid-gestation are shown diagrammatically in Fig. 2.8. The allantoic placenta fails to develop in culture and, as explained on p. 29, growth of explanted fetuses much beyond the 20-somite stage requires raised oxygen concentrations in the culture system, so that more oxygen passes to the fetus via the yolk-sac. Is it necessary also to compensate for nutritive and/or excretory functions of the allantoic placenta? In late pregnancy these functions of the allantoic placenta are probably of vital importance, but whether they are essential during organogenesis is unknown. No systematic studies have been made with raised nutrient concentrations comparable to those with raised oxygen. Removal of excretory products by renewal of the culture medium, which might be expected to assist excretory processes, has not significantly improved fetal growth (nor has reduction of carbon dioxide in the gas phase of the culture system from 5% to 2%—see Table 2.5). Such modifications of the medium would, of course, compensate for the failure of the allantoic placenta only if the yolk-sac could provide a substitute transport system; they would have no effect if the required nutrients or harmful excretory products could pass only through the allantoic placenta (e.g., the histo-chemical study of Wislocki, Deane & Dempsey, 1946, suggested that while the yolk-sac serves as the route of iron transport the chorio-allantois may be specific for calcium transfer). At present the role, if any, of the rodent allantoic placenta in fetal nutrition and excretion during organogenesis remains uncertain.

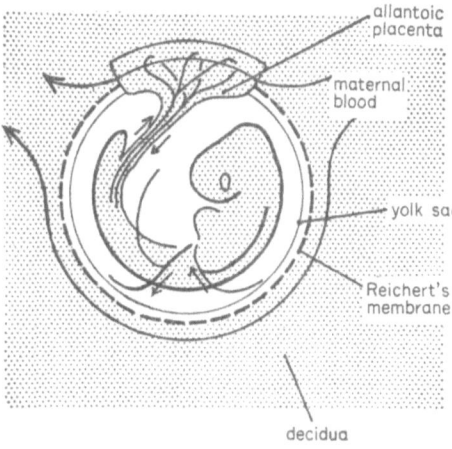

Fig. 2.8. Diagram of rat fetal blood circulations, and the relationship of the placentae to the surrounding maternal blood flow. The rat fetus develops two placentae. The first to form is the yolk-sac placenta and consists entirely of fetal tissue. The allantoic placenta, which forms later, is a more elaborate structure consisting of both fetal and maternal tissue

Fetuses explanted after the 20-somite stage maintain the allantoic blood circulation in culture although the allantoic placenta does not grow. There is no maternal circulation flowing through the placenta, and it must be almost or completely useless as an organ of nutrition, excretion or respiratory exchange. Of interest, therefore, is the fact that such embryos grow much further than those explanted just before the 20-somite stage, which have no allantoic blood circulation and in culture develop an abnormal vessel bypassing the allantois (Fig. 2.9). This suggests that the tissues of the allantois itself may at this stage be producing hormonal substances required by the fetus and conveyed by the allantoic circulation. Such an interpretation is at present only speculative but would be worth testing experimentally.

The yolk-sac would appear to be the main nutritive organ of the rodent embryo during much of the period of organogenesis. Several studies in recent years (e.g. Padykula, Deren & Wilson, 1966; Beck, Lloyd & Griffiths, 1967) have emphasized the capacity of the visceral endoderm layer of the yolk-sac to absorb macromolecules, particularly protein, and after digestion these are probably transported, together with other nutrient substances, to the fetus via the yolk-sac blood circulation. But not all proteins are digested. Brambell *et al.* (1954) discovered that antibodies are transferred selectively across the yolk-sac

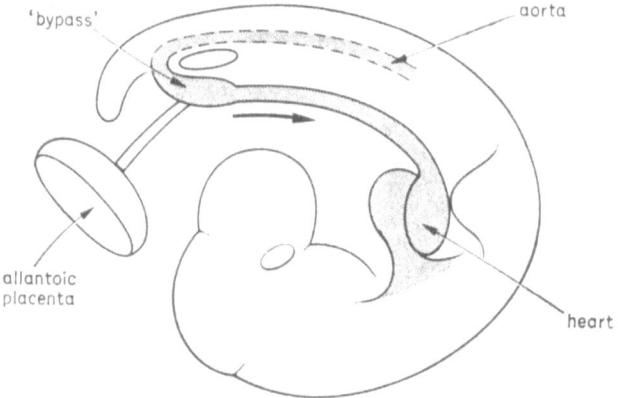

Fig. 2.9. Development of an abnormal blood vessel bypassing the blood flow to the allantois in rat fetuses explanted before the 20-somite stage

in the rabbit, and later the same was shown to be true for the rat (Brambell & Halliday, 1956). Wild (1970) has confirmed by fluorescent labelling methods that the endoderm cells of the visceral yolk-sac are the main site of this protein selection, and the electron-microscope studies of Slade (1970), Wild, Stauber and Slade (1972) and others, have revealed some of the intracellular mechanisms involved.

In view of these important functions of the rodent visceral yolk-sac, any agents harmful to this membrane can be expected to affect fetal development. New & Brent (1972) have recently demonstrated that yolk-sac antibody (sheep anti-rat yolk-sac gamma-globulin) retards fetal growth when added to the medium of rat fetuses in culture. The antibody was known to be teratogenic *in vivo* (Brent, Johnson & Jensen, 1971), but the culture experiments showed that it had a direct effect on the fetus or fetal membranes, independently of any possible reactions of the mother. Added to the culture medium in concentrations of 0·1 mg/ml or more it caused gross retardation of growth of both fetus and membranes (Fig. 2.10). That the action was antigenic in nature was shown by the absence of any effect when control gamma-globulin was added to the medium.

A particular advantage of growing these small fetuses in culture is that

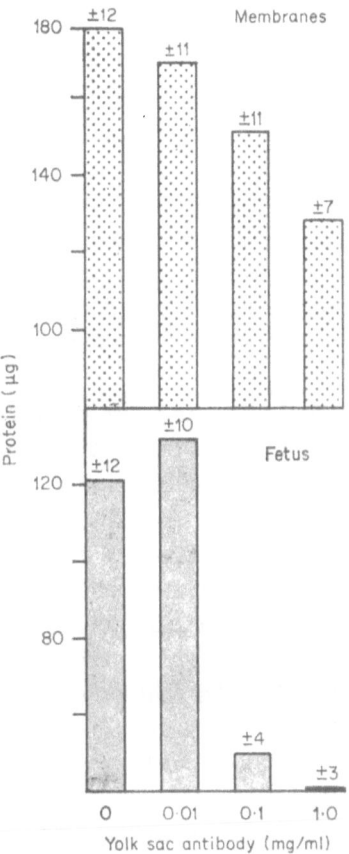

Fig. 2.10. Increase of protein of fetus (lower histogram) and fetal membranes (upper histogram) after growth in culture in different concentrations of yolk-sac antibody. Figures above each rectangle are standard errors. The fetuses were explanted at 10½ d gestation and initial protein content was about 40 µg in the fetus and 80 µg in the membranes

injections can be made into precisely determined regions of the explant (Plate 2.8). New & Brent were able to compare the effect of antibody added to the culture medium with antibody injected into the amniotic cavity or extra-embryonic coelom. In contrast to the externally applied antibody, the injections had little or no effect, suggesting that probably only the visceral endoderm (in direct contact with the culture serum) is sensitive to the antibody. Injection methods of this kind were also used by Turbow (1966) in studying the teratogenic action of trypan blue on rat fetuses in culture. In some of the cultures the trypan blue was merely added to the culture medium; in others it was injected inside the yolk-sac. At low concentrations of the dye, only the injected explants developed malformed embryos, and from this result Turbow concluded that trypan blue has a direct teratogenic effect on the rat fetus, from which it can be protected by the yolk-sac. However, trypan blue and other azo dyes are concentrated in the yolk-sac epithelium, and Beck & Lloyd (1966) have argued that the teratogenic effect *in vivo* could result from inhibition of the lysosomal enzymes of the visceral yolk-sac at a time when these are needed for the breakdown of macromolecules essential for embryonic nutrition. (See Shepard *et al.* (1971) for further discussion on the use of *in-vitro* methods for analysing teratogenic events.)

Transport of an antibody by the yolk-sac of the rat fetus in culture was shown by some experiments of Berry (1971). He explanted fetuses at $10\frac{1}{2}$ and $11\frac{1}{2}$ d gestation, and cultured them in antiserum against the contractile proteins of the rat heart. The transport and binding of the antibody in the fetal heart was traced by electron microscopy and ferritin labelling. The treated fetuses showed much poorer survival in culture than controls, and frequently developed malformations. Berry considered that this resulted from interference with fetal nutrition, caused by the reduced blood flow through the yolk-sac which followed damage to the developing myocardium.

Payne & Deuchar (1972), in an *in-vitro* study of the functions of embryonic membranes in the rat, found that after removal of the yolk-sac the amnion collapsed and the amniotic cavity disappeared. They have proposed that in addition to its nutritive and immunological functions, the yolk-sac may be important in regulating the volume of extra-embryonic fluids.

The Reichert membrane is usually torn open or removed when fetuses are explanted, because in culture it prevents expansion of the yolk-sac. This membrane is peculiar to rodents and insectivores, and its function is obscure, but Payne & Deuchar (1972) make the interesting suggestion that it may supplement the nutritive function of the yolk-sac. They obtained enhanced protein synthesis by $10\frac{1}{2}$-d rat fetuses explanted with the Reichert membrane intact. But protein synthesis by their control fetuses (Table 2.3) was unusually low, and the results

48

could usefully be confirmed by further experiments. If the Reichert membrane does increase fetal protein synthesis, an important point to clarify would be whether the membrane (with its attached trophoblast and parietal endoderm layers) actively transports nutrients itself or exerts its effect in some other way, e.g. by protection of the yolk-sac.

Growth in different culture media

Mouse and rat embryos of early somite stages develop well on the surface of plasma clots (New & Stein, 1964). But liquid culture media are more convenient and, for the older fetuses, provide the mechanical support needed for the enlarged yolk-sac. Most *in-vitro* studies of rat fetuses have depended on homologous serum as culture medium, usually with the addition of antibiotics.

In a study of heterologous sera, New (1966b) found that rabbit serum was rapidly lethal to rat fetuses in culture; it became less harmful if pre-heated to 56°C, but was still inferior to homologous serum as a culture medium. Some samples of fowl serum gave similar results, but others (possibly affected by longer storage) supported limited development of rat fetuses without pre-heating. Some commercially prepared samples of calf, horse and sheep serum, although not as rapidly lethal to rat embryos as fresh rabbit serum, were very inferior to rat serum.

Heterologous sera have commonly been found to be harmful to tissues in culture and often the harmful effects can be reduced or eliminated by pre-heating. The damage to the tissue appears to be due to an immune reaction which requires the complement present in the serum, and the heat treatment destroys the complement. Some of the results with rat fetuses in heterologous sera suggest similar antigenic mechanisms. But heterologous sera are not always inimical to fetal growth. Clarkson *et al.* (1969) obtained good development of mouse fetuses in rat serum, and Tanimura & Shepard (1970) found human serum to be as good a culture medium as rat serum for rat fetuses. It is interesting to compare these results with the studies by Chang (1949) on growth of rabbit eggs in different sera: most of the eggs developed normally in rabbit, horse, rat or guinea-pig serum; they were killed within 10 min by immersion in human, sheep, cattle, goat or fowl serum, but these lethal sera became harmless if pre-heated.

Although the rat fetus appears sensitive to many heterologous sera, it is remarkably tolerant of different kinds of homologous serum. It grows equally well in autologous serum, or serum from other pregnant females, from males, or from rats of a different strain (New, 1966b, 1967).

There is no indication from these results that any of the changes in composition of the maternal blood associated with pregnancy are critical for development of the fetus in culture.

The suggestion is sometimes made that better fetal growth might be obtained in circulator cultures if the oxygen-carrying capacity of the nutrient serum were increased by the addition of erythrocytes. But this is unlikely. The amount of oxygen carried in solution by the serum in the circulator cultures (without erythrocytes) is far in excess of that required by the fetuses. More critical would seem to be the oxygen pressure in the layer of serum flowing past the yolk-sac, and this can be controlled by varying the oxygen concentration of the gas with which the serum is equilibrated. A few trials with erythrocytes added to the serum have proved unpromising. Besides the inconvenience of working with opaque medium, avoiding haemolysis and sedimentation of the erythrocytes has proved difficult, and fetuses in such cultures have usually grown less than in control cultures.

Analysis of the nutritive requirements of fetuses would be greatly assisted if the culture serum could be replaced by a chemically defined medium. At present no such medium is available. Some of the standard tissue culture media have been tested for their capacity to support fetal growth but the results have been very poor. Rather better results have been obtained with mixtures of serum and chemically defined media. Rat fetuses can be grown in homologous serum diluted with up to 75% '199' or Waymouth's medium, or even Tyrode's saline solution. The fetuses often develop well in such mixtures, and differentiation may proceed as far as in whole serum, but protein synthesis is significantly reduced (Table 2.6).

This reduced growth might be the result of dilution of the serum proteins, which could be harmful in two ways. It could change the osmotic relationship between the culture serum and fetal blood in the yolk-sac, with possible effects on the transfer of nutrient materials, or it might reduce the rate of uptake of protein from the serum by the visceral endoderm. But at present any explanation is speculative and the few available facts do not form a clear picture. Tissue culture media made osmotically equivalent to serum by the addition of dextran have not shown improved capacity to support rat fetal growth. More confusing, studies with other species have sometimes indicated that diluted serum is preferable to whole serum. Givelber & di Paolo (1968) found whole homologous serum unsatisfactory for hamster fetuses, and obtained better results by diluting it with McCoy's medium. Daniel (1971) recommends dilution of rabbit serum for growing rabbit embryos. Clarkson *et al.* (1969) showed that mouse fetuses in Waymouth's medium with 20% rat serum developed better than in whole rat serum (but no comparison was made with growth in homologous serum).

Table 2.6. *Reduced growth of rat fetuses in culture resulting from dilution of the nutrient serum with '199' or Tyrode's solution. (The fetuses in the '199' experiments were cultured for 45 h, those in Tyrode for 24 h)*

	Explanted at 10½ d			Explanted at 11½ d		
Culture medium	Number of fetuses	Final protein (μg)	Final somites	Number of fetuses	Final protein (μg)	Final somites
100% serum	8	225±40	28·1±2·8	6	668±60	39·5±0·6
50% serum in 199	8	129±19	25·5±1·4	6	531±33	38·3±0·5
P for difference	$P=0.05$		$P=0.10$	$P=0.05$		$P=0.10$
100% serum	12	279±17	25·8±0·6	8	875±118	40·0±0·9
25% serum in 199	10	202±17	24·6±0·5	8	550±35	36·9±0·9
P for difference	$P=0.002$		$P=0.10$	$P=0.02$		$P=0.02$
100% serum				16	578±29	33·7±0·3
25% serum in Tyrode				15	455±26	33·6±0·4
P for difference				$P=0.002$		–

The available culture methods provide excellent opportunities for testing the capacity of different media to support fetal growth during the period of organogenesis. Particularly valuable would be work directed towards the synthesis of chemically defined 'fetal culture media' comparable with the available tissue culture media. Studies in this area would undoubtedly yield more certain information about the requirements of the fetus at its most critical stages of development. At present we have many questions and few answers.

Organogenesis and torsion

Until recently the lack of adequate culture methods severely impeded experimental work *in vitro* on organogenesis in mammalian fetuses. Only a few studies are to be found scattered through the literature of the last 40 years. Now that better culture methods are available we may hope that mammals will at last begin to rival frogs and chicks as profitable organisms for the analysis of developmental mechanisms.

Daniel & Olson (1966) examined the morphogenetic movements in the rabbit embryonic shield. By staining different regions of the shield with neutral red they were able to demonstrate a general pattern of cell movement, including convergence towards the primitive streak in the midline, similar to that in other amniotes. Waddington (1937) made *in-vitro* studies on rabbit embryos and examined the inductive capacity of the primitive streak. He was not able to demonstrate inductions by grafting fragments of rabbit streak to rabbit blastodisc, but obtained inductions by chick streak grafted to rabbit, or rabbit to chick. Törö (1938) transplanted pieces of neural tissue between rat embryos in culture, and reported inductions of secondary neural grooves; but the conclusions have been criticized by Waddington (1956). Smith (1964) studied mouse embryos in culture at slightly older (early somite) stages and concluded from transection experiments that there is a caudal extension during normal development of the notochord and neural tube. The results from all these studies suggest broad similarities with early chick development.

Advanced tissue differentiation can be obtained from mouse or rat embryos, either whole or in parts, isolated at primitive-streak stages and grown in the anterior chamber of the adult eye (Grobstein, 1951, 1952; Levak-Svajger & Skreb, 1965). Another favourable site is underneath the kidney capsule of the adult animal, and Levak-Svajger *et al.* (1969, 1971) have used this in studying the differentiation of rat germ layers separated by enzyme treatment. Professor Skreb has recently made similar studies on isolated rat embryonic shields grown *in vitro* (p. 21). The explants in these various studies are allowed to continue developing for two weeks or more, and the degree of tissue differentiation obtained is remarkable. In addition to neural tissue, muscle, adipose tissue, cartilage and epithelia of various types, bone, glands and hair are also sometimes found. But the tissues are usually irregularly arranged and do not develop as a normal fetus.

In-vitro methods are particularly good for studying the development of the heart and blood circulation. Goss (1938) described the beginning of contractile activity in the hearts of $9\frac{1}{2}$-d rat embryos observed in hanging drop cultures: 'The first contractions have a regular rhythm and a rate of 37–42/min. They are confined to a small area, three or four cells, in the lateral ventricular myocardium of the left primitive heart tube. The area is near a constriction which marks the junction of the primitive atrium and ventricle. The right myocardial tube begins contracting in a similar fashion 2 h after the left. Its rhythm is regular, but independent of and slower than the left. The contractile activity gradually extends over the ventricular myocardium of each side until all the portion surrounding the endocardial lumen is involved. The two lateral hearts become united into a single saccular ventricle by the progressive differentiation of the median splanchnic mesoderm. After this median myocardium becomes active, the independent activity of the right heart is suppressed and the left side becomes the pace-maker for the whole ventricle.' Earlier, Goss (1935) had shown that double hearts formed in rat fetuses if fusion of the two heart primordia were prevented. Jolly & Lieure (1938) found that if one of the heart primordia were destroyed, the other enlarged, united with the vitelline vessels of the operated side and supplied a circulation to the complete area vasculosa. Destruction of both primordia resulted in a fetus without a heart, but this did not hinder the formation of vascular rudiments in the yolk-sac, allantois and along the line of the aortae. Recently, Steele (1972) has shown that the formation of double or single hearts in culture is strongly affected by the nature of the culture serum (p. 28).

The curious 'inverted' yolk-sac of rodents such as the rat and mouse causes the curvature of the fetus, when first formed, to be concave dorsally and convex ventrally. After the body of the fetus has become folded off from the yolk-sac, it is free to twist round into the more characteristic 'fetal position' with the ventral surface concave (Fig. 2.1). This process occurs between the 10- and 20-somite stages—about $10\frac{1}{2}$– 11 d of gestation in the rat. The rotation of the fetus begins with the head end only, and the torsion then passes tailwards until the whole axis has reversed its curvature. The mechanism is little understood, but some interesting transection experiments by Deuchar (1969, 1971) have recently begun to throw some light on the problem. The effect of transection of the $9\frac{1}{2}$- to $10\frac{1}{2}$-d fetus is to inhibit rotation, particularly in the posterior part. The proportion of fetuses affected is greater after transection at cervical than at mid-trunk levels. Deuchar suggests that the cervical region has special contractile properties that initiate axial rotation; also, that the normal twisting of the cardiac tube in this region, together with the rhythmic movements of the head to the right and downward that result from the heart beat, may play a part.

After the fetus has turned, it is completely surrounded by the amnion and visceral yolk-sac and is less accessible for surgical operations. Fortunately, however, both these membranes are fairly transparent and heal rapidly from small wounds. By careful illumination the fetus can still be closely observed, and operations carried out with instruments inserted through the membranes. Provided only small holes are made in the yolk-sac and the major blood vessels are undamaged, recovery is good. In a study of nerve regeneration, A. Hughes & New (unpublished data) were able to make precisely located transections of the spinal cord of 25-somite rat fetuses in culture. The transections were made with fine glass needles. The punctured yolk-sac collapsed during the operation but rapidly healed and re-expanded afterwards, and the blood circulation was resumed; the operated fetuses continued developing almost as well as controls.

The present and the future

In this chapter reference has been made to many different studies of fetal development in culture. These studies have included various aspects of respiration and energy metabolism, of nutrition and growth, of normal and abnormal organogenesis. Such studies have led to some significant conclusions and have opened up numerous new lines of inquiry. The extent of the work provides some indication of the importance of *in-vitro* methods for studying fetal development during the period of organogenesis. In this last section we shall briefly assess the present achievements and limitations of these methods, and consider some possibilities for improving them in future.

How successful are the present culture methods?

Of the various species examined, rat and mouse fetuses have so far proved the most successful in culture during the period of organogenesis. Figure 2.11, which summarizes the performance of rat fetuses, can therefore be regarded as indicating the maximum that is at present obtainable. It shows that these fetuses can now be grown in culture at any time during the second week (i.e. middle third) of pregnancy, while all the major events of organogenesis are occurring. A valuable feature of the culture methods for experimental purposes is that most of the explanted fetuses continue developing for the greater part of the culture period, and closely resemble equivalent fetuses *in vivo*. The survival curves in Fig. 2.12 show the characteristic pattern of development in culture—continued growth and differentiation of a high proportion of the explanted fetuses up to a certain stage of development, followed by a fairly rapid decline.

Figure 2.11, besides indicating the successes, also shows the limitations of the present culture system. Although rat fetuses can be grown between the eighth and fifteenth days of gestation, the same fetus cannot be grown right through from the eighth to the fifteenth day. Fetuses explanted at Day 8 continue developing for about 3 d in culture; those

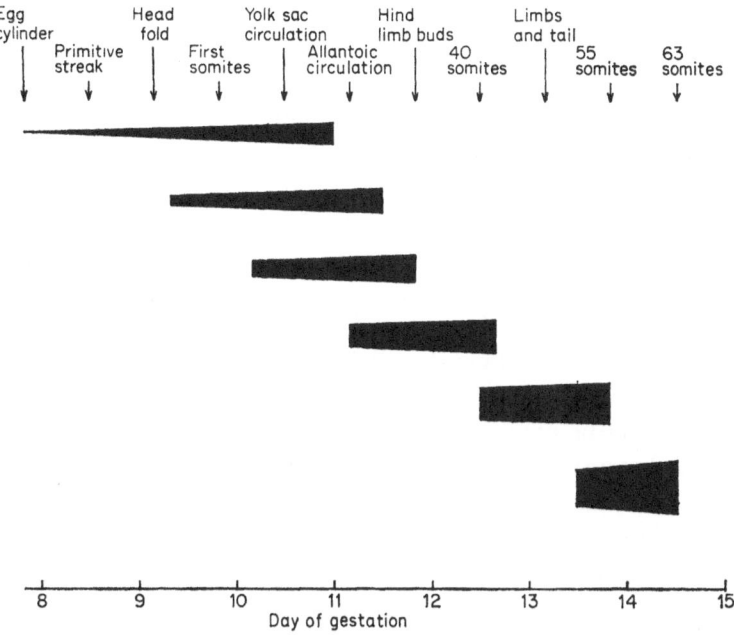

Fig. 2.11. Periods of development (represented by black areas) of rat fetuses in culture

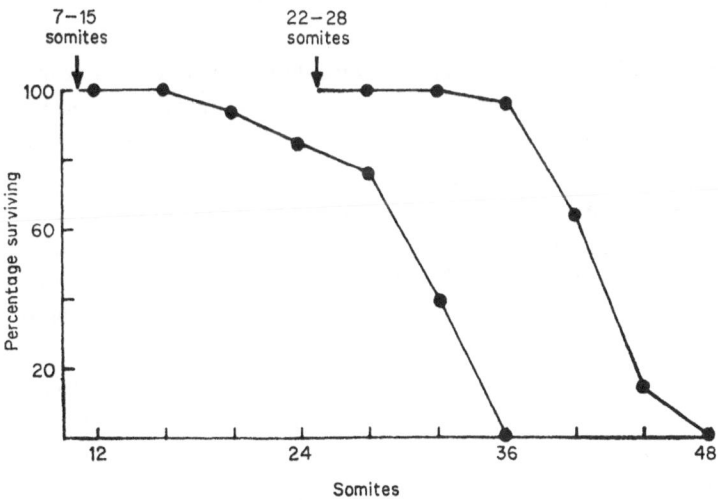

Fig. 2.12. Percentage surviving to different stages of development, as indicated by number of somites, of rat fetuses explanted at 7–15 somites (10½ d gestation) and 22–28 somites (11½ d gestation) and grown in circulating serum

explanted at later stages grow for progressively shorter periods. Earlier in this chapter various factors that may retard or restrict growth in culture have been discussed. On present evidence, the most important of these is the inadequate oxygen supply to the fetus resulting from the failure of the explanted allantoic placenta. For short periods this can be overcome by raising the oxygen concentration in the aqueous medium, so that more oxygen reaches the fetus via the yolk-sac. But this method is limited by the rapidly damaging effect of the increased oxygen on the yolk-sac itself. To obtain more extended development of rodent fetuses in culture, particularly of the older fetuses, a likely first requirement is a method either for growing the allantoic placenta or for perfusing the fetus through the placental blood vessels. Perfusion methods and their problems will be described in later chapters; they have been used in many studies on the late fetuses of larger mammals, but have not yet been applied on the miniaturized scale that would be required to maintain rat and mouse organogenesis. For these very small fetuses, the development of a culture system that would support growth of the allantoic placenta as part of a 'feto-placental' unit, although presenting formidable problems, at present seems the more promising line of attack.

Could the allantoic placenta be grown in culture?

The rodent allantoic placenta when fully developed is a highly complex structure composed of both fetal and maternal tissues and containing elaborately interlocked networks of blood vessels to promote maximum exchange of materials between the circulating fetal and maternal blood. To attempt to grow such a structure *in vitro* involves the problems of providing a substrate to replace the maternal tissue, together with the problems of a perfusion system to replace the maternal circulation. In all, a formidable task. There are, however, some indications that the developing placenta may not be too exacting in its substrate requirements. The evidence for this is the occurrence in ectopic pregnancies of advanced fetal and placental development at sites in the body very different from those in the uterus. In women the commonest site for ectopic pregnancies is the oviduct and such pregnancies are occasionally found also in animals. Eales (1933) describes one instance of a pregnant rabbit with advanced fetuses—at stages up to and beyond normal term— in the abdominal cavity. Apparently implantation had occurred in the oviduct which had then ruptured, releasing the fetuses but retaining the placentae, to which the fetuses had for a time remained connected. Abdominal pregnancies can sometimes occur in women (Baldwin, 1954), but are more uncommon under the conditions of modern gynaecology than in the past. 'The older medical literature—of the days when abdominal surgery was undertaken only as a last desperate hope—reports many well-authenticated cases of abdominal pregnancies in which the embryo went to several months gestation before causing the death of the mother by placental perforation of the viscera or by the intra-abdominal haemorrhage incident to trophoblast invasion. There are even occasional cases on record of abdominal pregnancies in which a viable fetus was removed by laparotomy' (Patten, 1946). Injection of blastocysts into extrauterine sites in experimental animals frequently results in massive invasion of the surrounding tissue by the blastocyst trophoblast, and occasionally to extensive development of the fetus. Kirby (1963) obtained development of a 12-d fetus from a mouse blastocyst injected into the testis. Nicholas (1934) separated the oviducts from the uterus in rats 2–3 d pregnant so that the cleaving eggs fell into the abdominal cavity; in 35 female rats so treated five fetuses developed to term after implanting on the mesenteries. But such advanced ectopic development is rare in rodents (Jollie, 1961; McLaren & Tarkowski, 1963).

In culture, the tendency of the trophoblast to spread over surfaces (Cole, Edwards & Paul, 1966; New & Daniel, 1969; Hsu, 1971) or into a plasma clot (New & Stein, 1964) is well known. When rabbit blastocysts are explanted with endometrium, the trophoblast attaches to and invades the endometrium; this invasion and the subsequent differentiation of the trophoblast and uterine tissue show several features resembling the earliest stages of placental formation *in vivo*, as demonstrated in the remarkable experiments of Glenister (1961, 1963, 1967). This is still a long way from a functional allantoic placenta in culture, but, together with the evidence from ectopic pregnancies, suggests that the development of such a placenta is not an impossibility.

Possibilities of the marsupial fetus

The rodent fetus is unusual among eutherian mammals in that, during early organogenesis, respiration and nutrition are mediated largely through a yolk-sac placenta; only later in development does the allantoic placenta become important. The yolk-sac placenta is a relatively simple structure, and when explanted with the early fetus it grows and develops a blood circulation in a manner very similar to that *in vivo*. Hence the rodent fetus can grow and differentiate in culture until such time as the allantoic placenta becomes indispensable.

As the yolk-sac placenta grows in culture and the allantoic placenta does not, would better results be achieved with a mammal that has no allantoic placenta and relies entirely on a yolk-sac? Many of the marsupials approach closely to this condition and provide an exciting challenge for studies on fetal development *in vitro*. The main difficulty at present is obtaining an adequate supply of live embryos; few marsupials are in any sense 'laboratory animals' and none have yet been bred in captivity with the ease and reliability of rats and mice. But one study (New & Mizell, 1972) has been made with the opossum *Didelphis marsupialis virginiana*, a marsupial common in North America.

Although the opossum breeds erratically in captivity and is much more difficult to maintain in self-perpetuating colonies than the common laboratory animals, a reliable method has been developed by Professor Mizell for obtaining embryos of predetermined age from recently captured opossums. The animals are trapped during the breeding season, at which time practically all the females either have young in the pouch, or within a few days give birth to litters sired in the wild. If the young are removed from the pouch, a post-lactational oestrus follows a week later. This oestrus seems of exceptional intensity, and when the females are caged with males a high proportion of fertile matings can usually be achieved; the females that fail to mate at this oestrus may do so at the end of the next cycle 28 d later. Because the young can be removed from the pouch at any desired time, the method is very convenient for initiating timed pregnancies as required.

The embryonic development of the opossum has been described in detail by McCrady (1938). The gestation period is $12\frac{3}{4}$ d, and the young are born at a stage of development corresponding about to that of a rat fetus of 14 d, except that the fore-limbs, lungs, pancreas and a few other organs are precociously developed. The allantois is present only as a simple sac that enlarges with urine during the last 3 d of gestation, and probably has no placental function. But the yolk-sac, which in rats is about 7 mm in diameter on the thirteenth day of gestation, develops in the opossum to over 20 mm in diameter (Fig. 2.13), with folds closely adhering to those of the endometrial surface.

The results obtained by New & Mizell with opossum fetuses in culture have been described in a previous section (p. 20 and Plate 2.1). In this preliminary study the number of fetuses used was fairly small (33) and the culture apparatus was very simple. Only a few variations in culture conditions could be tried. Nevertheless, the older fetuses

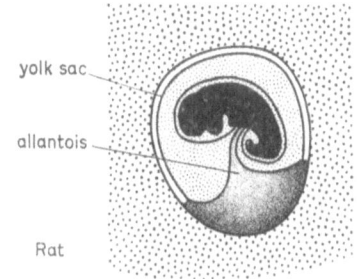

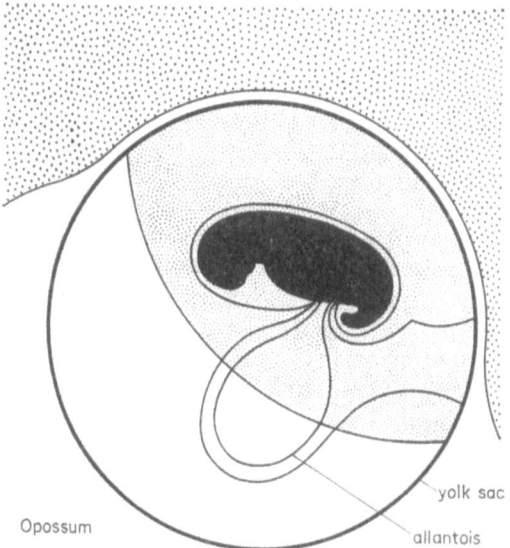

Fig. 2.13. Comparison of rat and opossum fetuses and fetal membranes. Both diagrams are of fetuses at 12 d gestation and drawn to same magnification × 3. Light stipple, area vasculosa

survived and developed well for up to 20 h, and there seems little doubt that with further work this period could be extended. Already the most advanced stage attained in culture is within 24 h of the end of normal gestation, suggesting the possibility of rearing opossums from cultured fetuses; it is known that newborn young can be placed by hand in the pouch to attach to the nipples and continue development. As a source of embryos for studying fetal development in culture, and perhaps for following the post-natal results of experiments on the fetus, the opossum has considerable promise.

References

AKSU, O., MACKLER, B., SHEPARD, T. H. & LEMIRE, R. J. (1968) Studies of the development of congenital anomalies in embryos of riboflavin-deficient, galactoflavin fed rats. II. Role of the terminal electron transport systems. *Teratology*, 1, 93–102.

BALDWIN, W. F. (1954) Abdominal pregnancy. *Obstetrics and Gynecology*, 4, 435–439.

BECK, F. & LLOYD, J. B. (1966) The teratogenic effects of azo dyes. *Advances in Teratology*, 1, 131–193.

BECK, F., LLOYD, J. B. & GRIFFITHS, A. (1967) A histochemical and bio-chemical study of some aspects of placental function in the rat, using maternal injection of horseradish peroxidase. *Journal of Anatomy*, 101, 461–478.

BERRY, C. L. (1968) Comparison of *in vivo* and *in vitro* growth of the rat foetus. *Nature*, 219, 92–93.

BERRY, C. L. (1971) The effects of an antiserum to the contractile proteins of the heart on the developing rat embryo. *Journal of Embryology and Experimental Morphology*, 25, 203–212.

BOELL, E. J. (1958) In *Physiology of Prematurity*, ed. Lanman, J. T. In chapter by Villee, C. A., pp. 9–76. New York: Josiah Macy Jr. Foundation.

BOELL, E. J. & NICHOLAS, J. S. (1939) Respiratory metabolism of mammalian eggs and embryos. *Anatomical Record (Supplement)*, 75, 66.

BRAMBELL, W. R., BRIERLY, J., HALLIDAY, R. & HEMMINGS, W. A. (1954) Transfer of passive immunity from mother to young. *Lancet*, i, 964–965.

BRAMBELL, F. W. R. & HALLIDAY, R. (1956) The route by which passive immunity is transmitted from mother to foetus in the rat. *Proceedings of the Royal Society, Series B*, 145, 170–178.

BRENT, R. L., JOHNSON, A. J. & JENSEN, M. (1971) The production of congenital malformations using tissue antisera. VII. Yolk-sac antiserum. *Teratology*, 4, 255–276.

CHANG, M. C. (1949) Effects of heterologous sera on fertilized rabbit ova. *Journal of General Physiology*, 32, 291–300.

CLARK, C. M. (1971) Carbohydrate metabolism in the isolated fetal rat heart. *American Journal of Physiology*, 220, 583–588.

CLARKSON, S. G., DOERING, J. V. & RUNNER, M. N. (1969) Growth of postimplantation mouse embryos cultured in a serum-supplemented, chemically defined medium. *Teratology*, 2, 181–186.

COCKROFT, D. L. (1973) Development in culture of rat foetuses explanted at 12½ and 13½ days of gestation. *Journal of Embryology and Experimental Morphology*. 29, 473–483.

COLE, R. J., EDWARDS, R. G. & PAUL, J. (1966) Cytodifferentiation and embryogenesis in cell colonies and tissue cultures derived from ova and blastocysts of the rabbit. *Developmental Biology*, 13, 385–407.

COX, S. J. & GUNBERG, D. L. (1972a) Metabolite utilization by isolated embryonic rat hearts *in vitro*. *Journal of Embryology and Experimental Morphology*, 28, 235–245.

Cox, S. J. & Gunberg, D. L. (1972b) Energy metabolism in isolated rat embryo hearts: effect of metabolic inhibitors. *Journal of Embryology and Experimental Morphology*, **28**, 591–599.

Daniel, J. C. (1970) Culture of rabbit embryo in circulating medium. *Nature*, **225**, 193–194.

Daniel, J. C. (1971) In *Methods in Mammalian Embryology*, ed. Daniel, J. C. Ch. 20, pp. 284–289. San Francisco: W. H. Freeman.

Daniel, J. C. Jr., & Olson, J. D. (1966) Cell movement, proliferation and death in the formation of the embryonic axis of the rabbit. *Anatomical Record*, **156**, 123–128.

De Plaen, J. L. (1970) In *Aspects dynamiques du métabolisme glucidique chez l'embryon de rat*, Part I. pp. 17–138. Brussels: Editions Arscia S.A.

Deuchar, E. M. (1969) Effects of transecting early rat embryos on axial movements and differentiation in culture. *Acta Embryologiae Experimentalis*, 1969, 157–167.

Deuchar, E. M. (1971) The mechanism of axial rotation in the rat embryo: an experimental study in vitro. *Journal of Embryology and Experimental Morphology*, **25**, 189–201.

Eales, M. B. (1933) Abdominal pregnancy in mammals with an account of multiple ectopic gestation in a rabbit. *Journal of Anatomy*, **67**, 108–117.

Folstad, L., Bennett, J. P. & Dorfman, R. I. (1969) The *in vitro* culture of rat ova. *Journal of Reproduction and Fertility*, **18**, 145–146.

Givelber, H. M. & DiPaolo, J. A. (1968) Growth of explanted eight day hamster embryos in circulating medium. *Nature*, **220**, 1131–1132.

Glenister, T. W. (1961) Organ culture as a new method for studying the implantation of mammalian blastocysts. *Proceedings of the Royal Society, Series B*, **154**, 428–431.

Glenister, T. W. (1963) In *Delayed Implantation*, ed. Enders, A. C. pp. 171–182. Chicago: University of Chicago Press.

Glenister, T. W. (1967) In *Fertility and Sterility*, ed. Westin, B. & Wiquist, N. International Congress Series No. 133, pp. 385–394. Amsterdam: Excerpta Medica Foundation.

Glenister, T. W. (1971) In *Methods in Mammalian Embryology*, ed. Daniel, J. C. Jr. Ch. 23, pp. 320–333. San Francisco: W. H. Freeman.

Goss, C. M. (1935) Double hearts produced experimentally in rat embryos. *Journal of Experimental Zoology*, **72**, 33–50.

Goss, C. M. (1938) The first contractions of the heart in rat embryos. *Anatomical Record*, **70**, 505–524.

Grobstein, C. (1950) Behaviour of the mouse embryonic shield in plasma clot culture. *Journal of Experimental Zoology*, **115**, 297–314.

Grobstein, C. (1951) Intra-ocular growth and differentiation of the mouse embryonic shield implanted directly and following *in vitro* cultivation. *Journal of Experimental Zoology*, **116**, 501–526.

Grobstein, C. (1952) Intra-ocular growth and differentiation of clusters of mouse embryonic shields cultured with and without primitive endoderm and in the presence of possible inductors. *Journal of Experimental Zoology*, **119**, 355–379.

Hsu, Yu-Chih. (1971) Post blastocyst differentiation *in vitro*. *Nature*, **231**, 100–102.

Jenkinson, E. J. & Wilson, I. B. (1970) *In vitro* support system for the study of blastocyst differentiation in the mouse. *Nature*, **228**, 776–778.

Jollie, W. P. (1961) The incidence of experimentally produced abdominal implantations in the rat. *Anatomical Record*, **141**, 159.

Jolly, J. & Lieure, C. (1938) Recherches sur la culture des œufs des mammifères. *Archives d'Anatomie microscopique*, **34**, 307–374.

Kirby, D. R. S. (1963) The development of mouse blastocysts transplanted to the scrotal and crytorchid testis. *Journal of Anatomy*, **97**, 119–130.

Kleiber, M., Cole, H. H. & Smith, A. H. (1943) Metabolic rate of rat fetuses in vitro. *Journal of Cellular and Comparative Physiology*, **22**, 167–176.

Le Goascogne, C., & Brun, J. L. (1969) Développment in vitro de l'embryon de rat. *Comptes rendus des seances de la Academie des sciences*. Paris, **268**, 3195–3198.

Levak-Svajger, B. & Skreb, N. (1965) Intraocular differentiation of rat egg cylinders. *Journal of Embryology and Experimental Morphology*, **13**, 243–253.

Levak-Svajger, B. & Svajger, A. (1971) Differentiation of endodermal tissues in homografts of primitive ectoderm from two-layered rat embryonic shields. *Experientia*, **27**, 683–684.

Levak-Svajger, B., Svajger, B. & Skreb, N. (1969) Separation of germ layers in presomite rat embryos. *Experientia*, **25**, 1311–1312.

Lowry, O. H., Rosebrough, N. J., Farr, A. L. & Randall, R. J. (1951) Protein measurement with the folin phenol reagent. *Journal of Biological Chemistry*, **193**, 265–275.

McCrady, E. Jr. (1938) *The Embryology of the Opossum*. American Anatomical Memoirs No. 16. Philadelphia: Wistar Institute.

McLaren, A. & Tarkowski, A. K. (1963) Implantation of mouse eggs in the peritoneal cavity. *Journal of Reproduction and Fertility*, **6**, 385–392.

Moscona, A., Trowell, O. A. & Willmer, E. N. (1965) In *Cells and Tissues in Culture*, ed. Willmer, E. N. Ch. 2, pp. 19–98. London & New York: Academic Press.

Netzloff, M. L., Chepenik, K. P., Johnson, E. M. & Kaplan, S. (1968a) Respiration of rat embryos in culture. *Life Sciences*, **7**, 401–405.

Netzloff, M. L., Johnson, E. M. & Kaplan, S. (1968b) Respiratory changes observed in abnormally developing rat embryos. *Teratology*, **1**, 375–386.

New, D. A. T. (1966a) *The Culture of Vertebrate Embryos*. London: Logos Press.

New, D. A. T. (1966b) Development of rat embryos cultured in blood sera. *Journal of Reproduction and Fertility*, **12**, 509–524.

New, D. A. T. (1967) Development of explanted rat embryos in circulating medium. *Journal of Embryology and Experimental Morphology*, **17**, 513–525.

New, D. A. T. (1971) Methods for the culture of post-implantation rodents. In *Methods in Mammalian Embryology*, ed. Daniel, J. C. pp. 305–319. San Francisco: W. H. Freeman.

NEW, D. A. T. & BRENT, R. L. (1972) Effect of yolk-sac antibody on rat embryos grown in culture. *Journal of Embryology and Experimental Morphology*, **27**, 543–553.

NEW, D. A. T. & COPPOLA, P. T. (1970a) Effects of different oxygen concentrations on the development of rat embryos in culture. *Journal of Reproduction and Fertility*, **21**, 109–118.

NEW, D. A. T. & COPPOLA, P. T. (1970b) Development of explanted rat fetuses in hyperbaric oxygen. *Teratology*, **3**, 153–162.

NEW, D. A. T., COPPOLA, P. T. & TERRY, S. (1973) Culture of explanted rat embryos in rotating tubes. *Journal of Reproduction and Fertility*, in press.

NEW, D. A. T. & DANIEL, J. C. (1969) Cultivation of rat embryos explanted at 7·5 to 8·5 days of gestation. *Nature*, **223**, 515–516.

NEW, D. A. T. & MIZELL, M. (1972) Opossum fetuses grown in culture. *Science*, **175**, 533–536.

NEW, D. A. T. & STEIN, K. F. (1964) Cultivation of post-implantation mouse and rat embryos on plasma clots. *Journal of Embryology and Experimental Morphology*, **12**, 101–111.

NICHOLAS, J. S. (1934) Experiments on developing rats. I. Limits of foetal regeneration; behavior of embryonic material in abnormal environments. *Anatomical Record*, **58**, 387–413.

NICHOLAS, J. S. (1938) The development of rat embryos in a circulating medium. *Anatomical Record*, **70**, 199–210.

NICHOLAS, J. S. & RUDNICK, D. (1934) The development of rat embryos in tissue culture. *Proceedings of the National Academy of Sciences*, **20**, 656–658.

NICHOLAS, J. S. & RUDNICK, D. (1938) Development of rat embryos of egg cylinder to head-fold stages in plasma cultures. *Journal of Experimental Zoology*, **78**, 205–232.

PADYKULA, H. A., DEREN, J. J. & WILSON, T. H. (1966) Development of structure and function in the mammalian yolk-sac. I. Development, morphology and vitamin B_{12} uptake of the rat yolk-sac. *Developmental Biology*, **13**, 311–348.

PATTEN, B. M. (1946) *Human Embryology*, p. 176. Philadelphia: Blakiston.

PAUL, J. (1970) *Cell and Tissue Culture*, 4th Edition. Edinburgh & London: E. & S. Livingstone.

PAYNE, G. S. & DEUCHAR, E. M. (1972) An *in vitro* study of functions of embryonic membranes in rat. *Journal of Embryology and Experimental Morphology*, **27**, 533–542.

PHILIPS, F. S. (1941) The oxygen consumption of the early chick embryo at various stages of development. *Journal of Experimental Zoology*, **86**, 257–289.

ROBKIN, M. A., SHEPARD, T. H. & TANIMURA, T. (1972) A new *in vitro* culture technique for rat embryos. *Teratology*, **5**, 367–376.

ROMANOFF, A. L. (1941) The study of the respiratory behaviour of individual chicken embryos. *Journal of Cellular and Comparative Physiology*, **18**, 199–214.

SHEPARD, T. H., TANIMURA, T. & ROBKIN, M. (1969) *In vitro* study of rat embryos. I. Effects of decreased oxygen on embryonic heart rate. *Teratology*, **2**, 107–110.

SHEPARD, T. H., TANIMURA, T. & ROBKIN, M. A. (1970) Energy metabolism in early mammalian embryos. *Developmental Biology Supplement*, **4**, 42–58.

SHEPARD, T. H., TANIMURA, T. & ROBKIN, M. A. (1971) In *Malformations Congénitales des Mammifères*, ed. Tuchmann-Duplessis, H. pp. 51–66. Paris: Masson.

SLADE, B. S. (1970) An attempt to visualize protein transmission across the rabbit visceral yolk-sac. *Journal of Anatomy*, **107**, 531–545.

SMITH, L. J. (1964) The effects of transection and extirpation on axis formation and elongation in the young mouse embryo. *Journal of Embryology and Experimental Morphology*, **12**, 787–803.

STEELE, C. E. (1972) Improved development of rat 'egg-cylinders' *in vitro* as a result of fusion of the heart primordia. *Nature New Biology*, **237**, 150–151.

TAMARIN, A. & JONES, K. W. (1968) A circulating medium system permitting manipulation during culture of postimplantation embryos. *Acta Embryologiae et Morphologiae Experimentalis*, **10**, 288–301.

TANIMURA, T. & SHEPARD, T. H. (1970) Glucose metabolism by rat embryos *in vitro*. *Proceedings of the Society for Experimental Biology and Medicine*, **135**, 51–53.

TÖRÖ, EMERIC. (1938) The homeogenetic induction of neural folds in rat embryos. *Journal of Experimental Zoology*, **79**, 213–236.

TROWELL, O. A. (1961) Problems in the maintenance of mature organs *in vitro*. *Colloques Internationaux du Centre National de la Recherche Scientifique*, **101**, 237–254.

TURBOW, M. M. (1966) Trypan blue induced teratogenesis of rat embryos cultivated *in vitro*. *Journal of Embryology and Experimental Morphology*, **15**, 387–395.

WADDINGTON, C. H. (1937) Experiments on determination in the rabbit embryo. *Archives de Biologie*, **48**, 273.

WADDINGTON, C. H. (1956) *Principles of Embryology*, p. 240. London: Allen & Unwin.

WADDINGTON, C. H. & WATERMAN, A. J. (1933) The development *in vitro* of young rabbit embryos. *Journal of Anatomy*, **67**, 355–370.

WARBURG, O., POSENER, K. & NEGELEIN, E. (1924) Uber den Stoffwechsel der Carcinomzelle. *Biochemische Zeitschrift*, **152**, 309–344.

WHITTINGHAM, D. G. (1971) Culture of mouse ova. *Journal of Reproduction and Fertility*, Supplement 14, 7–21.

WILD, A. E. (1970) Protein transmission across the rabbit foetal membranes. *Journal of Embryology and Experimental Morphology*, **24**, 313–330.

WILD, A. E., STAUBER, V. V. & SLADE, B. S. (1972) Simultaneous localisation of human y-Globulin I[125] and ferritin during transport across the rabbit yolk-sac splanchnopleur. *Zeitschrift Zellforschung*, **123**, 168–177.

WILDENTHAL, K. (1971) Substrate requirements of foetal mouse hearts maintained for long periods in organ culture. *Journal of Physiology*, **217**, 56–57P.

WILLMER, E. N. (ed.) (1965) *Cells and Tissues in Culture*. London & New York: Academic Press.

WISLOCKI, G. B., DEANE, H. W. & DEMPSEY, E. W. (1946) The histochemistry of the rodent's placenta. *American Journal of Anatomy*, **78**, 281–345.

WITSCHI, E. (1962) In *Growth. VII. Pre-natal Vertebrate Development*, ed. Altman, P. L. & Dittmer, D. S. Biological Handbooks of the Federation of American Societies for Experimental Biology, Washington, D.C.

3 Adaptations of Marsupial Pouch Young for Extra-uterine Existence

G. B. SHARMAN

The primary reproductive adaptation for terrestrial existence was the development of the land (cleidoic) egg. The evolution of this closed egg with its stored nutrients, fluid-filled chamber (amnion) and associated embryonic membranes (yolk-sac, allantois and chorion), all of which have been retained and adapted for new functions in viviparous amniotes, was as important in the colonization of the land as was the evolution of legs in place of fins.

Mammals that lay eggs (monotremes), mammals that give birth to living but incompletely developed young (marsupials), and mammals that give birth to comparatively well-developed young (eutherians) are all descended from ancestors that once reproduced by means of the closed egg. In eutherians the embryonic development that formerly took place in a nutrient-filled egg now takes place in the uterus where fetal nutrition, respiration and excretion are mediated via an extensive and complex placenta. Marsupial embryos are nourished via a placental connection in the early stages of development, but much of fetal growth occurs in the marsupium, or pouch, where milk-feeding provides for the nourishment of very small young while they undergo the development and differentiation that, in birds and oviparous reptiles, occurs in the egg and, in eutherian mammals, in the uterus. The adoption of milk feeding was probably thus of paramount importance in the evolution of marsupial viviparity, and milk production by the mother may have evolved, as in monotremes, while an ancestral oviparous method of reproduction prevailed (Sharman, 1965).

There have been many separate evolutions of viviparity in both vertebrate and invertebrate animals and the unique features of marsupial reproductive physiology may indicate that viviparity evolved separately in marsupial and eutherian groups after their derivation from a common oviparous ancestor (Sharman, 1970). Viewed in this light the marsupial neonate ought not to be considered a premature eutherian young, but the young of an amniote animal that is adapted to complete development, in the absence of a placental connection or food supply stored in an egg, while kept under incubation conditions and with a continuous milk supply in the mother's pouch. The adaptations that marsupial young have evolved for survival are therefore unlikely to be present in the prematurely born eutherian young, but the marsupial pouch young may exhibit some of the adjustments that have to be made by, or for, the prematurely born eutherian if it is to survive.

There are four distinct phases during the development of marsupial young:

(1) A period of intrauterine development varying, in different species, from 12 to about 38 d.
(2) A period of free existence after birth, of several minutes duration only, during which the newborn young reaches the pouch by independent effort.
(3) An extended period of development in the pouch varying, in different species, from about 2 to 11 months.
(4) A period of independence of the pouch during which milk feeding continues. This is equivalent to the suckling period of eutherians such as the horse and ruminants, which are born in an advanced state of development.

Phase 1 of marsupial development is essentially similar to that of eutherians, and roughly corresponds to that period of eutherian embryogenesis during which a functional yolk-sac placenta is present. Phase 2, the first period of free existence, has no counterpart in eutherians which complete normal embryonic development *in utero*. It is of relevance to this review only for a description of the newborn, and of the structures that enable it to reach the pouch and attach to a teat. Phase 3 is the period during which continuous or almost continuous milk feeding, in the relatively stable environment of the pouch, is essential for life. An attempt is made to describe the pouch environment, the composition of the milk and its changes during pouch suckling, and the changes that occur in the pouch young itself during phase 3.

This chapter is not concerned with the intrauterine marsupial embryo or fetus, on which there is a considerable literature. Recent work (e.g. New & Mizell, 1972, and see Chapter 2) has also been directed to the maintenance and growth of intrauterine stages of marsupials in culture.

Development of melanocytes, hair follicles and epidermis in pouch

68

young of the brush possum from birth was described in papers by Lyne (1970a, b) and by Lyne, Henrikson & Hollis (1970).

The neonatal marsupial

Adaptations for reaching the pouch and teat

Observations have been made on the newborn young of a number of marsupial species. A representative series of these is listed in Table 3.1 together with their gestation periods and birth weights.

According to Block (1964) the newborn American opossum closely resembles a 10-d rat or mouse embryo, a 12-d guinea-pig embryo or an 8-week human embryo. Comparisons such as these well illustrate the

Table 3.1. *Gestation periods and birth weights of marsupials mentioned in the text**

Species	Uterine gestation (copulation to birth) (d)	Weight at birth (mg)
Didelphis marsupialis virginiana (Virginian opossum)	13	161
Dasyurus viverrinus (native cat)	8–20	13
Antechinus flavipes (marsupial mouse)	32	16
Perameles nasuta (long-nosed bandicoot)	12	200
Trichosurus vulpecula (brush possum)	17·5	200
Setonix brachyurus (quokka)	27	343
Macropus eugenii (tammar)	29	ca. 400
Macropus robustus (wallaroo and euro)	32	?
Macropus giganteus (eastern grey kangaroo)	37	828
Megaleia rufa (red kangaroo)	33	817

* References are given by Sharman (1965) except for the gestation periods of *Macropus robustus* and *M. giganteus* which are cited from Calaby & Poole (1971).

small size (Table 3.1) and general lack of development of the newborn marsupial, but fail to emphasize the numerous specialized features, of an adaptive nature, which occur and which enable the neonate to reach the pouch where it completes embryonic and fetal development.

At birth the mouth opening of the newborn native cat is triangular, bounded laterally by two curved thickenings, which converge dorsally to form the tip of the snout and below by a slightly arc-shaped transverse thickening (Fig. 3.1). This is the oral shield which Hill & Hill (1955)

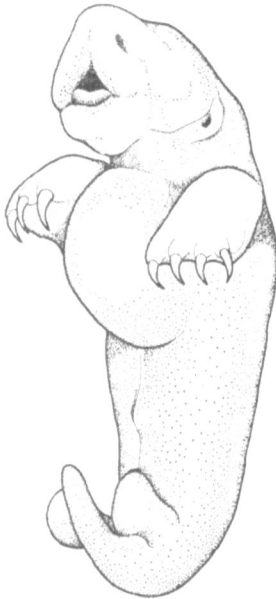

Fig. 3.1. Newborn pouch young of the native cat *Dasyurus viverrinus* ($\times$ 11·8). The mouth opening is surrounded by the oral shield, the structure between the forelimbs is the cervical swelling and the forelimb digits have deciduous claws. (Reproduced from Hill & Hill, 1955)

suggested acted as a 'more or less air-tight washer', when the mouth was applied to the teat by coming into contact with the surrounding skin. A similar structure is found in the newborn Virginian opossum, but is apparently absent in neonatal bandicoots, brush possums, koalas and kangaroos which are more advanced at birth.

The remarkable bladder-like cervical swelling (Fig. 3.1) which lies on the thoracic wall between the forelimbs has been described only for the neonatal native cat. Its suggested function is to act as a supporting cushion which keeps the head at right angles to the trunk, thus facilitating application of the mouth to the teat (Hill & Hill, 1955).

Plate 3.1. Newborn young of the red kangaroo *Megaleia rufa* (×5) attached to a teat in the pouch. The nostrils are open and large but the eyes are closed and there is no external ear opening. The young is gripping the hairs surrounding the teats with the clawed forelimb digits. (Reproduced from Sharman & Calaby, 1964)

Plate 3.2. Newborn young of the tammar (*Macropus eugenii*) and young aged 4 and 8 d (×2·5). (Photograph taken by Mr. R. J. Oldfield from specimens supplied by Miss Lynn Day)

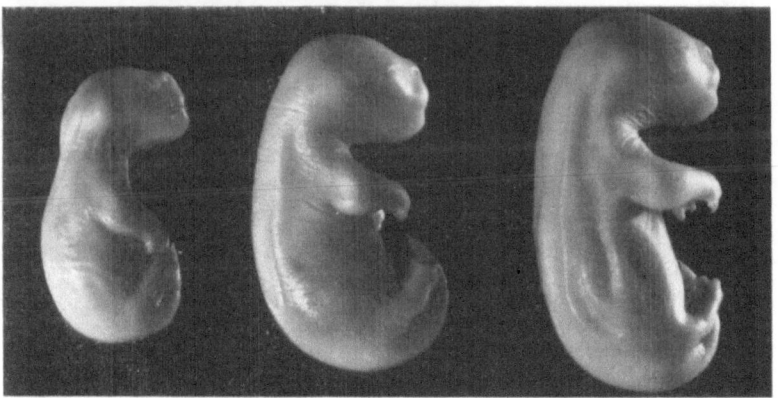

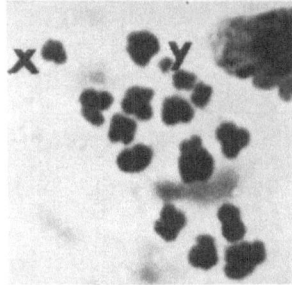

Plate 3.3. The diploid chromosome complement (2n =16) from a liver cell of a 5-d-old, sexually undifferentiated, pouch young of the tammar (×1500). There are 16 chromosomes including one X sex chromosome (X) and the Y sex chromosome (Y). Female pouch young have two X chromosomes and no Y. (Photograph supplied by Miss Lynn Day)

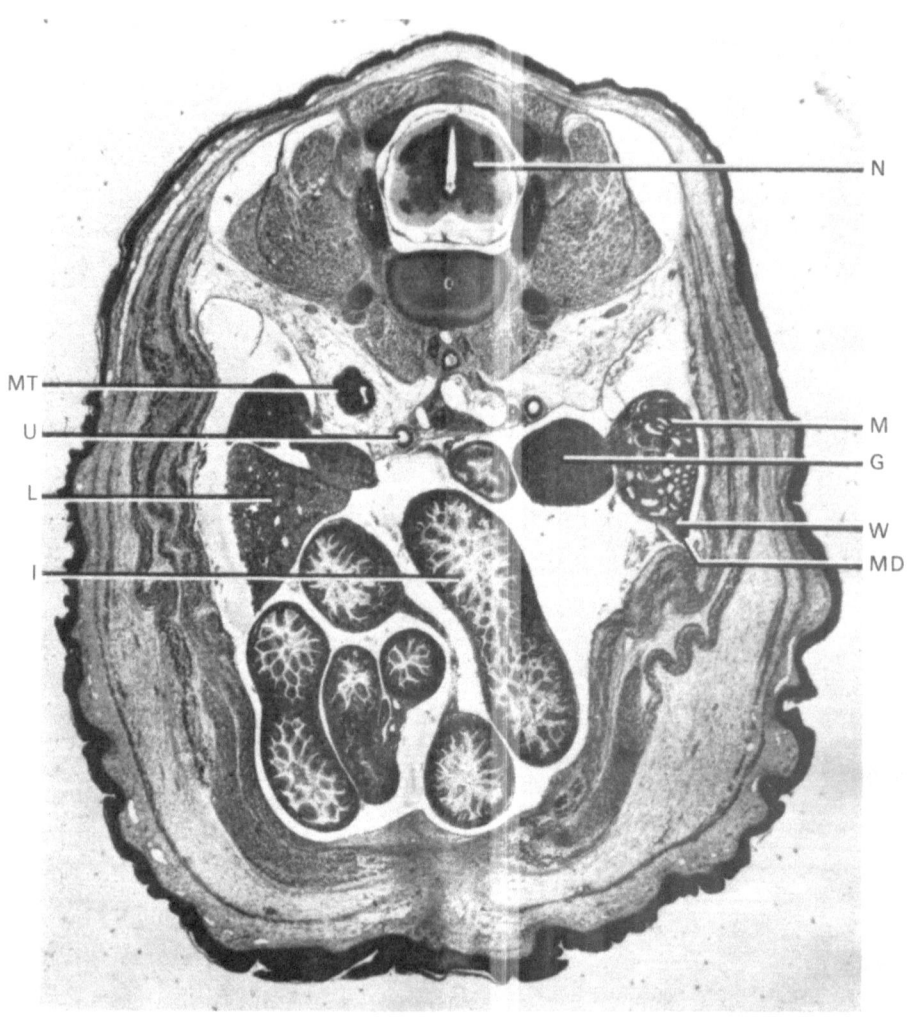

Plate 3.4. Transverse section of a male pouch young of the tammar aged 7 d
(×20). G =gonadal ridge; I =intestine; L =liver; M =mesonephros; MD =
Müllerian duct; MT =metanephros; N =nerve cord; U =ureter; W =Wolffian
duct. (Photograph supplied by Miss Lynn Day)

All marsupials are born with claws on their forelimbs which are used to grasp the mother's fur as the young climbs to the pouch, and to grip the fur around the teats after attachment (Plate 3.1). Claws are absent from the hindlimb at birth and for a variable period thereafter, according to species, until they develop during pouch life. The claws present at birth in American opossum, native cat (Hill & Hill, 1955) and bandicoots (Lyne, 1952) are deciduous, being shed and replaced by the permanent claws some time after the young reaches the pouch. The claws present on the forelimb digits at birth in the brush possum (Dunnet, 1956) and in at least several kangaroo species (Sharman, 1957; Sharman & Pilton, 1964) are precociously developed permanent claws not shed during pouch life.

At birth the forelimbs have a well-developed muscular system (Plate 3.2) and cartilaginous skeleton. The claws are controlled by functional flexor and extensor muscles. Shoulder-girdle elements including scapula and coracoid are recognizable, and the ribs are well formed and becoming cartilaginous in the native cat. By contrast the future hindlimb skeleton is represented only by tracts of condensed mesenchyme, and no muscle primordia are recognizable in the hindlimb region (Hill & Hill, 1955).

An interesting property of the undeveloped hindlimbs was investigated by Mizell (1968) and Mizell & Isaacs (1970) in the Virginian opossum, and this relates to the capacity for regeneration. An earlier author (McCrady, 1938) had found that amputation of neonatal limbs did not lead to regeneration. Mizell and his colleague confirmed this finding, but, knowing that regeneration could be induced in adult frogs and in lizards by augmenting the nerve supply (Singer, 1951, 1961), applied this procedure to newborn opossums with some success. When hindlimbs were amputated after implantation of nervous tissue (developing forebrain), partial regeneration was obtained. No regeneration occurred following forelimb amputation. In newborn mice, regeneration was not induced even after implantation of forebrain.

The buccal cavity and pharynx of the newborn native cat and other marsupials show structural adaptations that permit attachment to the teat, and allow breathing and feeding to proceed simultaneously. The tongue is large with its intrinsic and chief extrinsic muscles fully established, with large hypoglossal nerves innervating them. The surface of the tongue is convex in the native cat and grooved in the brush possum and kangaroos. When the teat is drawn into the buccal cavity the tongue assumes a semi-tubular form, and the ridged hard palate

71

becomes indented so that a tubular structure of pouch-young tissues surrounds the teat. A bulbous swelling on the end of the teat, within the buccal cavity, is formed during the first few days of pouch life, so that the tube surrounding the teat terminates in an almost spherical body (Merchant & Sharman, 1966).

Hill & Hill (1955) traced the history of the belief that, in pouch young of marsupials, the epiglottis is intranarial. However, in the newborn and later pouch young of the native cat, the epiglottis is not intranarial but intrapharyngeal, as it is also in the bandicoot *Isoodon obesulus*. In the brush possum the epiglottis is intranasopharyngeal. Merchant & Sharman (1966) showed that air and food passages do not cross in the pharynx of the pouch young of the red kangaroo. The epiglottis just pierces the soft palate so that the glottis opens into the nasopharynx. The red kangaroo epiglottis is thus intranasopharyngeal rather than intranarial, but it achieves the same purpose as would an intranarial epiglottis. Milk apparently passes from the back of the tongue in a divided stream around the upper portion of the trachea at the same time as air passes through the trachea.

The optic and auditory organs of the newborn marsupial are in an embryonic condition, but in all newborn marsupials the nostrils are open and large (Fig. 3.1, Plate 3.1), and the olfactory organs appear capable of functioning in newborn marsupials of at least two species. Olfactory sense cells are present in the nasal epithelium and nerve fibrillae enter the olfactory bulbs in the newborn native cat (Hill & Hill, 1955). McCrady (1938) inferred that the olfactory organs of the neonatal Virginian opossum could be functional because of the presence of a ciliated olfactory epithelium and of nerve fibres passing from epithelium to olfactory lobe. There is no evidence that the female marsupial assists the passage of the young from urogenital opening to teat other than by adopting a position that does not hinder the progress of the young by its own efforts (Sharman & Calaby, 1964). The presence of large open nostrils and of an apparent sense of smell suggest that newborn marsupials may be guided to the pouch by olfactory stimuli.

The newborn young of the brush possum is semi-transparent, pinkish in colour and with a conspicuous network of blood vessels visible in the skin (Lyne, Pilton & Sharman, 1959). The lungs are partly developed at birth (see below) and respiratory movements have been observed in the newborn marsupial, but cutaneous respiration may supplement normal respiration during the climb to the pouch.

Internal anatomy

The lungs of the newborn native cat are thin-walled sacs lined by an epithelial layer with capillaries beneath. They are devoid of bronchioles, alveolar ducts and alveoli—the normal constituents of the adult lung of the native cat and other mammals. Lung development is, however, precocious in marsupials, as the pharynx develops pharyngeal pouches and lung rudiments at an earlier stage than in other amniotes (Renfree, 1972). Renfree's illustrations show better developed lungs in the near-term tammar wallaby embryo than are found in the newborn native cat (Hill & Hill, 1955) or American opossum (McCrady, 1938).

The alimentary tract of the newborn marsupial is complete but relatively undifferentiated. In newborn American opossums (Heuser, 1921) and native cats (Hill & Hill, 1955) the stomach consists of fundic and pyloric regions lined by a low columnar-to-cuboidal epithelium, with a wall of undifferentiated mesenchyme. It has no glands. The duodenum is a relatively large thin-walled structure with its lining produced into prominent villous folds. This is apparently the absorptive region of the gut since the epithelium increases in thickness after the beginning of milk feeding and becomes functional in the absorption of mammary secretion. The large intestine is of relatively small diameter compared to the small intestine.

Hill & Hill (1955) describe the nervous system of the newborn native cat as being 'at the lowest possible grade of structural differentiation compatible with a considerable degree of reflex activity'. The forebrain is largely in an embryonic condition apart from the olfactory bulbs which, in keeping with the development of the nostrils and olfactory epithelium, are much in advance of the remainder. The floor of the mid-brain exhibits some degree of differentiation, but differentiation reaches its highest level in medulla oblongata, cerebellum and cervical and thoracic regions of the spinal cord—areas concerned with co-ordination of the crawling movements exhibited by the newborn young as it makes its way to the pouch.

The sex of the newborn marsupial cannot readily be determined except by examination of its chromosomes (Plate 3.3). The gonadal ridge (Plate 3.4) remains undifferentiated in the tammar until it is about 7 days old at which time primary sex cords appear in the male gonad (Lynn Day, personal communication, 1972).

Further observations on the anatomy of the newborn marsupial are to be found in succeeding sections of this chapter.

73

The pouch environment

Thermal and gaseous components

In the Virginian opossum the pouch temperature increases slightly during pregnancy from about 34·5 to 35·5°C. There is a slight drop after birth of the young but during development in the pouch the temperature is maintained fairly constantly at about 34·8°C with a variation of about 1·7°C—the smallest variation during any stage of the reproductive cycle (Reynolds, 1952). The pouch temperature is maintained, during the period while the young are incapable of regulating their own body temperature, at a point very near the thermal neutrality of the young when they do begin to control their body temperature.

In the quokka (*Setonix brachyurus*) pouch temperatures varied from 36·5 to 37·8°C (Bartholomew, 1956). Shield (1966) in an extensive series of measurements on quokkas with pouch young aged 34–171 d found the pouch temperature to vary from 36·0 to 38·3°C (average 36·9°C). The average rectal temperature of the mothers was less than one degree higher (37·49°C) while the rectal temperatures of the pouch young varied from 35·4°C at 59 d to 36·8°C at 181 d (average 36·47°C). In this species a thermal neutral zone is established in the range 32–36°C the age of about 150 d, and Shield concluded that 'incubation in the pouch keeps the body temperature of quokka joeys (i.e. pouch young) at about 36·5°C (1°C lower than the deep rectal temperature of the mother) and, like placental mammals before birth, the joey's thermal stability is not threatened'.

The carbon dioxide content of the pouch of the Virginian opossum did not exceed 2% when the young were 26 d old. Higher carbon dioxide contents were found in the pouches of females with later pouch young but never exceeding 6%, even in females with young 78 d old just before they left the pouch to ride the mother's back (Reynolds, 1952). In the brush possum (*Trichosurus vulpecula*), which bears but a single young, the carbon dioxide content of the pouch rises from 2% during the first 40 d of pouch suckling to a maximum of about 5·3% at 70 d (Bailey & Dunnet, 1960). These authors experimentally increased the oxygen content of the pouch but found no deleterious effects of higher oxygen concentration on the pouch young.

A commonplace laboratory observation is that naked marsupial young, taken from the pouch, survive better if placed in a humid environment. Reynolds (1952) placed pouch young of the Virginian opossum, aged 6 and 16 d, in a stream of dry air maintained at pouch temperatures. Movement of the young did not cease for 7 or 8 h, by which time they had lost nearly 30% of their initial body weights. In view of the relatively slow rate of dehydration of the young and the amount of dehydration they could stand, Reynolds concluded that, under normal circumstances, dehydration is not a great danger to the marsupial pouch young because suckling is continuous and water in the milk more than counterbalances that lost under normal circumstances.

Marsupial milk

True milk is not found in the alveoli and ducts of the mammary glands of recently post-parturient marsupials. Numerous observers, including O'Donoghue (1911) and Sharman (1962), have remarked that only a clear serous fluid may be expressed from the teats at the time of parturition. The developmental changes that occur in the mammary glands of marsupials during pregnancy are exactly paralleled by changes during the oestrous cycle following ovulations that are unaccompanied by mating and fertilization; the non-mated, parous or virgin female marsupial will suckle young placed in the pouch at a time after ovulation corresponding to that at which mated females give birth. Such foster-young undergo normal growth and development, and normal lactational development of the suckled teat and mammary gland occurs (Sharman, 1962; Merchant & Sharman, 1966).

Owing to the small amount of milk available from the female marsupial suckling early young, most investigators have confined their quantitative analyses of marsupial milk to late stages of lactation. However, Griffiths, McIntosh & Leckie (1972) made measurements of crude lipid content, and of the component fatty acids of the triglycerides of the lipid, in milk taken by red kangaroo pouch young aged from one to 360 d. Crude lipid varied from 0·9 g/100 g milk at 1–4 d to 11·9 g at 226 d. In general the crude lipid content increased about ninefold during 360 days of suckling. Bulked samples of 80- to 120-d, 170-d and 210- to 280-d milks showed that phospholipids, cholesterol and free fatty acids were present in detectable amounts, but that the triglyceride fraction was by far the major component of the milk lipids. Up to and including the seventy-second day of lactation, the milk triglyceride contained an average of 44·2% palmitic acid and 24·1% oleic acid. From 100 to 360 d the triglyceride contained 24·8% palmitic acid and 48·1% oleic acid. Stearic acid also increased in concentration throughout lactation, from values of 3·3% for early milk to 9·0% for mature milk.

Lemon & Barker (1967) showed that the fat content of red kangaroo milk increased from about 1·5% at 120 d of lactation to as much as 14% at 290 d of lactation, while protein content increased from about 4% at 90 d to one-and-a-half times or double this amount at 200 to 300 d of lactation. At the same time the reducing sugars showed only slight random fluctuations around 2%. In general the compositions of other marsupial milks appear to be similar to that of the red kangaroo, and the only typical marsupial feature seems to be the very low fat content of the milk taken by the newborn young. Vitamin assays show no clear lactational trends during pouch suckling (Ford & Thompson, 1965).

There are, however, qualitative differences between the milks of marsupials and eutherians, which may reflect phylogenetic differences or adaptations of marsupials for the feeding of 'prematurely born' off-spring. Gross & Bolliger (1958) pointed out that the milks of the brush possum and wallaroo (*Macropus robustus*) contained carbohydrates other than lactose. Jenness, Regehr & Sloan (1964) showed that lactose, the principal milk sugar of placental mammals, is not the predominant milk sugar of the Virginian opossum, quokka or red kangaroo. Varying quantities of glucose, galactose and oligosaccharides were detected, the latter being characteristically composed of glucose and galactose. The data presented by these authors were consistent in implicating stage of lactation as a major factor influencing composition of the carbohydrate mixture. In none of the three species was there any evidence for the occurrence of pentoses, although pentoses were reported in wallaroo milk by Bolliger & Pascoe (1953).

There is apparently no transfer of iron and copper to the fetal mar-supial during gestation, and the young obtain these minerals from the milk during pouch suckling. In human and bovine milk the iron concentration is less than that of the maternal plasma, whereas in the quokka the milk assumes iron values considerably higher than those of the plasma, and the haemoglobin concentration of the blood of the pouch young rises throughout pouch life. The amount of non-haemo-globin storage iron accumulated in the liver of the six-month-old animal is greater than the corresponding fraction in maternal livers (Kaldor & Ezekiel, 1962). Similarly copper levels in quokka milk were usually higher than blood copper levels in the same animals (Barker, 1962), and copper levels in the livers of marsupial pouch young are usually about ten times higher than in the maternal livers (Beck, 1961).

Red kangaroos, and some other monotocous marsupials, have over-lapping periods of lactation (Sharman & Calaby, 1964). In such species an advanced young, equivalent to the early suckling young of animals like the horse and ruminants, is suckled outside the pouch from one of the four teats, while a very much younger offspring is suckled in the pouch from another teat. Under such circumstances the two suckled mammary glands concurrently produce milks of different compositions, each characteristic of the lactational stage of the suckled young. Lemon & Bailey (1966) showed that in the red kangaroo whey proteins, which appear at late stages of suckling, were present in milk taken by a 330-d-old young, but not in the milk taken by a 77-d-old young suckled simultaneously. The two lactating glands differ anatomically and, in addition to the specific protein difference, the larger gland produces milk with little palmitic and much oleic acid, whereas the smaller gland produces milk with a large amount of palmitic and relatively little oleic acid (Griffiths et al., 1972). This is a remarkable adaptation achieved in spite of the common endocrine environment of the two active mammary glands. Its occurrence strongly suggests that production of a special milk for the 'prematurely born' is a necessity.

In man, goat and cow, eutherian species that give birth to comparatively well-developed young, the Na:K ratio is always less than one. In the quokka it is initially 1·8 and remains fairly steady until the young is about 160 d old, at which time the Na:K ratio decreases to about 0·7. As Bentley & Shield (1962) point out, the significance of these changing ratios of sodium to potassium is open to conjecture, but may reflect some altered metabolic need such as change in the relative growth of intra- and extracellular fluid compartments.

The belief that the young marsupial in the pouch is fed with milk forced into its mouth by the compressing action of muscles in the mother's pouch area has been questioned by numerous authors (Merchant & Sharman, 1966). It appears to have been finally disproved by Enders (1966).

Physiology of pouch young

Growth

There are numerous studies on the growth of young in the pouch of various species of marsupials. These include the native cat (Hill & Hill, 1955), quokka (Shield & Woolley, 1961), euro (Sadleir, 1963), red kangaroo and other kangaroos (Kirkpatrick, 1965), a rat kangaroo (Tyndale-Biscoe, 1968) and the tammar wallaby (Murphy & Smith, 1970). However, only Lyne & Verhagen (1957) have compared growth in a marsupial (the brush possum) with that of eutherian mammals. These authors used linear equivalence (cube root of body weight) as an overall measure of size for studying differences between species. Evidence of the relatively early birth of the brush possum was provided by the postnatal part of its growth curve which was large compared to that of the eutherians—man, mouse, cow and sheep. Up to 40 d from conception the mouse is larger (i.e. it has a faster rate of growth) than the remaining species including the brush possum. Initially the brush possum is larger than man, but at about 80 d after conception intrauterine man comes to exceed the brush possum pouch young in size.

In larger marsupial species, such as the red kangaroo, the growth rate of the young in the pouch exceeds that of intrauterine man until vacation of the pouch, which occurs at approximately the same time after conception as birth occurs in man. Among the numerous members of the kangaroo family, there undoubtedly occurs a monovular species in which the intrauterine and postnatal growth rates in the pouch very nearly approximate to the intrauterine growth rate of man. A week-by-week comparison of embryogenesis in the two species would undoubtedly reveal many similarities and perhaps some group-specific differences. For example sex-chromosome anomalies resulting in intersexuality are known both in marsupials (Sharman *et al.*, 1970) and man, but whether there is a selection against sex chromosome aneuploids in the marsupial pouch, such as undoubtedly occurs during intrauterine gestation in man, remains to be determined.

Müller (1969) concluded that the first viviparous mammals had gestation periods of some 10–12 d, that similar developmental times are still evident in primitive marsupials, and that in those marsupials with longer gestation periods there is no corresponding increase in neonatal organization. The postnatal development of the Virginian opossum in the pouch is, however, extremely retarded immediately after birth, and Müller surmises that all marsupials, including the bandicoots which complete pouch development in as little as 60 d, have an extended postnatal development period.

Developmental physiology

Digestion

In the adult red kangaroo the stomach is elongated and coiled spirally on itself, the inside of the helix being modified to form the spiral groove, which is lined by stratified cornified epithelium. The spiral groove is found only in kangaroos, which have ruminant-like digestive processes, and not in other marsupials. In the newborn red kangaroo some cells are differentiated to form the anlage of the spiral groove, but otherwise the cells lining the inside of the stomach constitute a simple columnar epithelium. In red kangaroo pouch young 3 d old, the lining epithelium contains recognizable glands, and at 6 d the glandular structure is well developed. Glands of considerable depth and complexity are achieved at 35 d, but no differentiated chief or parietal cells are detectable by light microscopy until the young is 200 d old, or about 35 d before it vacates the pouch. A proteinase, which has an optimum activity at pH 3·5, is present from birth onwards, and its activity increases in a roughly linear manner with age. Griffiths & Barton (1966) concluded that the whole of the stomach of red kangaroo young, up to 200 d old, possesses peptic activity due to the secretion of acid and pepsin, by cells of one type only. This cell type exhibits characteristics of both chief and parietal cells when viewed by electron microscopy, and is apparently analogous to the bipotent cell in the stomach mucosa of frogs. The marsupial bipotent gland cell possibly occurs in the stomachs of all marsupial pouch young as an adaptation to allow milk feeding at an early stage of ontogeny.

Kidney function

Marsupials are born with a functional mesonephros (Plate 3.4) and, for a period, the mesonephros continues as an organ of excretion until the metanephros assumes its full function. In the native cat (Hill & Hill, 1955) and brush possum (Buchanan & Fraser, 1918), the demonstration of a functional mesonephros during early pouch life rests on anatomical evidence. The retention of an apparently active mesonephros for some time after birth in the Virginian opossum was noted by early investigators, and Gersh (1937) has presented experimental evidence for its functional status, which apparently lasts until a week after birth (Müller, 1969, Table V). According to Müller the newborn grey kangaroo, one of the largest of newborn marsupials and one with a relatively long gestation period, has a mesonephros filling (in diameter) more than one-half of the coelom. The allantois of the red kangaroo, although not taking part in the formation of a placenta, is large and filled with fluid at the termination of intrauterine development. At birth the allantoic sac, with a diameter about 2 cm, is delivered intact, and the allantoic fluid contains 180 mg/100 ml urea. It appears to be almost impermeable to the diffusion of urea from interior to exterior (Sharman & Calaby, 1964). Müller (1969, Table V) states that the grey kangaroo is also born with an intact allantoic sac.

For the purposes of studying kidney function in quokka pouch young, Bentley & Shield (1962) divided the young into two groups aged 59–119 d and 120–200 d. The choice of age for this division was influenced by the degree of development of the young. The eyes open and there is shivering in response to cold at 115 d, and hair starts to grow at 125 d. At the age of 100–120 d the quokka is at about the same developmental stage as the neonatal rat, and at this time the body weights of both species represent roughly the same proportion of their respective mother's weight. The brain-weight-to-body-weight ratio shows a sharp decrease in the 10-d-old rat, and a similar decrease occurred in the 120-d-old quokka (Shield, 1961). A nephrogenic zone persisted at the periphery of the quokka kidney (metanephros) until the age of about 100 d, and the ratio of kidney weight to body weight was nearly constant during this period. With the disappearance of the nephrogenic zone, kidney weight developed progressively, indicating that new nephrons are differentiated during the first 100 d of pouch life and that kidney growth thereafter consists of enlargement of these. The glomerular diameter did not start to increase in size until the young were 140 d old.

The quokka less than 120 d old has a low plasma sodium concentration, as have newborn and fetal human beings and piglets; on the other hand plasma potassium concentration is elevated in the quokka but low in newborn eutherians. Plasma urea concentrations in the quokka during the first 120 d of pouch life are similar to those of adults, and subsequently decrease. Elevated plasma urea levels are also found in newborn eutherians but the decrease comes sooner, namely about 3 d after birth. The decrease in plasma urea concentration in the quokka is probably related to a more efficient clearance of urea by the kidney and a better utilization of milk proteins for growth (Bentley & Shield, 1962).

Haematopoiesis and fetal haemoglobin

Block (1964) gives an exhaustive account of the blood-forming tissues and blood of the Virginian opossum from birth to the one hundredth day of life (i.e. until two to three weeks after leaving the pouch). The liver of the newborn opossum is 50% haematopoietic and there is an increased concentration of haematopoietic tissue at 2 d. Liver haematopoiesis becomes more focalized at 5 d and begins to decline at 8–9 d after birth.

At birth the thymus is a pure epithelial sheet with large lymphocytes bordering capillaries between lobules. Mitotic activity becomes pronounced in the thymus at 2 d, and does not begin to decrease until 13–16 d. Small lymphocytes begin to appear in the centre of the developing thymus at 5 d, and predominate there on days 6 and 7. Medullary lymphatic tissue appears at 13–16 d and there is an increase of medullary cortical tissue at 17–22 d. Between 23 and 32 d the thymus achieves full development.

The spleen of the newborn Virginian opossum is avascular, but contains rare transitional forms between mesenchyme stem cells and large lymphocytes. Haematopoietic foci become prominent between 6 and 7 d after birth, and at 13–16 d there is rapid growth of the spleen, at which time it has more mitoses than any other organ. At 17–22 d the spleen is 'solidly filled with erythroblasts' (Block, 1964), and at 23–32 d white pulp becomes evident. There is thereafter an increase in the ratio of white to red pulp, and a decrease in myeloid haematopoiesis in the white pulp, until differentiation of lymphatic nodules becomes evident at 65–100 d.

At birth the endochondral bones of the Virginian opossum are solidly cartilagenous, but at between 6 and 7 d of pouch life most of the cartilage in the diaphyses is replaced by marrow. The marrow, unlike other myeloid organs at the onset of haematopoiesis, has primarily mature rather than immature cells. Endochondral marrow continues to increase at the expense of cartilage until, by days 23–32, endochondral bones are filled with marrow except at the epiphyses. At 65–100 d the marrow of the developing young in the pouch is similar to that of mature animals in which the marrow remains the main centre of erythropoiesis throughout life. In the opossum, however, 'unlike in eutheria, erythropoiesis is intravascular, reflecting a phylogenetic relationship to birds' (Block, 1964).

The blood of such newborn marsupials as have been investigated has a high percentage of nucleated red blood cells. These are primarily primitive erythroblasts or megaloblasts (Block, 1964), and are formed in the yolk-sac, which persists (as the yolk-sac placenta) until birth in marsupials. In the Virginian opossum practically every cell in the heart blood of the newborn is a nucleated eosinophil megaloblast, but since the yolk-sac—the source of the megaloblasts—is lost at birth their frequency rapidly declines. Nucleated megaloblasts disappear from the blood by the thirteenth to sixteenth day of pouch development in the Virginian opossum. At the tenth day of pouch development in the red kangaroo, normoblasts and leucocytes become increasingly apparent, thus reducing the proportion of megaloblasts in the nucleated cell population, and by day 25 very few megaloblasts are left (Richardson & Russell, 1969). Fewer data are available for the grey kangaroo and euro pouch young, but a nucleated cell picture very like that of the red kangaroo emerges.

Block (1964) points out that as mammalian fetal haemoglobin has a greater affinity for oxygen than adult haemoglobin its presence would tend to compensate for the inadequacies of the opossum lung at birth. The presence of a fetal (or embryonic) haemoglobin (HbF) in the pouch young of various kangaroos was discovered by Richardson & Russell (1969). HbF was present, alone, in a number of red and grey kangaroo young up to about 40 d of age, while the youngest wallaroo young studied had a mixture of HbF and HbA (adult haemoglobin). Apparently in kangaroo pouch young the switch from HbF to HbA can occur as early as 7 d or as late as 40 d, varying from animal to animal, and there is no obvious correlation between haemoglobin type and changes in the population of nucleated red blood cells. The situation in marsupials appears unlike that in man, if the suggestion that HbF is contained only in megaloblasts in the human subject is correct, because HbF is found in definite marsupial erythrocytes. Richardson & Russell, assuming that cutaneous respiration may be important until some time after birth, suggest that the transfer from HbF to HbA probably occurs when the lungs take over all respiratory functions in marsupials. In eutherians the change from fetal to adult haemoglobin is seen to be related to the change from placental to lung respiration.

Whether the haemoglobin of young marsupials should be termed fetal or embryonic haemoglobin is open to doubt. An embryonic haemoglobin occurs in man, being found in embryos of less than 10 cm crown-rump (CR) length, and in greatest quantity in embryos of less than 6·5 cm CR length. HbF is found in kangaroo pouch young of up to 6·5 cm CR length, and these are approximately 70 d after conception, as are human embryos of the same CR length. Richardson & Russell comment that kangaroo HbF could actually be an embryonic haemoglobin, adapted to respiration by diffusion, which is immediately replaced by adult haemoglobin adapted to lung respiration. If this is true kangaroos do not develop an equivalent of eutherian fetal haemoglobin, which is apparently adapted for placental respiration.

Block (1964) showed that the development of blood-forming organs in the Virginian opossum was very similar to that in eutherians. The absence of a fetal haemoglobin at a similar developmental stage to that at which fetal haemoglobin is present in eutherians perhaps reflects differences between the intrauterine environment of eutherians and the pouch environment of marsupials. The occurrence of HbF at around the time the marsupial young is born may be owing to the need for a haemoglobin adapted for respiration by diffusion, which is important during the journey from uterus to pouch and at the period during which full lung development is attained.

Oxygen consumption and temperature regulation
Shield (1966) stated that, while no newborn eutherian yet studied is poikilothermic, quokka pouch young less than 23 d old show a linear dependence of oxygen consumption on ambient temperature and must be regarded as poikilothermic. Rowlatt, Mrosovsky & English (1971) studied the neck and axilla of 285 perinatal mammals from 17 orders, and found that brown fat was absent only in young marsupial pouch young. Since brown fat is thermogenic its absence presumably indicates that the perinatal marsupial is incapable of exercising effective control of its body temperature. Virginian opossum young removed from the pouch and maintained at 13·2°C showed a rapid drop in body temperature when 69 d old, and a not-so-rapid drop when 75 d old. At 81 d they took 10 h to reach approximate environmental temperature, and at 94 d they maintained their body temperature at around 32°C for at least 12 h (Reynolds, 1952).

Quokka pouch young over 100 d old show an increase in oxygen consumption at 20°C, whereas those between 120 and 150 d old show a further increase but not sufficient to establish a thermal neutral zone. At 150–180 d the rate of oxygen consumption is approximately 12 times greater than at ages less than 100 d, and a thermal neutral zone is established in the range 32–36°C (Shield, 1966).

Immunological competence

Important studies have been initiated on the early development of the immune response in marsupial pouch young. Both in terms of cellular differentiation and capacity to respond to antigenic challenge, the establishment of the immune system is entirely postnatal, in the species investigated so far, and takes place as a succession of events. In *Didelphis marsupialis virginiana* the thymus, cf. section on haemopoiesis and fetal haemoglobin, receives its first lymphocytes on the second day of life; lymph nodes become distinguishable on the third day (Rowlands *et al.*, 1964). Lymphocytes begin to appear in the blood at about the end of the first week (Block, 1964). The spleen gains recognizable lymphoid tissue between the seventeenth and twenty-fourth days, plasma cells and secondary nodules appear at about the sixtieth day, and later on tonsils and Peyer's patches can be seen (Rowlands *et al.*, 1964).

Although the thymus is generally held to be important for the development of immunological competence, it is evidently inadequate alone and the formation of antibodies seems to require also the existence of lymph nodes. In addition reactivity differs for different antigens. The earliest antibody response obtained in *Didelphis* was at 5 d, and this was to bacteriophage f2; from 8 d onwards *S. typhosa* flagellar antigens evoked a response, and at 15 d or more a hapten determinant, DNP, was effective (Rowlands & Dudley, 1969). Mizell (1968) and Mizell & Isaacs (1970) found that homografts of kidney tissue made at 3 d after birth became accepted and when examined 23 d later there was no sign of lymphocytic infiltration. Even xenografts of tadpole brain were tolerated.

Examination of the antibody response in pouch young and adult Virginian ópossums, and in adult specimens of the quokka (*Setonix brachyurus*), revealed the production of immunoglobulins corresponding to the IgM and IgG of eutherian mammals, as well as a low-molecular-weight antibody (Rowlands, 1970; Rowlands *et al.*, 1972; Thomas *et al.*, 1972). The slow conversion from IgM to IgG placed the marsupials in an intermediate position between eutherians and lower vertebrates.

Conclusions

The postnatal development of the marsupial young in the pouch differs from fetal development in the uterus of eutherians in several ways that reflect the requirements of the 'prematurely born' young.

(1) Special features, of an adaptive nature, are found in the newborn.

    (i) The marsupial young is born with claws on the forelimb digits, well-developed forelimb musculature and development of the necessary co-ordination centres in the nervous system to aid the newborn in its climb to the pouch.

    (ii) The nostrils of the newborn marsupial are open and large, olfactory sense cells exist in the nasal epithelium and the olfactory bulbs of the brain are well developed, suggesting that the young finds its way to the pouch and teat by a precociously developed sense of olfaction.

(iii) In those marsupial young that are least developed at birth, there are modifications to the mouth (native cat and Virginian opossum) and cervical region (native cat only) which aid in attachment to the teats in the pouch.

(iv) In the newborn and in the developing pouch young, probably of all marsupials, there are modifications to the tongue, buccal cavity, pharynx and epiglottis that permit attachment to the teat, and allow breathing and swallowing of milk to proceed simultaneously.

(2) Special characteristics of the pouch environment compensate for the absence of the intrauterine environment in which the eutherian young completes differentiation and much of development.

(i) The pouch temperature is maintained within about 1 °C of the mother's body temperature, and with minimum fluctuations until the young develop homoiothermic mechanisms and begin to leave the pouch.

(ii) Marsupial milk differs in various ways from the milk of eutherians and has special characteristics that may be essential for the nursing of 'prematurely born' young. The early 'milk' is a clear serous fluid of low fat content; in later milks lactose, the principal milk sugar of eutherians, is not the predominant marsupial milk sugar, and the Na:K ratio of the milk remains fairly steady at 1·8 until the pouch young has reached an advanced state of development. In those marsupials that have overlapping periods of lactation involving two different mammary glands, the differences in the two milks, secreted simultaneously, strongly suggest that the production of a special 'early milk' is a necessity for proper development of the newborn.

(3) Physiological studies on the developing pouch young suggest that some developmental processes occur in an essentially similar manner to those of the intrauterine eutherian. In other systems, however, there are indications of special marsupial adaptations.

(i) No differentiated chief or parietal cells occur in the stomach of the developing red kangaroo until it is more than 200 d old. The single cell type present, which has the characteristics of both chief and parietal cells when viewed by electron micro-

85

scopy, may be analogous to the bipotent cell in the stomach mucosa of frogs.

(ii) Although the development of the blood-forming organs is essentially similar in intrauterine eutherians and in the pouch young of marsupials, there are indications that the marsupial 'fetal haemoglobin' (HbF) may be analogous to the embryonic haemoglobin of eutherians. Marsupial HbF may be adapted for respiration by diffusion, and may be important during the journey to the pouch and the period before full lung development is attained, perhaps while the newborn young is partly dependent on cutaneous respiration.

Acknowledgement

Support from the Australian Research Grants Committee is acknowledged.

References

BAILEY, S. W. & DUNNET, G. M. (1960) The gaseous environment of the pouch young of the brush-tailed possum, *Trichosurus vulpecula* Kerr. *C.S.I.R.O. Wildlife Research*, **5**, 149–151.

BARKER, S. (1962) Copper levels in the milk of a marsupial. *Nature*, **193**, 292.

BARTHOLOMEW, G. A. (1956) Temperature regulation in the macropod marsupial *Setonix brachyurus*. *Physiological Zoology*, **29**, 26–40.

BECK, A. B. (1961) The copper levels of foetal and newly born marsupials and whales. *The Australian Journal of Science*, **24**, 245–246.

BENTLEY, P. J. & SHIELD, J. W. (1962) Metabolism and kidney function in the pouch young of the macropod marsupial *Setonix brachyurus*. *Journal of Physiology*, **164**, 127–137.

BLOCK, M. (1964) The blood forming tissues and blood of the newborn opossum (*Didelphys virginiana*). *Reviews of Anatomy Embryology and Cell Biology*, **37**. Berlin: Springer-Verlag.

BOLLIGER, A. & PASCOE, J. V. (1953) Composition of kangaroos milk (wallaroo, *Macropus robustus*). *The Australian Journal of Science*, **15**, 215–217.

BUCHANAN, G. & FRASER, E. A. (1918) The development of the urogenital system in the Marsupialia, with special reference to *Trichosurus vulpecula*. Part 1. *Journal of Anatomy*, **53**, 35–95.

CALABY, J. H. & POOLE, W. E. (1971) Keeping kangaroos in captivity. *International Zoo Yearbook*, **11**, 5–12.

DUNNET, G. M. (1956) A live trapping study of the brush-tailed possum *Trichosurus vulpecula* Kerr (Marsupialia). *C.S.I.R.O. Wildlife Research*, **1**, 1–18.

ENDERS, R. K. (1966) Attachment, nursing and survival of young in some didelphids. *Symposium of the Zoological Society of London*, **15**, 195–203.

FORD, J. E. & THOMPSON, S. Y. (1965) Composition of kangaroos milk. *National Institute for Research in Dairying Report*, 153–155.

GERSH, I. (1937) The correlation of structure and function in the developing mesonephros and metanephros. *Contributions to Embryology of the Carnegie Institution of Washington*, **26**, 35–58.

GRIFFITHS, M. & BARTON, A. A. (1966) The ontogeny of the stomach in the pouch young of the red kangaroo. *C.S.I.R.O. Wildlife Research*, **11**, 169–185.

GRIFFITHS, M., MCINTOSH, D. L. & LECKIE, R. M. C. (1972) The mammary glands of the red kangaroo with observations on the fatty acid components of the milk triglycerides. *Journal of Zoology*, **166**, 265–275.

GROSS, R. & BOLLIGER, A. (1958) The occurrence of carbohydrates other than lactose in the milk of a marsupial *Trichosurus vulpecula*. *The Australian Journal of Science*, **20**, 184.

HEUSER, C. H. (1921) The early establishment of intestinal nutrition in the opossum—the digestive tract just before and soon after birth. *American Journal of Anatomy*, **23**, 341–356.

HILL, J. P. & HILL, W. C. O. (1955) The growth stages of the pouch young of the native cat (*Dasyurus viverrinus*) together with observations on the anatomy of the newborn young. *Transactions of the Zoological Society of London*, **28**, 349–452.

JENNESS, R., REGEHR, E. A. & SLOAN, R. E. (1964) Comparative biochemical studies of milks—II Dialyzable carbohydrates. *Comparative Biochemistry and Physiology*, **13**, 339–352.

KALDOR, I. & EZEKIEL, E. (1962) Iron content of mammalian breast milk: measurements in the rat and in a marsupial. *Nature*, **196**, 175.

KIRKPATRICK, T. H. (1965) Studies of Macropodidea in Queensland. 2. Age estimation in the grey kangaroo, the red kangaroo, the eastern wallaroo and the red-necked wallaby, with notes on dental abnormalities. *Queensland Journal of Agricultural and Animal Sciences*, **22**, 301–317.

LEMON, M. & BAILEY, L. F. (1966) A specific protein difference in the milk from two mammary glands of a red kangaroo. *The Australian Journal of Experimental Biology and Medical Science*, **44**, 705–708.

LEMON, M. & BARKER, S. (1967) Changes in milk composition of the red kangaroo, *Megaleia rufa* (Desmarest), during lactation. *The Australian Journal of Experimental Biology and Medical Science*, **45**, 213–219.

LYNE, A. G. (1952) Notes on the external characters of the pouch young of four species of bandicoot. *Proceedings of the Zoological Society of London*, **122**, 625–649.

LYNE, A. G. (1970a) The melanocyte population in the skin during development of the marsupial *Trichosurus vulpecula*. *Australian Journal of Biological Sciences*, **23**, 697–708.

LYNE, A. G. (1970b) The development of hair follicles in the marsupial *Trichosurus vulpecula*. *Australian Journal of Biological Sciences*, **23**, 1241–1253.

LYNE, A. G. & VERHAGEN, A. M. W. (1957) Growth of the marsupial *Trichosurus vulpecula* and a comparison with some higher mammals. *Growth*, **21**, 167–195.

LYNE, A. G., HENRIKSON, R. C. & HOLLIS, D. E. (1970) Development of the epidermis of the marsupial *Trichosurus vulpecula*. *Australian Journal of Biological Sciences*, **23**, 1067–1075.

LYNE, A. G., PILTON, P. E. & SHARMAN, G. B. (1959) Oestrous cycle gestation period and parturition in the marsupial *Trichosurus vulpecula*. *Nature*, **183**, 622–623.

McCRADY, E. (1938) The embryology of the opossum. *American Anatomical Memoir*, **16**, 1–234.

MERCHANT, J. C. & SHARMAN, G. B. (1966) Observations on the attachment of marsupial pouch young to the teats and on the rearing of pouch young by foster-mothers of the same or different species. *Australian Journal of Zoology*, **14**, 593–609.

MIZELL, M. (1968) Limb regeneration: induction in the newborn opossum. *Science*, **161**, 283–286.

MIZELL, M. & ISAACS, J. J. (1970) Induced regeneration of hindlimbs in the newborn opossum. *American Zoologist*, **10**, 141–155.

MÜLLER, F. (1969) Zur fruhen Evolution der Sauger-Ontogenesetypen. Versuch einer Rekonstruktion aufgrund der Ontogenese-Verhaltnisse ber den *Marsupialia*. *Acta Anatomica*, **74**, 297–404.

MURPHY, C. R. & SMITH, J. R. (1970) Age determination of pouch young and juvenile Kangaroo Island wallabies. *Transactions of the Royal Society of South Australia*, **94**, 15–20.

NEW, D. A. T. & MIZELL, M. (1972) Opossum foetuses grown in culture. *Science*, **175**, 533–536.

O'DONOGHUE, C. H. (1911) The growth changes in the mammary apparatus of *Dasyurus* and the relation of the corpora lutea thereto. *Quarterly Journal of Microscopical Science*, **57**, 187–234.

RENFREE, M. B. (1972) Ph.D. Thesis, Australian National University.

REYNOLDS, H. C. (1952) Studies on reproduction in the opossum (*Didelphis virginiana virginiana*). *University of California Publications in Zoology*, **52**, 223–284.

RICHARDSON, B. J. & RUSSELL, E. M. (1969) Changes with ages in the proportion of nucleated red blood cell types and in the type of haemoglobin in kangaroo pouch young. *The Australian Journal of Experimental Biology and Medical Science*, **47**, 573–580.

ROWLANDS, D. T., BLAKESLEE, D. & LIN, H-H. (1972) The early immune response and immunoglobulins of opossum embryos. *Journal of Immunology*, **108**, 941–964.

ROWLANDS, D. T., Jr. (1970) The immune response of adult opossums (*Didelphys virginia*) to the bacteriophage f$_2$. *Immunology*, **18**, 149.

ROWLANDS, D. T., Jr., LA VIA, M. F. & BLOCK, M. H. (1964) The blood forming tissues and blood of the newborn opossum (*Didelphys virginiana*).

II. Ontogenesis of antibody formation to flagella of *Salmonella typhi*. *Journal of Immunology*, **93**, 157–164.

ROWLANDS, D. T., Jr. & DUDLEY, M. A. (1969) The development of serum proteins and humoral immunity in opossum 'embryos'. *Immunology*, **17**, 969.

ROWLATT, V., MROSOVSKY, N. & ENGLISH, A. (1971) A comparative survey of brown fat in the neck and axilla of mammals at birth. *Biology of the Neonate*, **17**, 53–83 (seen in abstract only).

SADLEIR, R. M. F. S. (1963) Age estimation by measurement of joeys of the euro *Macropus robustus* Gould in Western Australia. *Australian Journal of Zoology*, **11**, 241–249.

SHARMAN, G. B. (1957) The quokka. In *The UFAW Handbook on the Care and Management of Laboratory Animals*, ed. Worden, A. N. & Lane-Petter, W. Ch. 57, Second Edition. London: The Universities Federation for Animal Welfare.

SHARMAN, G. B. (1962) The initiation and maintenance of lactation in the marsupial, *Trichosurus vulpecula*. *Journal of Endocrinology*, **25**, 375–385.

SHARMAN, G. B. (1965) Marsupials and the evolution of viviparity. In *Viewpoints in Biology*, ed. Carthy, J. D. & Duddington, C. L. Vol. 4, Ch. 1. London: Butterworth.

SHARMAN, G. B. (1970) Reproductive physiology of marsupials. *Science*, **167**, 1221–1228.

SHARMAN, G. B. & CALABY, J. H. (1964) Reproductive behaviour in the red kangaroo, *Megaleia rufa*, in captivity. *C.S.I.R.O. Wildlife Research*, **9**, 58–85.

SHARMAN, G. B. & PILTON, P. E. (1964) The life history and reproduction of the red kangaroo (*Megaleia rufa*). *Proceedings of the Zoological Society of London*, **142**, 29–48.

SHARMAN, G. B., ROBINSON, E. S., WALTON, S. M. & BERGER, P. J. (1970) Sex chromosomes and reproductive anatomy of some intersexual marsupials. *Journal of Reproduction and Fertility*, **21**, 57–68.

SHIELD, J. W. (1961) The development of certain external characters in the young of the macropod marsupial *Setonix brachyurus*. *Anatomical Record*, **140**, 289–294.

SHIELD, J. (1966) Oxygen consumption during pouch development of the macropod marsupial *Setonix brachyurus*. *Journal of Physiology*, **187**, 257–270.

SHIELD, J. W. & WOOLLEY, P. (1961) Age estimation and measurement of the pouch young of the quokka (*Setonix brachyurus*). *Australian Journal of Zoology*, **9**, 14–23.

SINGER, M. (1951) Induction of regeneration of forelimb of the frog by augmentation of the nerve supply. *Proceedings of the Society of Experimental Biology and Medicine*, **76**, 413–416.

SINGER, M. (1961) Induction of regeneration of body parts in the lizard, *Anolis*. *Proceedings of the Society of Experimental Biology and Medicine*, **107**, 106–108.

THOMAS, W. R., TURNER, K. J., EADIE, M. E. & YADEV, M. (1972) The immune response of the quokka (*Setonix brachyurus*): the production of a low molecular weight antibody. *Immunology*, **22**, 401–406.

TYNDALE-BISCOE, C. H. (1968) Reproduction and post-natal development in the marsupial *Bettongia lesueur* (Quoy & Gaimard). *Australian Journal of Zoology*, **16**, 577–602.

4 Experimental Techniques in Fetal and Placental Physiology

The late D. A. NIXON

The fetus, lying within the intact uterus, presents many problems in physiological investigations. Without some operative technique, only limited and often indirect observations are possible. This article deals with experimental procedures whereby access to the fetus and its placenta may be gained, particularly for metabolic investigations in the sheep. Techniques applicable to the blastocyst and early embryo will not be discussed.

Animal experimentation forms an essential feature of many aspects of medical research. If the investigations are to have relevance to man the animal model should obviously bear as close a physiological resemblance to him as possible. On a phylogenetic basis the anthropoid apes should be most suitable; however, their numbers in the wild are rapidly decreasing, and in captivity their rate of reproduction, at least at the present time, is too slow, so that economics alone impose a limitation on the demands of experimentalists. Furthermore, for the fetal physiologist there arises the prospect of sacrificing not one animal but two! These considerations also apply, although to a lesser extent, to the Old World monkeys such as the rhesus monkey (*Macaca mulatta*) and the baboon (*Papio anubis*). Particularly valuable contributions to the use of primates in research are to be found in the monograph edited by Goldsmith & Moor-Jankowski (1969). That there are dangers in the extrapolation of results from one species to another is obvious, yet in biology it is often necessary to make generalizations. Thus we know more about the physiology and biochemistry of fetal life in general from the study of the sheep than for any other species.

Since the sheep has been, and no doubt will continue to be, a favourite experimental animal of the fetal physiologist, no apology is made for the bias towards this species in the present chapter. This animal offers a number of desirable features to the experimentalist. It can tolerate a wide range of climatic conditions and does not require expensive food. As it has long been domesticated much is known of its management, and it is reared in vast numbers for meat and wool. These factors make it a relatively inexpensive animal. Furthermore, it is an easy animal to handle as it does not bite or scratch, and the females of many breeds are hornless. The dating of conception is easy and reasonably accurate. In addition, it has a quiescent uterus bearing a single fetus or occasionally twins. But the principal advantage lies in the relatively large size of its fetus (indeed the weight at birth is similar to that of the human infant). With a fetus of this size surgical and analytical procedures are possible which would prove very taxing in more conventional laboratory animals.

It is perhaps appropriate here to mention a few aspects of sheep reproductive physiology and the means for acquiring pregnant animals. The sheep is a seasonal polyoestrous animal with cycles repeated at 16-d intervals. The breeding season begins in the autumn and lasts until the end of the year. However, in some breeds the season is extended to the spring and in a few the oestrous cycles continue throughout the year; in this country the best example is the Dorset Shorthorn. Sheep of the Welsh Mountain breed are particularly suitable for experimental purposes as they are small hardy animals weighing around 30 kg. The actual duration of oestrus is about 24–48 h with ovulation occurring spontaneously towards the end of this phase of the cycle. Spermatogenesis takes place throughout the year although there is some decline during the summer months. The dating of pregnancies can be indicated by applying some suitable colouring material to the underside of the ram who is then turned out with a portion of the flock. Every day the ewes, each of which bears a numbered ear tag, are inspected for the dye, this being evidence of having been mounted by the ram. By presenting portions of the flock to the ram at intervals a spread of tupping dates over a period of two or three months is possible. The gestation period is about 148 d and with arrangements, such as those outlined above, an experimental season of six to eight months can be achieved. The fetal ages of undated animals may be calculated from a consideration of their body weights or preferably their lengths (Huggett & Widdas, 1951; Meschia et al., 1965a). It is not usually practical to mate the ewes in animal houses of research departments; fortunately, however, sheep farmers are often willing to provide ewes bearing fetuses of known age, to maintain them on the farm and to shear them as they are required to be dispatched for experimentation.

The uterus of the sheep is bicornate. The placenta is of the cotyledonary type and some 80–100 cotyledons are spread over both uterine horns. Maximum placental weight is attained at about the ninetieth day and remains approximately constant for the rest of gestation. The maternal circulation is separated from that of the fetus by the endothelium and connective tissue of the mother and by the trophoblast, connective tissue and endothelium of the fetus. Paired arteries and veins are present in the long umbilical cord, which is relatively free from tight coils. Along the length of the cord runs the allantoic duct connecting the bladder with

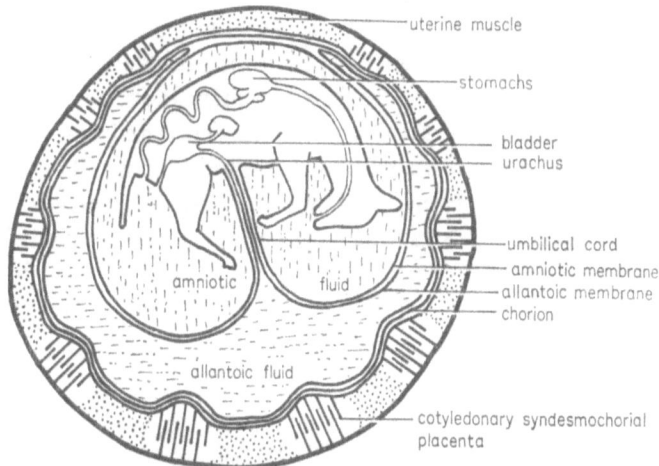

Fig. 4.1. Schematic diagram showing the relationship of the fetal sheep to its amniotic and allantoic fluids

the allantoic sac. As the fetus grows older the volume of fluid contained within this sac increases and at term averages 834 ml. An increase in amniotic fluid volume also occurs but reaches a mean maximum of 686 ml at the 120th day, to fall to a mean of 369 ml at term (Arthur, 1969). The general arrangement of the sheep fetus within the uterus is shown schematically in Fig. 4.1.

The works of Barcroft (1946), Asdell (1946), Barclay, Franklin & Prichard (1944) and Dawes (1968) together with those edited by Parkes (1952, 1956, 1960, 1966) give extensive coverage of reproduction in the sheep, and of the physiology and biochemistry of its fetus. A general account of sheep anatomy has been produced by May (1964), while Bassett (1965) has paid particular attention to the pelvic and perineal regions of the ewe.

The easiest experimental approach to the fetus is through a Caesarian section in an anaesthetized mother, leaving the umbilical circulation intact. This may be considered as the classical preparation of the fetal physiologist and has its origins in the work of Cohnstein & Zunty (1884). Much of our knowledge of the physiology and biochemistry of fetal life has and will continue to be derived from this preparation. However, to augment and to complement the observations made upon the acute preparation, further approaches have been developed in recent times. It was appreciated that factors such as anaesthesia, posture and the removal of the fetus from the uterine cavity may limit the accuracy of the observation or give misleading results. Such considerations have led to the use of chronic as opposed to acute preparations.

A certain degree of reluctance in performing recovery operations upon fetuses was evident. Among the fears were that the stress of surgery on the mother would prove detrimental to her fetus, and that interference with the uterus and its contents would result in abortion. The early work in this field has been reviewed by Swenson (1925), Nicholas (1925) and Hess (1957). These fears have been shown in the main to be unfounded.

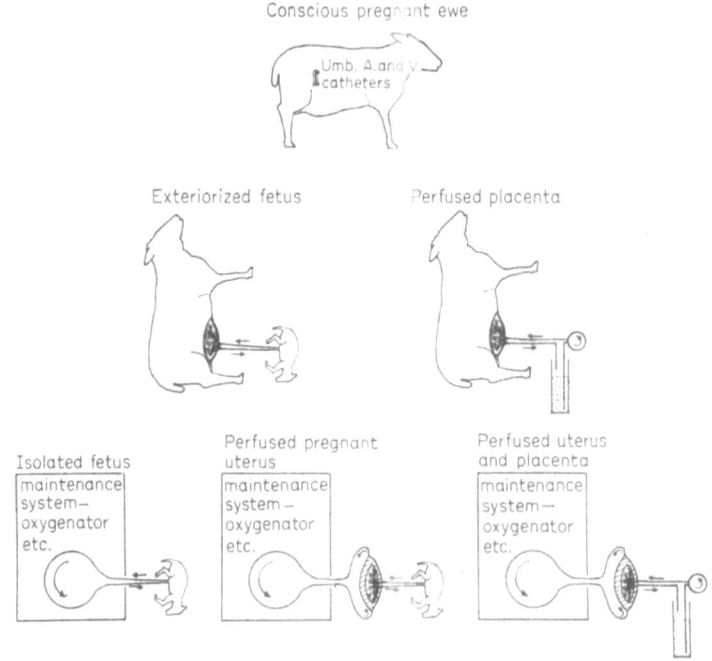

Fig. 4.2. Schematic diagram of several preparations used in metabolic investigations upon the fetus and placenta of the sheep. (Nixon, 1972a, reproduced by courtesy of the publisher)

It is now possible to insert catheters into the fetus at a prior operation, and after recovery to make observations upon the fetus in the unanaesthetized mother who is in a normal standing posture. Such an approach offers the possibility of repeated observations or of conducting different experiments at intervals over the remainder of the gestation period. Longitudinal studies of this nature are valuable since, by following one animal over a period of time, the biological variations seen in cross-sectional observations are eliminated and the maturation of a response more clearly seen. Surgery upon the fetus has also revealed the effects brought about by ablation of organs in chronic preparations.

Another approach has been to examine the maternal-placental-fetal unit, not as a whole, but rather in its component parts with perfusion techniques. Such techniques are a traditional feature of many investigations since a more direct assessment of the physiology of an organ may be achieved through its isolation from the rest of the body and artificial maintenance.

A schematic diagram of several preparations referred to in this article is shown in Fig. 4.2.

Exteriorized fetus

The exteriorization of the fetus has already been referred to as the classical approach of the fetal physiologist. Basically here the mother is anaesthetized either by a general or a spinal anaesthetic, the uterus exposed by laparotomy and access to the fetus gained by hysterotomy. With this acute preparation investigations extending over many hours are often possible.

Ewes undergoing general anaesthesia should be starved for 12–24 h, since pulmonary ventilation will be embarrassed by the large stomachs if the animal is in the supine position and to a lesser extent if placed laterally. Again in animals under general anaesthesia an endotracheal tube should be inserted or in acute experiments the trachea can be cannulated to circumvent the inhalation of regurgitated stomach contents and saliva.

The two most widely used general anaesthetics have been choralose and sodium pentobarbitone (Nembutal) administered intravenously at about 40 mg and 30 mg/kg, respectively (Dawes, 1968). These may be injected into the external jugular vein with a wide-bore needle. This vein is usually readily palpable, particularly if distension is produced by

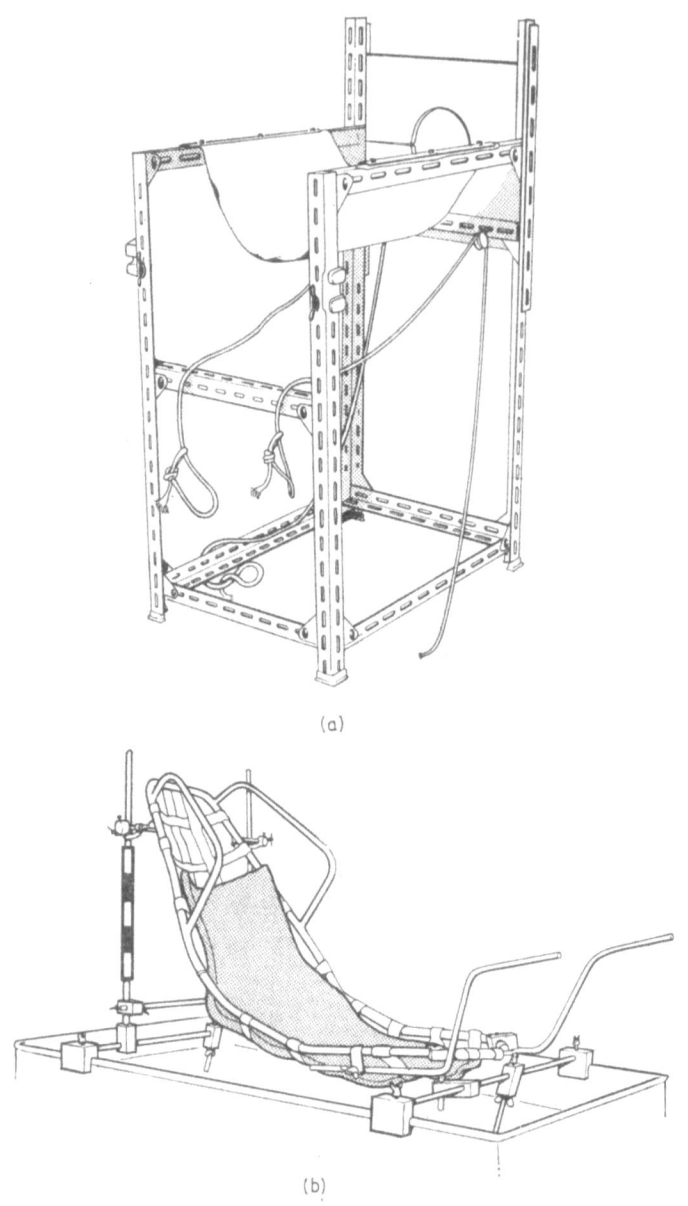

(a)

(b)

Fig. 4.3. (a) The pillory-like frame for positioning the ewe for the administration of a spinal anaesthetic. (Constructed from 4-cm 'Handy-Angle')

(b) The cradle that secures the ewe during the experiment. The head is elevated at an angle of 45° to the horizontal. The fore-limbs are tied to the upper supports while the hind limbs are attached to adjustable rods. The fetus is delivered on to a small table placed between the hind limbs. (Scale attached to the upright support is in 5-cm units)

compression of the vein at the base of the neck with the hand. After the initial injection the syringe may be detached from the needle and a thin catheter passed along the bore into the blood vessel. The needle can then be withdrawn from the vein leaving the catheter within the vessel so that further injections may be made to maintain the desired plane of anaesthesia. It is also possible to cut down on to a superficial vein for catheterization, after infiltration of the area with a local anaesthetic (2% procaine hydrochloride). A particularly useful vessel is the dorsalis pedis vein, which may be exposed through a 3-cm incision on the anterior surface of a hind limb about midway along the metatarsus. The dorsalis pedis artery lies deeper, beneath and slightly medial to the tendon.

An alternative approach is to subject the ewe to spinal anaesthesia. This procedure avoids the possibility of adverse effects upon the fetus through a passage of drugs across the placenta. In order to inject the anaesthetic the hind legs of the ewe need to be drawn forward as far as possible to give a good exposure of the lower lumbar spaces. Adequate exposure may be obtained by an assistant kneeling in front of the ewe, tucking the animal's head under his arm, and gripping firmly the hind limbs which are then drawn towards him. A better method is to construct a rectangular frame which has a pillory-like attachment at one end and a canvas sling (Fig. 4.3a). The animal is lifted onto the sling and its head placed in the pillory. Cords are attached to the hind legs and pass over pulleys mounted on the front of the frame; these can be drawn to the rear and secured, thus arching the lumbar region. An area roughly at the level of the iliac crests is infiltrated with a local anaesthetic (2% procaine hydrochloride). A spinal needle is then advanced through the tough intervertebral disc until resistance disappears. The stilette of the needle is removed and for acute experiments 2 ml of 20% procaine hydrochloride is injected. Sometimes the needle enters or touches the cord resulting in a convulsive kick of the hind limbs. (The injection of alcohol into the spinal cord of pregnant rabbits to produce a 'chemical transection' has been employed by Basmajian & Ranney (1961) in their investigations upon the fetuses.) Where subsequent recovery of the ewe is required, particular care must be taken to ensure that the needle tip is sited satisfactorily, and this is indicated by the passage of cerebrospinal fluid through the needle. A suitable spinal anaesthetic for use under these conditions is 10–15 mg amethocaine in 2·5 ml of 4·5% glucose (Alexander & Nixon, 1963). With spinal anaesthesia the head and upper portions of the trunk must be elevated during the experiment. It is convenient to place such animals on a frame like that shown in Fig. 4.3b. A pronounced tranquillizing effect on the ewe can be achieved by masking her eyes before the administration of the spinal anaesthetic.

The laparotomy is performed by making a skin incision along the midline from just short of the umbilicus to the mammary glands. On either side of the incision for a few centimetres the skin is freed from the abdominal wall, care being taken not to rupture the large abdominal veins. The abdominal cavity is then entered through a midline incision; prolongation caudally, below the mammary glands, allows the uterus to be partially delivered through the incision. Extrusion of the gut is prevented by partial closure of the abdominal cavity with artery forceps or by suturing the uterus to the abdominal wall.

The uterus is wrapped in lint swabs, soaked in warm physiological saline. A longitudinal incision in the uterine horn containing the fetus is made through the myometrium, with care to avoid the cotyledons, starting from about the mid-point on the horn and running for a few centimetres towards its upper pole. The chorion and amnion are cut and the fetus withdrawn onto a platform fixed at the mother's side or between her legs. In fetuses older than about 120 d the head should be enclosed in a bag of warm physiological saline (a rubber glove serves as an admirable container); such a procedure prevents the onset of breathing. The fetus and its umbilical cord are covered by swabs, and kept warm and moist by the frequent application of physiological saline. Prevention of drying is of great importance in sheep fetuses up to about 130 d of age since it is only at this time that the wool has grown sufficiently to afford some protection to the skin.

An alternative method is to place the lower half of the ewe in a bath of physiological saline maintained at body temperature. The fetus can then be delivered into the bath and tethered by lint slings to prevent tension on the cord. This approach was introduced by Huggett in 1927 and used by him in experiments upon the transfer of gases across the placenta of the goat. The work is of historical interest since it perhaps marks the start of the extensive use of the ruminant as an experimental model in fetal physiology. In addition to goats and sheep the technique has been applied to guinea-pigs (Karvonen & Räihä, 1954). While the problems of temperature control and drying of the fetus are of course eliminated by this method it does have some disadvantages—the mother's peritoneal cavity is exposed to a large volume of saline, usually none too clean because of excreta adhering to the fleece, and physiological parameters are difficult to measure on the less static fetus. The delivery of fetal sheep into a saline bath was recently used by Dawes, Fox & Richards (1972) in studies upon the duration of gasping when the umbilical cord was occluded.

Fetal blood can be obtained by withdrawing samples from the umbilical vessels with a syringe. However, a better method is to establish constant sampling sites by inserting polythene catheters. These should be of an appropriate external diameter and primed with heparinized saline (5000 u/100 ml 0·9% NaCl); the free end is fitted with a three-way tap. Suitable tributaries of the umbilical vessels are located; those of the umbilical arteries can be distinguished from the veins by their somewhat whiter colour, owing to their thicker walls. The tributary is carefully freed along a few centimetres of its length. Two cotton ligatures, preferably of different colours, are placed around the vessel; the distal ligature is then tied to occlude the vessel. An assistant applies traction to the other ligature, which has been loosely tied around the vessel, to provide temporary occlusion; and with a finger placed under the vessel to give support a small incision is made. The catheter tip is advanced along the tributary until it lies within the cord vessel, and it is then tied in place. A useful safeguard against possible displacement of the catheter is to tie a second knot, using the ends of the distal ligature, over the first knot. Care should be taken when cutting the bevel of the catheter tip that the angle is not too acute since difficulty may be encountered in withdrawing blood or it may even puncture the wall of the vessel. A possible disadvantage of this procedure may lie in the effective loss of a few cotyledons.

The cord vessels and their tributaries constrict readily on manipulation. This may be prevented by fixing the tissue through the application of a few drops of formol-saline (4% formaldehyde in 0·9% NaCl solution) to the region selected for catheterization (Barcroft, 1946). Another method to prevent a spasm in these vessels is to apply a solution of hexylcaine and phenoxybenzamine hydrochloride (Rudolph & Heymann, 1967). Vessels that have constricted may often be persuaded to relax by covering the region with saline swabs at 2–3°C above body temperature, or by applying procaine externally. Faber & Hart (1966) found that nitroglycerine was also an effective vasodilator in this situation.

Many metabolic and related investigations have been conducted on the exteriorized fetal sheep. In considering carbohydrate metabolism it was shown for example that glucose crossed the placenta by facilitated diffusion (Widdas, 1952), that maternal glucose was the precursor of fetal fructose (Alexander, Andrews, Huggett, Nixon & Widdas, 1955), that the fetal kidney could reabsorb glucose (Alexander & Nixon, 1963), and that the fetal gut could absorb fructose (Nixon & Wright, 1964).

Acute *in-utero* preparation

In this type of preparation the fetus remains within the uterus of the anaesthetized mother. Reynolds & Paul (1955) showed that in the sheep satisfactory catheterization of tributaries of the umbilical vessels could be achieved through a small uterine incision. Since these blood vessels are external to the amnion the intra-amniotic pressure is preserved. Another approach is to withdraw a fetal limb through a small uterine incision for vascular catheterization, after which it is returned to the uterus and the incision repaired around the catheter. Lack of distension might be expected to reduce uterine and umbilical blood flows; indeed when Heymann & Rudolph (1967) compared the umbilical blood flow before and after the exteriorization of the sheep fetus they found a drop in flow of between 21 and 61%. In their studies on the secretion and function of the tracheal fluid of goats and sheep, Goodlin & Rudolph (1970) had as many as 15 catheters inserted in the same acute *in-utero* preparation.

Using this method van Duyne *et al.* (1960) showed that the concentration of free fatty acids in the plasma of the fetal sheep was 10% of that present in the maternal plasma.

This new method of approach appears to be essential in the monkey where exteriorization of the fetus brings about uterine contraction and separation of the placenta. Reynolds, Paul & Huggett (1954) exploited the fact that in about 80% of rhesus monkeys there is a primary placenta (from which the umbilical cord arises) which is connected to a smaller secondary placenta by blood vessels which lie between the chorion and amnion. It is possible to identify the site of the interplacental vessels by transillumination of the exposed uterus. A careful incision down to the vessels can then be made and their cannulation effected, leaving the amnion intact. Dawes *et al.* (1960), in their experiments on rhesus monkeys, favoured the direct catheterization of fetal vessels such as femoral, brachial or carotid arteries. In their colony, only about 60% of the animals presented a secondary placenta; furthermore the secondary placenta could form as much as 59% of the total placental mass. Ligation for cannulation of the interplacental vessels could thus exclude a considerable part of the placenta from the circulation. Recently Myers *et al.*

(1971) have produced placental insufficiency, as shown by a reduced body weight, by ligating the interplacental vessels in the rhesus monkey. However, circulation can be maintained if two catheters are inserted in a vessel in opposite directions and united through a three-way tap (Plentl & Friedman, 1962).

Acute *in-utero* preparations have been used in the rhesus monkey to study glucose transfer across the placenta (Chinard *et al.*, 1956; Little, Nasser & Spellacy, 1971).

Chronic *in-utero* preparation

While the acute *in-utero* preparation may be considered to be physiologically more normal than the exteriorized preparation, nevertheless anaesthesia coupled with the unnatural posture of the animal may place a limitation on the usefulness of the observations obtained. In the chronic *in-utero* preparation access to the fetus is gained through previously implanted catheters. The technique has therefore the advantage that the fetus can be examined in an unanaesthetized mother who maintains a normal posture, allowing 2–3 d recovery from surgical stress. Furthermore, the fetus may be examined repeatedly over the remainder of its gestation and indeed may even be delivered normally at term.

The technique has its origin in the successful implantation of catheters into the uterine vein of the pregnant sheep (Meschia, Wolkoff & Barron, 1959). They were able to show that blood samples could be removed in unstressed ewes over a number of weeks, and that the procedure need not be detrimental to either fetus or mother. Later Meschia *et al.* (1965b) introduced catheters into the umbilical vessels of fetal sheep and goats. Implantation of catheters directly into the fetal sheep has been made through exposure of the axillary artery and jugular vein (Quilligan *et al.*, 1968) and femoral artery and vein (Alexander, Britton, Mashiter, Nixon & Smith, 1970). Catheters have also been chronically implanted in other sites in the fetal sheep. The flow and composition of lymph from several regions was investigated by Smeaton *et al.* (1969). Catheters, inserted into the fetal urinary bladder and amniotic sac, and joined together through a three-way tap situated on the flank of the ewe, permitted Buddingh, Parker & Ishizaki (1969) to take intermittent urine samples while still preserving the general flow of urine into the amniotic fluid compartment. The fetal bladder has also been catheterized by Mellor, Williams &

Matheson (1972) and access to the allantoic fluid, in addition to the amniotic fluid, has been achieved by Mellor (1970).

Having succeeded in the chronic implantation of catheters into the fetal sheep in order to remove fluids, investigators are now becoming more ambitious. Quilligan *et al.* (1968) have introduced flow probes together with ECG and EEG electrodes to study cephalic metabolism. Pressure changes in the trachea, oesophagus and amniotic fluid compartment have been measured by Merlet, Hoerter, Devilleneuve & Tchobroutsky (1970). The multiplicity of recordings possible from the chronic fetal sheep preparation is now such that Dawes, Fox, Leduc, Liggins & Richards (1972) were recently able to report on carotid arterial, tracheal, oesophageal and amniotic sac pressures, tracheal and aortic flows, and on the electrocortical and transorbital potentials.

Undoubtedly the chronic *in-utero* preparation is expensive, both in terms of the number of animals required to obtain an acceptable series of observations and in the labour involved in the postoperative care of the animals. Many workers who have used the technique have refrained from commenting on their rate of success; this is often taken to mean that they have patent catheters for at least 3 d after surgery. Meschia, Makowski & Battaglia (1970) have reported a success rate of 75% (48 animals) and 100% (six animals); Comline & Silver (1970) 60% (17 animals); Soma, White & Kane (1971) 63% (46 animals), while the author's more limited experience with the technique initially gave 37% (11 animals) but in a current series 80% (five animals). Failure to establish a preparation can be due to infections of the mother or fetus, abortion, or the blocking of the catheter through kinking or blood coagulation.

In the method of Meschia *et al.* (1965b) the ewe was sedated with intravenous Nembutal followed by a spinal anaesthetic. The uterus was incised carefully down to the cotyledonary vessels, as in the acute *in-utero* technique of Reynolds & Paul (1955). Catheters were then threaded along the vessels until their tips lay in the umbilical vessels of the cord. The abdominal wound was closed and the free end of the catheter brought out through a skin tunnel to emerge on the flank of the mother. Closure of the catheter was effected by inserting a pin into the lumen, and protected by inclusion in a bag sutured to the skin. A somewhat similar approach to fetal vasculature was made by Comline & Silver (1970) in animals under intravenous sodium pentobarbitone anaesthesia.

The procedure of Alexander, Britton, Mashiter, Nixon & Smith

(1970) was to lightly anaesthetize the ewe with sodium thiopentone, given either via the jugular vein or a catheter in the dorsalis pedis vein inserted under local anaesthesia. An endotracheal tube was then passed with the aid of a sheep laryngoscope (Holborn Surgical Instrument Co., London) and general anaesthesia induced and maintained by the inhalation of an oxygen/Halothene gas mixture in a closed circuit. The wool over the abdomen, extending about 20 cm from the umbilicus and about 10 cm on each side of the midline, was closely clipped and finally shaved or a depilatory agent used. An area about 5 cm² on the flank was similarly prepared. Both sites were cleaned with Cetrimide and Hibitane (I.C.I.). Sterile surgical towels were placed around the site and the operative procedures carried out under strict aseptic conditions. The abdominal skin was incised along the midline, care being taken to avoid the large veins, and freed from the underlying tissue for about 5 cm either side of the incision, while bleeding was controlled by diathermy. An incision some 15 cm in length was made along the linea alba. The lie of the fetus within a uterine horn was ascertained by palpation. It was manipulated so as to bring the hoof of a hind limb up against the uterine wall in a region devoid of cotyledons. A small incision made into the uterus and membranes over the hoof allowed the leg to be drawn out to give access to the femoral triangle. By making the incision small and holding the leg vertical the loss of amniotic fluid can be minimized. After an injection of local anaesthetic the femoral artery and vein were exposed. The proximal ends of the vessels were ligated and some 8–12 cm of a catheter, having a total length of about 120 cm and filled with heparinized saline, was inserted along each vessel and tied in place. The catheter may be further secured by pulling forward the ends of the ligating ligature and using these to place a second tie centrally to the first. A catheter with an internal diameter of 0·5 mm and an external diameter of 1·5 mm (Portex, Portland Plastic Ltd.) has been found suitable in sheep fetuses from about 115 d. Tubing of these dimensions has the advantage that it is not occluded by bending. The free end of the catheter can be sealed very effectively by inserting a pin along the lumen. Closure of the thigh wound was effected by a continuous gut suture and the catheter anchored by a single tie on the outer aspect of the thigh. The leg was returned to the uterine cavity together with some 20 cm of catheter to allow for growth and movement of the fetus. The uterus was then closed by a continuous gut suture; particular attention was given to ensuring the inclusion of the amnion in the suture line. Excessive loss of amniotic fluid may be made good by introducing warm sterile physiological saline into the sac before finally closing the uterus. Again it may be wise to anchor the catheter by a single tie to the uterine wall. The free end of the catheter was led out of the abdominal cavity through a small incision made about 3 cm laterally to the midline. Some 30 cm of catheter should be left within the abdominal cavity to allow for the descent of the uterus when the animal regains the upright posture. A skin tunnel and exit site was made by pushing a metal tube, whose internal diameter was just a little wider than the external diameter of the catheter, through the subcutaneous tissue to the depilated area on the flank of the mother and cutting down upon it. The catheter was then threaded along the tube which was withdrawn outwards through the body wall of the ewe to leave the catheter protruding. The exit wound was closed around the catheter by a purse-string suture. Interrupted chromic suture lines closed the peritoneum and also joined the abdominal muscles. Finally the abdominal skin incision was closed by interrupted silk sutures and the area was dusted with antibiotics (Rikerospray, Riker Laboratories) and sealed with a plastic dressing (Nobecutane, B.D.H. Pharmaceuticals Ltd.). The site was covered by a lint pad held in place by adhesive tape and bandage. The protruding catheters were protected by being contained in a bag sutured to the flank of the ewe. The animal was then removed from the operating theatre to a small straw-filled pen containing hay and water. Recovery is rapid, and within a few hours the animal is capable of walking and eating.

It is important to ensure the patency of the catheters by a daily flushing with heparin solution (5000 u/ml). The sealing pin is removed and the heparin administered through a Number 20 syringe needle which can just be inserted along the lumen of the 0·5-mm internal diameter catheter. It is of course essential that sterility is maintained and the use of swabs soaked in Hibitane has been found to be effective when handling the catheter tips. As an alternative to the intermittent flushing of the catheter a continuous slow infusion of heparin may be given. Suzuki & Plentl (1969) achieved this by implanting a self-contained electrolytic pump and heparin reservoir beneath the maternal skin, while an external rotary pump was used by Dawes, Fox, Leduc *et al.* (1972). In addition to heparin a broad-spectrum antibiotic such as ampicillin should also be given to both mother (450 mg i.m.) and fetus (50 mg i.v. or i.a.) daily. It has been shown that this dose when administered to the fetus maintains an effective concentration for 24 h (Alexander *et al.*, unpublished). When the technique was initially introduced (Meschia *et al.*, 1965b) it was thought necessary to administer progesterone to the ewe for several days prior to the implantation of the catheters; however many workers using this experimental preparation, including Meschia, Makowski & Battaglia (1970), now consider that this premedication is not required.

Sheep fetuses equipped with such catheters continue to develop, and indeed an unassisted vaginal delivery of a live lamb can occur despite the catheters. Three animals in which catheters had been placed in the fetal femoral arteries some three to five weeks earlier delivered viable animals, one on the 147th day and the other two prematurely on the 137th and 138th days. The catheter in two of the animals had been pulled out of the vessels during birth, while in the third the mother had chewed through the catheter (Alexander, Britton & Nixon, unpublished).

Whilst skin incisions in the fetal sheep heal satisfactorily this may not be true for fetuses of other species, such as the rabbit (Somasundaram & Prathap, 1970).

Among the recent applications of the technique in the sheep have been studies on the transfer of amino acids (Hopkins, McFadyen & Young, 1971), on the changes in heart rate, blood pressure and gas tensions of the fetus when the mother was made to exercise on a tread-mill (Emmanouilides, Hobel, Yashiro & Klyman, 1972), on the composition of the fetal and maternal bloods before and after parturition (Comline & Silver, 1972), and on the glucose, fructose and oxygen uptake in fetuses of fed and starved ewes (Tsoulos, Colwill, Battaglia, Makowski & Meschia, 1971).

The chronic implantation of catheters into fetuses of other species has been performed; these include the dog (Hodari & Thomas, 1969) and the rhesus monkey (Bangham, Hobbs & Tee, 1960; Suzuki & Plentl, 1969).

Chronic ablations

The feasibility of performing operations *in utero* upon fetal rats was demonstrated by Nicholas (1925); in this work limbs, tails and eyes were removed, yet viable young were born at term. Reference to other early works involving surgery upon the fetus with survival are given by Hess (1957).

General details on the surgical approach for recovery operations on the fetus have been given for the rat (Wells, 1950), rabbit (Cowen & Laurenson, 1959; Thomasson & Ravitch, 1969), dog (Barnard, 1957;

Jackson & Egdahl, 1960), pig (Rosenkrantz, Simon & Carlisle, 1968), and for the rhesus monkey (Chez & Hutchinson, 1969).

The use of the sheep fetus as a model upon which the effects of severing or removal of tissue could be observed at a later stage appears to stem from the work of Barcroft & Barron (1937a,b). They studied the effect of transections of the central nervous system at different levels upon the reflexes seen when the fetus was exteriorized a number of days after the initial operation. Later they showed (Barron & Barcroft, 1938) that a fetus in which a portion of a forelimb had been amputated when it was 50 d old could survive and grow for a further 90 d *in utero*.

A number of papers has been published concerning the involvement of the fetal pituitary and adrenal glands in the initiation of parturition. As term approaches, the release of adrenocorticotrophin from the fetal pituitary increases. As a consequence of this elevation there is a stimulation of the adrenal cortex resulting in a rise in the output of corticosteroids. These hormones may bring about a suppression of the myometrial progesterone concentration with the result that uterine contractility increases (Liggins, 1969b). This picture has been built up by demonstrating that destruction of the pituitary in fetal sheep by electrocoagulation prolongs gestation (Liggins, Kennedy & Holm, 1967). A method for the adrenalectomy of sheep fetuses was introduced by Drost (1968) and in fetuses in which this operation had been carried out there was again a prolongation of gestation (Drost & Holm, 1968). Administration of adrenocorticotrophin or cortisol by continuous infusion into the fetuses, a month or more before term, could result in parturition 4–7 d after the start of the infusion; no such premature birth occurred however when these hormones were infused into the ewe (Liggins, 1968, 1969a). However, Lanman & Schaffer (1968) found that of three sheep fetuses successfully decapitated *in utero*, two delivered at term, while the third delivered a large dead fetus after a gestation of 168 d.

Methods have been devised for the hypophysectomy of fetuses of other species *in utero*; in rabbits by decapitation (Jost, 1947), in rats by electrocoagulation (Ostergard & Contopoulos, 1970) and in monkeys by implanting an yttrium[90] pellet in the hypophysis (Hutchinson, Westover & Will, 1962).

The fetal sheep has also featured in immunological investigations. It was shown by Schinckel & Ferguson (1953) that skin homografts could be rejected by the eightieth day of gestation. Fetal kidneys transplanted into the neck region of other fetuses showed a cessation of urine flow by the fourth day and a complete rejection of the graft 7–9 d after transplantation (Niederhuber *et al.*, 1971).

Thymectomy of the sheep fetus *in utero* was shown by Cole & Morris (1971a) to result in a reduction in the number of circulating lymphocytes; while splenectomy was without effect. Other species in which the fetuses have been subjected to the removal of the thymus include the rabbit (Kisken & Swenson, 1968), the dog (Fisher *et al.*, 1964; Dixit & Coppola, 1970) and rhesus monkey (Parshall & Silverstein, 1969).

The immunological response by the fetal sheep varies with the antigenic stimulus. By intramuscular injection into the fetus (made through an initially exposed but intact uterus) it was shown by Silverstein *et al.* (1963) for example that antibody could be produced towards bacteriophage as early as the sixty-sixth day of gestation; on the other hand there was no production of antibody throughout fetal life when diphtheria toxoid was given. Immunoglobins were also found after injecting fetuses with rubella vaccine virus (Osburn *et al.*, 1971). Cole & Morris (1971b) could detect no difference in the production of antibodies between sheep fetuses thymectomized *in utero* and control animals.

The effects of ligation of an umbilical artery in sheep have been studied by Emmanouilides, Townsend & Bauer (1968). They showed that survival was possible for more than 7 d after the operation in some of the fetuses despite signs of malnutrition.

Techniques for the *in-utero* ligation of the ureters or removal of the kidneys have been developed in several species: in the sheep (Beck, 1970), in the rabbit (Berton, 1970), and in the rhesus monkey (Hutchinson, Bashore & Will, 1962).

Perfused placenta

In studying the passage of substances across the placenta or of the synthesizing ability of the tissue it is often advantageous to remove the fetus and to substitute an *in-situ* artificial perfusion system.

The perfusion may be of two types—'recirculating' or 'through'. In the recirculating type of perfusion the venous drainage passes into a reservoir from where it is pumped continuously into the umbilical arteries. In the 'through' perfusion the pump draws upon a large reservoir volume and the umbilical venous outflow is collected in separate containers (Fig. 4.4).

The *in-situ* recirculating perfusion of the sheep placenta was first carried out by Huggett, Warren & Warren (1951) with physiological saline to prime the extracorporeal circuit. This medium proved to be of limited use owing to a rapid development of oedema; a somewhat longer life of the preparation was obtained with plasma. A further prolongation was achieved by using adult sheep blood obtained either from an abattoir or from the mother (Alexander, Huggett, Nixon & Widdas, 1955). Failure in attaining a satisfactory perfusion has never been attributable to the mixing of adult blood with the residual fetal blood in the placenta. Artificial

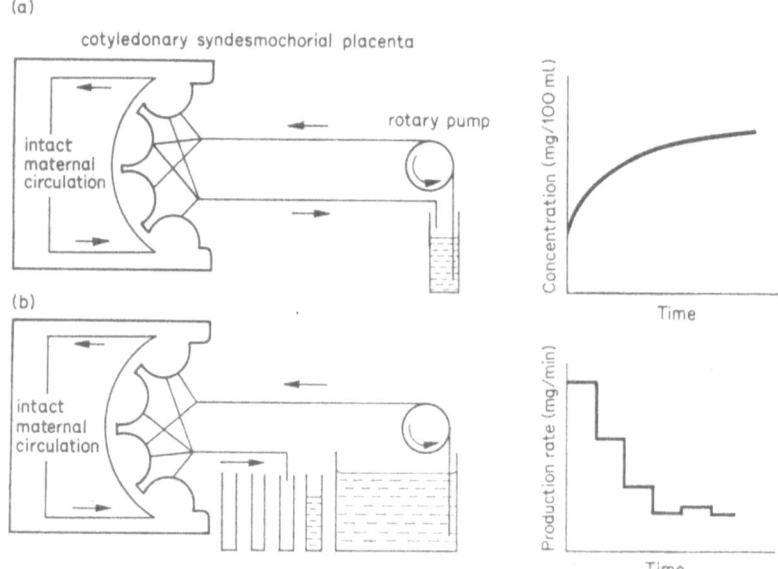

Fig. 4.4. Diagrams illustrating the two types of *in-situ* placental perfusion techniques, together with the changes in the fructose concentration of the perfusate with time. (a) Recirculating perfusion, (b) 'through' perfusion

perfusates containing Dextran (Metcalfe *et al.*, 1965) or polyvinylpyrrolidone (Nixon, unpublished) appear to be suitable and may play a particularly useful role when large volumes of perfusate are required as in the 'through' perfusion technique.

The perfusion apparatus consists essentially of a variable-speed rotary pump which draws through a filter (that found in a disposable blood administration set is ideal) upon a reservoir of blood, maintained at body temperature. The output of the pump is then distributed through a Y-shaped connector to the two glass cannulae inserted into the umbilical arteries. Venous drainage from the placenta is collected into a single outflow through a similar Y-shaped connector from the two umbilical vein cannulae. A pressure-limiting device situated between the pump and the placenta allows blood to pass back into the reservoir should the pressure exceed a value usually fixed at about 40 mm Hg. This limitation in pressure may be brought about by channelling the blood through a thin-walled latex tube (12-mm diameter Paul's drainage tubing) enclosed in a water-filled rigid tube upon which the desired external pressure can be applied.

The umbilical blood flow may be ascertained by passing the blood through an electromagnetic flow meter sited beyond the pressure-limiting device. An alternative and simpler method is to lead the outflow from the placenta first into a calibrated tube; the flow can then be measured by noting the time taken to fill the tube to the mark, when the silicone rubber tubing attached to the end is temporarily occluded by applying a pair of artery forceps. Also inserted along the inflow tubing are sites for temperature and pressure measurements. A schematic diagram of the perfusion circuit is depicted in the upper portion of Fig. 4.5.

All the glass used in the circuit should be treated with silicone ('Repelcote', Hopkins & Williams Ltd.), and the connections made with silicone rubber tubing (Esco Rubber Ltd.). In order to minimize the circuit volume the tubing should be as short as is possible without kinking.

Before use the circuit is cleaned by connecting the cannulae of the umbilical arteries and veins together and recirculating distilled water for 15–30 min. This is then discarded and is followed by three washes with physiological saline for the same duration.

A range of cannulae should be available to cater for umbilical vessels of different sizes. Those selected should have a diameter as large as is commensurate with their rapid insertion. The cannulae may be constructed by tapering glass tubing (5-mm internal diameter, 7-mm external diameter) and forming a neck some 1–2 cm from the tip whose angle should not be too acute.

The fetus is exposed by Caesarean section in a ewe that has received spinal anaesthesia, with supplements of 5% sodium thiopentone where necessary. About 200 ml of blood are withdrawn from the maternal dorsalis pedis artery in the presence of heparin to prime the perfusion circuit. While the umbilical vessels are prepared for cannulation saline is removed from the circuit and replaced with blood. This is then recirculated in the perfusion system and the speed of the pump is adjusted so that a delivery of about 100 ml/min is attained; the efficiency

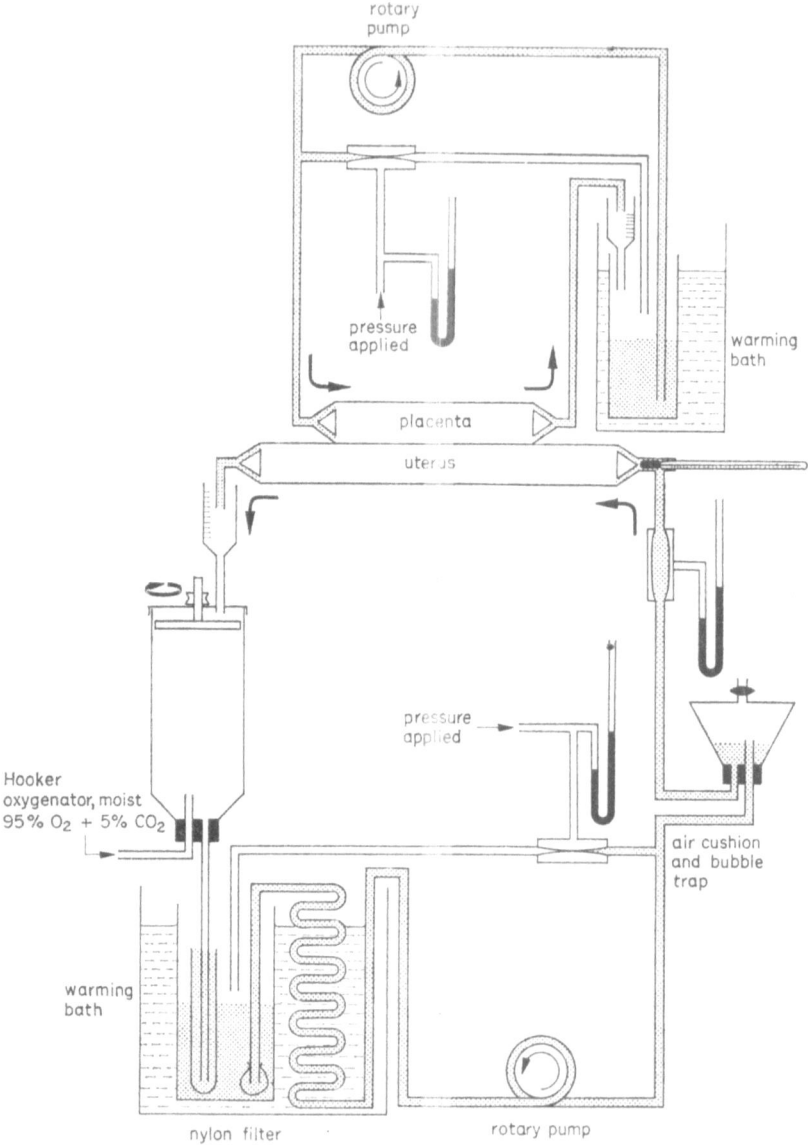

Fig. 4.5. Schematic diagram of the circuits for perfusing the placenta and uterus (Nixon, 1963; reproduced by courtesy of the *Journal of Physiology*)

of the bypass is checked and any air bubbles in the circuit are removed.

A small incision is made into the Wharton's jelly some 10 cm from the abdominal wall of the fetus. Into this is inserted a spinal needle, whose tip is guarded by a small sheath of polythene tubing, attached to

110

a syringe filled with formol saline (4% formaldehyde in 0·9% NaCl solution). As the tip is advanced the jelly is infiltrated with the formol saline. The region so treated extends for about 5 cm on either side of the incision point. After 5 min the coating of the umbilical cord is stripped off to leave the umbilical vessels free for cannulation. Around each of the four vessels is placed a loosely tied braided silk surgical suture (B.P.C. 3 Pearsall).

Arterial forceps are applied to the silicone rubber tubing (which joins pairs of cannulae to the Y-shaped connectors) of the four cannulae, simultaneously with switching off the pump. Tubing that temporarily connected the cannulae of the umbilical arteries with the veins is then removed.

The umbilical cord is occluded by applying two large arterial forceps and the fetus is removed by cutting between them. An assistant occludes the vessels in turn, by traction on the ligature, while they are cannulated and then ties the knot so that the ligature lies in the neck of the cannula. When all four vessels have been cannulated the arterial forceps are released simultaneously and at the same time the pump is switched on.

A prolonged stasis of placental blood flow during the period of cannulation should be avoided. With practice the time from the initial application of the forceps to the umbilical cord and the establishment of perfusion can be as short as 1–2 min. At this stage the lie of the cannulae and vessels should be checked to ensure optimum outflow of blood from the placenta.

Under favourable conditions the *in-situ* perfusion of the sheep's placenta may continue for at least 4 h. Useful parameters for gauging the success of the perfusion are: constancy in the volume of blood in the reservoir, stability of the haemoglobin concentration, a rise in the fructose concentration, and no change in the concentration of myo-inositol in the perfusate on its intravenous injection to the mother (Campling & Nixon, 1954). Nevertheless examination of placental cotyledons before and after a satisfactory *in-situ* perfusion does reveal some swelling of the fetal trophoblasts, but with minor changes in the other elements (Hoyes & Nixon, in preparation). An example of these changes, as observed by the electron microscope following a recirculating perfusion of 89 min in a 122-d-old sheep placenta, is seen in Plate 4.1. Such an oedematous state may have some influence upon the trans-placental passage of substances in both directions.

The presence of fructose in the fetal blood is a characteristic of those species that possess either an epitheliochorial or a syndesmochorial type of placenta (Huggett & Nixon, 1961; Nixon, Huggett & Amoroso, 1966). In the sheep the plasma concentration of fructose always exceeds that of the glucose, although there is a decline as term approaches (Fig. 4.6). Extensive investigations upon the production of this carbohydrate by the syndesmochorial placenta of the sheep have been carried out with the *in-situ* perfusion techniques (Huggett, Warren & Warren, 1951; Alexander, Huggett, Nixon & Widdas, 1955; Nixon, Alexander & Huggett, 1966; Britton, Huggett & Nixon, 1967).

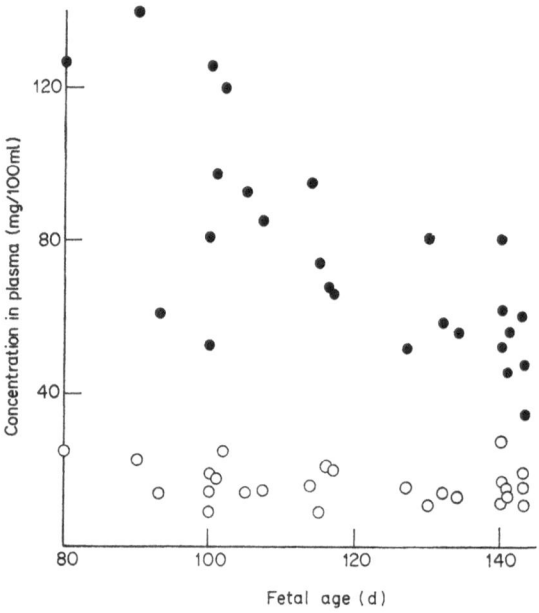

Fig. 4.6. Plasma concentrations of glucose (o) and fructose (●) in twenty-seven fetal sheep, in whom blood samples were obtained immediately on exposure, plotted against age

The technique has also demonstrated the accumulation of *a* amino nitrogen in the perfusate of the sheep's placenta (Nixon, unpublished). An example of the rise in perfusate blood concentrations of *a* amino nitrogen and fructose, together with the stability of the glucose concentration, is shown in Fig. 4.7; the data were obtained from the same animal whose placental histology is shown in Plate 4.1.

Perfusion studies in the sheep have revealed the absence of effective placental transfer of acetoacetate (Alexander, Britton & Nixon, 1966b), insulin (Alexander *et al.*, 1969b), glucagon (Alexander *et al.*, 1971) and conjugated bilirubin (Alexander, Andrews, Britton & Nixon, 1970).

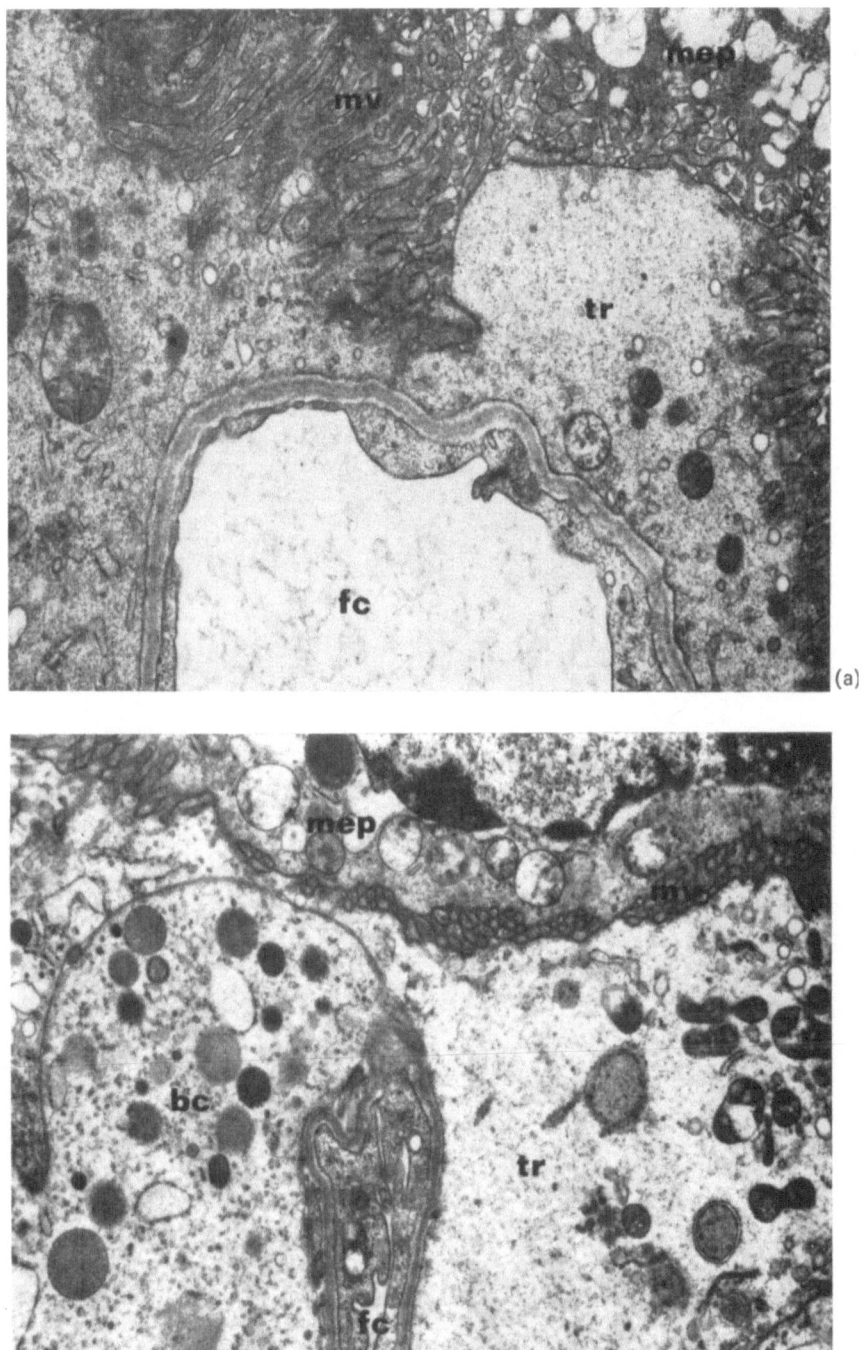

Plate 4.1. Electron micrographs of sheep placental cotyledons (a) before and (b) after 89 min *in-situ* recirculating perfusion. Magnification in (a) ×19700 and (b) ×13000. Gestational age 122 d. fc, fetal capillary; tr, trophoblast; mep, maternal epithelium; mv, microvilli; bc, binucleate cell

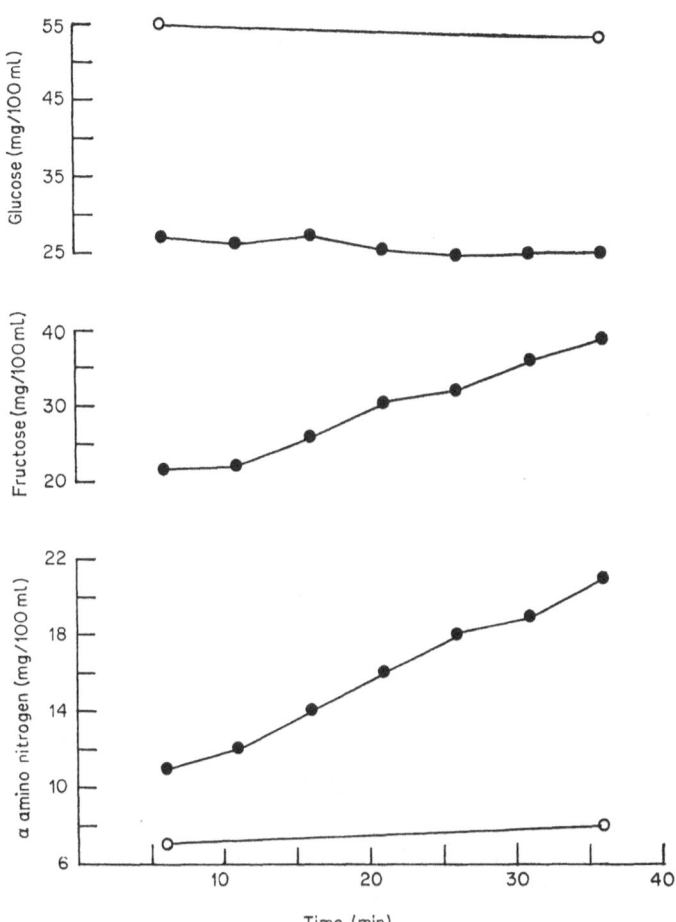

Fig. 4.7. Changes occurring in the concentration of glucose, fructose and α amino nitrogen in the plasma of the blood perfusing a sheep placenta *in situ* by the recirculating technique (●). Maternal values (o). Gestational age 122 days

The perfused sheep placenta has also been used by Metcalfe *et al.* (1965) to study the transfer of carbon monoxide and nitrous oxide. They also examined the transfer of these gases in retrograde perfusion; that is, with the perfusate passing to the placenta along the umbilical veins. Studies on the transfer of argon, ether and water have been made by Bissonnette & Gurtner (1970). The technique has been used by Campbell *et al.* (1966) to measure the oxygen consumption of the tissue and fetal membranes.

In-situ perfusions of placentae have been carried out in other species:

goat (Neil, Walker & Warren, 1961), rabbit (Faber & Hart, 1966; Baker & Morgan, 1970), cats and dogs (Dawson & Robson, 1940) and for short periods in the human subject (Mikhail, Wiqvist & Diczfalusy, 1963; Jaffe & Ledger, 1966). However the species that has received considerable attention is the guinea-pig. Money & Dancis (1960) developed a technique whereby a recirculating or a 'through' type of perfusion could be carried out in this species. With these approaches the transfer across the guinea-pig placenta has been investigated for carbohydrates (Folkart, Dancis & Money, 1960; Ely, 1966), amino acids (Dancis *et al.*, 1968), lipids (Kayden, Dancis & Money, 1969), oestrogens (Levitz *et al.*, 1960) and inorganic ions (Dancis & Money, 1960; London, Money & Rawson, 1964; Twardock & Austin, 1970). A 'through' perfusion technique with a solution of electrolytes and Dextran has been used by Reynolds & Young (1971) to study the transfer of amino acids.

Perfused pregnant uterus

Investigations on the transfer of a substance across the placenta from the fetus to the mother may be hampered by the dilution of the material in the larger maternal equilibrating space, and by the efficiency of removal mechanisms. These difficulties may be overcome by perfusing the pregnant uterus.

In the method introduced by Nixon (1963) blood from a reservoir was drawn through a nylon filter and heat exchanger by a variable-speed rotary pump. A pressure-limiting device allowed blood to pass back to the reservoir should the perfusion pressure exceed the elected value. Along this arterial perfusion line was situated an air cushion, to dampen excessive pulsatile flow imparted by the pump, and also to serve as a bubble trap. The approximate pressure of the perfusing blood was measured together with its temperature further along the line. A Y-shaped connection distributed the blood to cannulae inserted into the uterine arteries on each side of the uterus. Catheters inserted in the two uterine veins led blood through a Y-shaped connector to the top of a vertically held spinning-disc oxygenator, gassed with moist 95% oxygen and 5% carbon dioxide. If the venous outflow is led first through a graduated tube, then a measure of the flow may be obtained by temporarily occluding the flow to the oxygenator. The uterine perfusion circuit is schematically represented in the lower half of Fig. 4.5. The apparatus was assembled, with the arterial and venous catheters joined together, and washed by recirculating first distilled water and then several changes of physiological saline. Some 700–800 ml of defibrinated heparinized blood, obtained from a donor animal or from an abattoir, was used to prime the circuit.

General anaesthesia was induced in the ewe by the intravenous administration of sodium thiopentone. The ewe was placed in a supine position with the rear of the body elevated at an angle of about 15° to the horizontal. A midline abdominal incision was then made extending from the xiphoid process to the pubis. The skin was reflected from the underlying tissue for some 15–20 cm on each side of the incision. The abdominal cavity was entered and the uterine arteries and veins exposed as deeply as possible to allow unbranched vessels to be cannulated. Great care is required in preparing these vessels for cannulation owing to their delicate nature and to the scarcity of connective tissue.

In the early stages of the development of the preparation, difficulties were encountered in the catheterization of the uterine veins. The catheter could only be advanced a short way before resistance was felt, this was found to be owing to the presence of numerous valves which may be distributed at 1-cm intervals along the uterine vein (Nixon, 1963). Valves have, incidentally, been observed in the pregnant uteri of dromedary (*Camelus dromedarius*) and hog deer (*Axis porcinus*) and were similar to those encountered in the pregnant sheep (Nixon, unpublished).

To secure an adequate insertion the tip of the catheter must be advanced for about 5 cm in a retrograde direction to the flow of blood through the valves of the veins. Another difficulty encountered was the tendency of the veins to be occluded by twisting due to the weight and torsion of the catheters and tubing.

Just before catheterization 25,000 u heparin were injected intravenously into the ewe. The catheters were then inserted and tied into the veins first since they lie more dorsal than the arteries. The extracorporeal perfusion circuit was then started. A strong tape was tied firmly around the uterus in the cervical region. Finally the remainder of the broad ligament was occluded by applying large arterial forceps thus isolating the uterus from the rest of the body. A continuous spray of warm physiological saline kept the tissues moist. No attempt was made to detach the uterus from its position *in situ*.

The uterus was opened and access to the fetal circulation gained through the catheterization of tributaries of the umbilical vessels.

With this preparation it was possible to demonstrate that fructose present in the fetal circulation, contrary to what was originally thought, could pass across the placenta into the maternal circulation (Nixon, 1963).

Perfusion of the pregnant sheep uterus near term has also been carried out by Peirce *et al.* (1970). In this technique the uterus was transferred to a large container filled with maternal blood. The uterine arteries were cannulated and a recirculating perfusion established with the pump drawing upon the pool of blood. Venous drainage from the organ into the pool was from uncannulated cut vessels. Gaseous exchange was achieved with a membrane oxygenator. The perfusion time varied from 30 to 390 min and live fetuses were obtained at the end of some experiments.

A preparation of the *in-situ* type has been developed for the dog by Benzi, Berté, Crema & Arrigoni (1968), and used by them to investigate the distribution in fetal tissue of metabolites of aminopyrine after the introduction of this substance into the perfusion circuit.

Combined uterine–placental perfusion

The survival of the fetus in the perfused-pregnant-uterus preparation was limited to about 60 min. While the cause of fetal death was not ascertained it is probable that the perfusion system was inadequate for maintaining the requirements of the uterus, placenta and fetus. By eliminating the fetus and substituting for it a perfusing system, the mass of metabolizing tissue was brought more within the capacity of the extracorporeal supporting system. The combined uterine–placental perfusion preparation therefore consists of two perfusion systems and is depicted schematically in Fig. 4.5 (Nixon, 1963).

With the combined uterine–placental preparation in the sheep it was shown that the rate of transfer of fructose from the blood perfusing the uterus was similar to that observed in the perfused pregnant uterus preparation. In addition the behaviour of urea and inositol, when added to the placental perfusate, was observed. The results (Fig. 4.8) show that during the first 60 min of the combined perfusion the concentrations of urea and inositol in the uterine perfusate remain stable whilst that of fructose rose. When fructose, urea and inositol were later added to the placental perfusate, the concentration of urea rose as equilibration across the placenta occurred. The rate of transfer of fructose was now

116

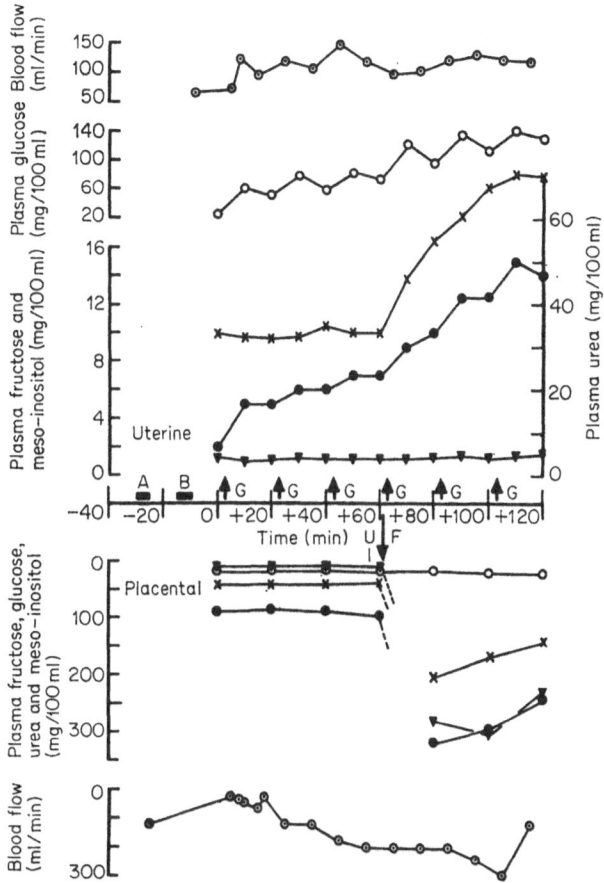

Fig. 4.8. Concentrations of fructose (●), urea (X), glucose (o) and meso-inositol (▼) in the plasma of the perfused uterus (above the abscissa) and of the perfused placenta (below the abscissa) of the sheep. At F, U, I 1 g each of fructose, urea and meso-inositol were added to the placental circuit. G represents the addition of 0·5 g of glucose to the uterine circuit. Establishment of the placental and uterine perfusions is indicated by A and B respectively. Volumes of the uterine and placental circulations were approximately 800 and 300 ml, respectively. Gestational age 118 days. (Nixon, 1963; reproduced by courtesy of the *Journal of Physiology*)

somewhat faster than in the previous phase of the experiment. On the other hand there was no change in the concentration of inositol, indicating the functional integrity of the placental barrier.

Fischer (1967) has described a method whereby the uterus and placenta of the guinea-pig were perfused simultaneously.

117

Perfused feto-placental unit

In this preparation introduced by Lerner & Diczfalusy (1968) the isolated fetus with its placenta is perfused by a 'through' perfusion technique. Its application has been confined to the investigation of steroid metabolism in the human fetus and placenta.

At the elective termination of pregnancy, between the seventeenth and twentieth week, the pre-viable human fetus was placed in a saline bath. The placenta, still connected to the fetus, was transferred to a separate bath filled with blood which represents the blood on the maternal side of the placenta. A catheter inserted into an umbilical artery perfused the preparation with oxygenated compatible blood. The oxygenated blood

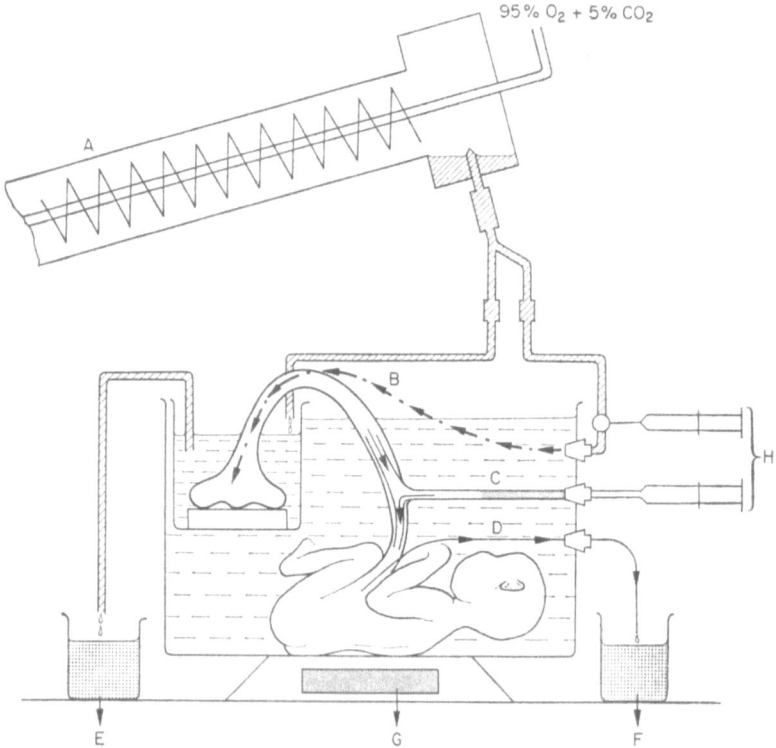

95% O_2 + 5% CO_2

Fig. 4.9. System for the *in-vitro* perfusion of the complete feto-placental unit at mid-pregnancy. A, oxygenator; B, placental perfusion via an umbilical artery; C, 'T'-tube in an umbilical vein for fetal perfusion; D, catheter in the other umbilical artery to collect perfusate; E, perfusate from the 'maternal' side; F, container with fetal perfusate; G, heater; H, infusion pumps. (Reproduced by courtesy of Excerpta Medica Foundation)

also fed the bath in which the placenta was immersed. The outflow from this bath was collected and analysed for substances which may have passed from the feto-placental unit across the placenta into the maternal circulation. Blood was also obtained for analysis from the preparation through a catheter inserted into the other umbilical artery. Substances under investigation may be introduced into the placental flow through the umbilical-artery catheter conveying the perfusate, or into the fetus through a T-shaped catheter inserted in the umbilical vein (Fig. 4.9).

The results obtained with this preparation, taken in conjunction with those derived from experiments where either the placenta or fetus were separately perfused, indicate a complementary role for the two units in the metabolism of steroid hormones. Their interdependence is explained by the necessary enzyme systems involved in steroid metabolism not being present in both the placenta and fetus. Thus some of the enzymes required are to be found functional only in the placenta while others operate only in the fetus.

The extensive studies on steroid metabolism by Diczfalusy and his colleagues involving the use of the preparation have been reviewed by Diczfalusy (1969, 1970).

Fetal perfusion

The maintenance of the fetus detached from its placenta results in a preparation upon which the nutritional requirements of the fetus may be studied directly. Thus the complications introduced by the trans-placental passage of material in both directions or by placental synthesis, storage and utilization are eliminated. A perfusion system that supports the isolated fetus has been called an artificial uterus by Greenberg (1951) or artificial placenta by Lawn & McCance (1961). It would however seem desirable to refer to the support of the isolated fetus as extra-corporeal maintenance, since perfusion circuits are substituted for only a few of the complex functions of the placenta (Nixon, 1972b). In addition to investigations upon the nutritional requirements of fetal life, the technique offers the prospect of an applied use. The possibility exists that, coupled with further knowledge of the physiology and bio-chemistry of fetal life, the technique could be of use in the management of the prematurely delivered human infant, or in the treatment of the acutely hypoxic infant at term.

The first experiments on the survival of detached fetuses were apparently made by Nicholas (1934) on rats. The only extracorporeal support that he gave to the young fetuses was to immerse them in various physiological media. Their survival was judged solely upon histological observation.

A more recent extension of this approach has been made by Goodlin (1962, 1963) in which the submerged fetuses were subjected to hyperbaric oxygen. With this procedure the fetal hearts of rabbits and mice were found to be still beating 30 h after detachment. Physiological and biochemical parameters are, however, not easily monitored under these conditions. An additional objection to the technique lies in the fact that many tissues are known to be damaged by high oxygen tensions (Davies & Davies, 1965). Goodlin & Perry (1964) concluded that while hyperbaric oxygen may be effective in maintaining cardiac activity in a 22-week human fetus the thicker skin and smaller ratio of the surface area to body mass of a full-term infant would in any case render the technique ineffective.

An extracorporeal perfusion system was designed by Thomas (1948). Using this system he and his colleagues were the first to maintain isolated fetuses by a perfusion technique (Thomas, Salomon & Salomon, 1948a, b; Thomas *et al.*, 1948). The fetuses they chose for these pioneering investigations were from cows and sheep.

In the system designed by Nixon, Britton & Alexander (1963) for the fetal sheep the arterial outflow from the fetus passes to the top of a vertically held spinning-disc oxygenator. After gassing with moist $3\% \ CO_2 + 97\% \ O_2$ the blood accumulates in a small reservoir situated at the bottom of the oxygenator. A variable-speed rotary pump conveys the blood from the reservoir through an air cushion and bubble trap, filter and electromagnetic flow meter back to the umbilical veins. The apparatus is shown schematically in Fig. 4.10; in this version of the apparatus gas absorbers for the collection of radioactive carbon dioxide are included.

The apparatus, assembled and carefully washed as with other perfusions, was primed with about 200 ml of heparinized maternal blood. The cord was infiltrated with formol-saline up the abdominal wall of the fetus and the vessels stripped of Wharton's jelly. An umbilical artery and vein were tied off and a temporary anastomosis established between the two vessels. The other pair of umbilical vessels were cannulated and incorporated into the perfusion circuit which was then started. The circuit was completed by the separation of the anastomosis and the inclusion of these vessels.

The speed of the pump was adjusted so that the blood flow was appropriate for the weight of the fetus (100–200 ml/min kg). A further consideration was the regulation, by means of a screw clip, of the outflow of blood from the fetus into the oxygenator, so as to maintain a

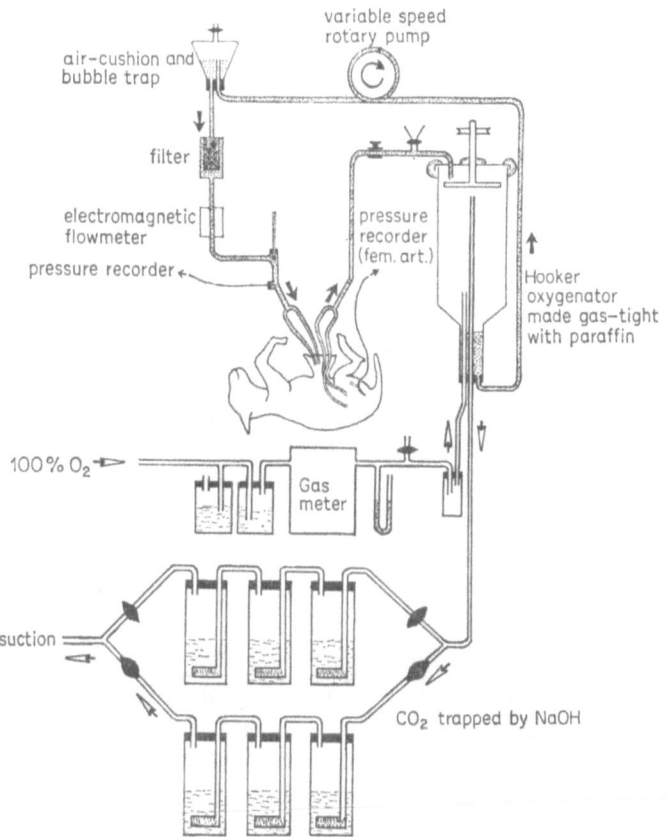

Fig. 4.10. Perfusion circuit for the isolated sheep fetus. The basic perfusion circuit has here been modified for the collection of radioactive carbon dioxide. (Alexander, Britton & Nixon, 1970; reproduced by courtesy of the *Quarterly Journal of Experimental Physiology*)

constant volume of blood in the reservoir. Stability of this blood level therefore indicated a balance between umbilical arterial outflow and umbilical venous inflow. An imbalance between these two flows often occurred about one hour after the start of the perfusion, so that inflow exceeded outflow. This situation necessitated the opening of the screw clip or a reduction in the speed of the pump in order to re-establish constancy of the corporeal and extracorporeal blood volumes. In a long perfusion successive corrections may eventually reduce the blood flow below an acceptable level. This was particularly common in the early experiments where cannulae were inserted just into the umbilical arteries of the cord. It seemed likely that vasoconstrictor agents were released and that the intra-abdominal portion of the umbilical arteries was particularly sensitive to their action. In later experiments acceptable blood flows were maintained for prolonged periods by passing catheters (adapted from umbilical feeding tubes) along the umbilical arteries so that their tips lay in or near the internal iliac arteries—a procedure that channelled the blood through the intra-abdominal arteries.

The fetus may either be submerged in a saline bath maintained at body temperature or placed upon an electric blanket with the head enclosed in a bag filled with saline. The use of the former approach gives a more satisfactory method of temperature control, prevents dehydration and by giving support reduces the bruising of the fetus by its own weight. It is necessary to secure the fetus to the bottom of the bath to restrict its movements which if unrestrained may occlude the catheters. Spontaneous natural movements of the head, trunk and legs occur occasionally; however, painful stimuli applied to the skin elicit withdrawal reflexes. On the other hand perfusion carried out on fetuses that are not submerged greatly facilitates experiments involving the collection of urine samples.

The technique outlined above has supported the isolated fetal sheep for a number of hours; the youngest animal perfused was 72 d and weighed 150 g, while the heaviest full-term fetus weighed 5 kg. The oxygenator used was however not very efficient in maintaining fetuses near term since it could not cope satisfactorily when blood flows of the order of 500 ml/min or more were required. One good feature of the vertically held spinning-disc oxygenator is that the volume of blood it contains is relatively small when compared with either horizontal multi-disc or membrane oxygenators. This enables the volume of blood in the extracorporeal circuit to be kept small, and so facilitates the detection of changes in concentration of plasma constituents. Furthermore, changes in the distribution of blood between the fetus and the extracorporeal circuit can be readily detected by looking for departures from constancy in the volume of blood in the reservoir. Walker, Rose & Simons (1962) have controlled this distribution of blood by using a servomechanism by which the weight of the fetus determines the speed of the perfusion pump.

Validation of the technique outlined above was obtained by subjecting sheep fetuses within a few days of term to a short period of whole-body perfusion and then detaching them from the circuit (Nixon, Britton & Alexander, 1963; Alexander, Britton & Nixon, 1964b). Since fetuses of this age survive premature delivery their behaviour over the following 24 hours of independent existence could be observed. Each animal appeared to be unaffected by the whole-body perfusion. Suckling and gait were good, they bleated, responded to light and sound, showed interest in their surroundings and had no retention of faeces or urine. These features were particularly clearly seen in one animal which was observed together with its non-perfused twin (Nixon, 1964). Although the observation period was limited to 24 h, chiefly because no aseptic techniques or antibiotic coverage had been used, it was felt that had the perfusion produced deleterious effects these would have become apparent during this period. At the end of this time the animals were killed and post-mortem examination showed no macroscopic changes. Survival of fetal sheep following perfusion has also been reported by Callaghan et al. (1963), Callaghan, Maynes & Hug (1965) and Tchobroutsky, Clauvel & Laurent (1966).

As a further consideration of an applied aspect of the technique a target duration of 24 h of whole-body perfusion of the sheep fetus was set and attained (Alexander, Britton & Nixon, 1968). A 2-d perfusion of a sheep fetus has been reported by Zapol et al. (1969). However, our technique was designed principally as a means for investigating aspects of metabolism directly in the fetal sheep. With this preparation it was shown that an avid utilization of glucose occurred. From measurement of the oxygen consumption glucose was concluded to be the main energy source in fetal life. Fructose on the other hand was only slowly utilized (Alexander, Britton & Nixon, 1964a, 1966a, 1970). Administration of anti-insulin serum suggested that fetal tissues were insensitive to endogenous insulin (Alexander, Britton, Cohen & Nixon, 1970).

Acetate and ketone bodies are capable of being rapidly removed from the perfusate of the isolated fetal sheep, but it is likely that the fetus does not have an opportunity to utilize these substances, since the placenta appears to be relatively impermeable to them (Alexander, Britton & Nixon, 1966b, 1967, 1969). Removal of amino acids has been demonstrated (Alexander, Britton, Nixon & Cox, 1970). Perfusion studies have indicated that the fetus produces inositol (Nixon, 1968).

Harned et al. (1957), Callaghan & Angeles (1961), Tchobroutsky et al.

(1966) and Zapol *et al.* (1969) have also designed perfusion circuits for the maintenance of the isolated sheep fetus.

Recently Pierrepoint *et al.* (1971) have introduced a technique that will prove of great importance in physiological and biochemical investigations on the isolated sheep fetus. In this method the animal is merely detached from its umbilical circulation and placed in a thermostatically controlled saline bath, and its lungs are artificially ventilated (Fig. 4.11). The limitation of the technique lies in its being only applicable to fetuses

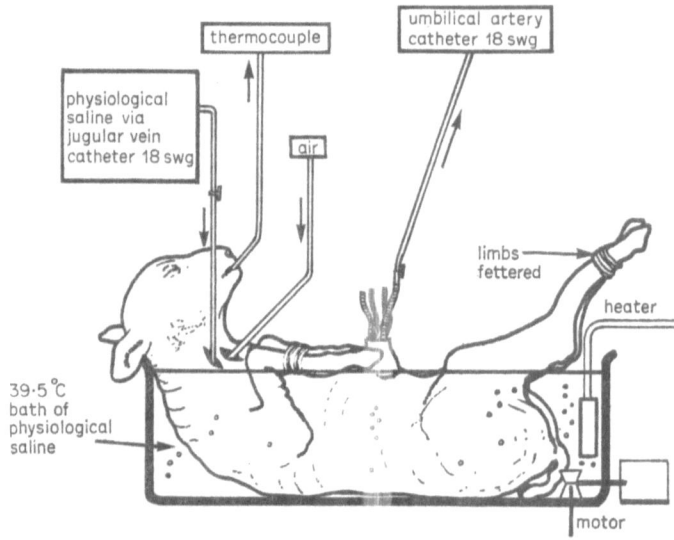

Fig. 4.11. Isolated fetal sheep preparation of Pierrepoint *et al.*, 1971. (Permission to use this diagram, made for the authors by Mrs J. Morgan, is gratefully acknowledged)

of an age at which sufficient lung maturation has occurred to permit adequate gaseous exchange across the alveolar membrane; however, a fetus as young as 119 d was satisfactorily maintained in this way. Thus by this simple and elegant method the need for an oxygenator, pump, priming blood, a complexity of tubing, to say nothing of the dexterity in connecting the fetus to and maintaining it on the perfusion circuit, is eliminated.

A method for the perfusion of the pre-viable human fetus obtained at therapeutic abortion was introduced by Westin, Nyberg & Enhörning (1958). Extensive use of a 'through' perfusion technique under hypothermic conditions has been made by Diczfalusy and his colleagues in their investigations upon steroid metabolism in human fetal life

(Diczfalusy, 1969). In addition to perfusing intact human fetuses those in whom adrenalectomy has been carried out have also been used (Wilson *et al.*, 1966). Human fetuses have also been maintained with a recirculating perfusion at normal body temperature by Chamberlain (1968).

The only other species upon which fetal perfusion has been carried out is the pig (Lawn & McCance, 1961, 1962, 1964, 1967). In these investigations a recirculating perfusion was used and in some experiments a dialyser was incorporated into the circuit. These workers observed that in the fetal pig, as with the fetal sheep, there was a removal of glucose but not fructose from the perfusate.

Premature delivery

Although the lamb is well developed at birth it can survive premature delivery only from about the 139th day (Andrews, Britton & Nixon, 1961; Dawes & Parry, 1965). Detached premature lambs have been used in investigations on the influence of birth upon the development of hepatic enzymes and on the alteration in plasma composition that occurs following birth (Andrews, Britton & Nixon, 1961; Alexander, Britton, Cohen & Nixon, 1969; Alexander, Assan & Nixon, unpublished).

The lambs were delivered by Caesarian section from ewes under a spinal anaesthesia. The umbilical cord was stretched to induce contraction of the blood vessels and then tied and cut. Usually respiratory movements occurred at once. Occasionally, however, some assistance was required, either by rocking or by mouth to mouth resuscitation after drainage of the nasopharynx. The lambs were quickly dried and stimulated generally by rough towelling. Within 1–2 h the lambs were capable of standing voluntarily and became quite active within 5–7 h. They were confined to small pens with an ambient temperature of 15–22°C. The animals were bottle-fed on warm cow's milk at intervals of 3–4 h throughout the 24 h. They remained in good condition on this feeding regimen; one animal which, at Caesarian section at 142 d fetal age, weighed 2·51 kg, had attained the weight of 4·4 kg 13 d later, when its liver was perfused.

Lambs have also been used in investigations into the features of germ-free life (Luckey, 1963). With strict aseptic technique the uterus was removed from the ewe and transferred to a sterile chamber where it was opened and the fetus detached from its placenta. Incorporated in the chamber were devices for the long-term maintenance of the animal.

Cardiovascular parameters have been measured in the calf before and after birth, through implanted catheters (Olsen & Allred, 1968).

Isolated organs and tissues

Fetal liver

The liver is an organ which, because of its vasculature, is suitable for extracorporeal maintenance. With a recirculating perfusion technique some aspects of hepatic metabolism have been investigated in the fetal sheep.

The fetus was exposed by Caesarian section and lightly anaesthetized with intravenous thiopentone, and the cord was divided. An abdominal incision was made so as to expose the common umbilical vein and to gain access to the vena cava. A schematic diagram of the perfusion circuit devised by Andrews *et al.* (1960) is shown in Fig. 4.12. Heparinized

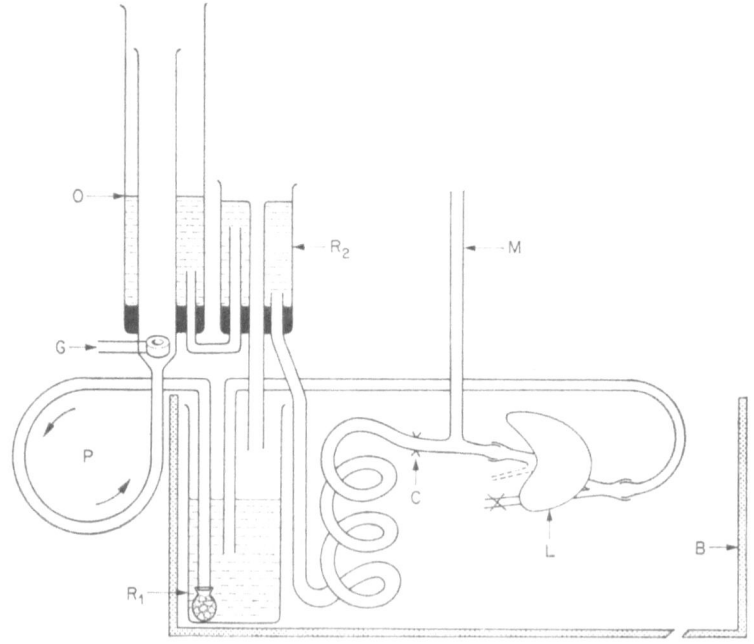

Fig. 4.12. A schematic diagram of the fetal liver perfusion circuit. B, saline bath; C, screw clip; G, 95 % O_2 + 5 % CO_2; L, liver; M, manometer; O, oxygenator; P, pump; R_1, reservoir; R_2, constant-level reservoir and bubble trap. (Andrews *et al.*, 1960; reproduced by courtesy of the *Journal of Physiology*)

maternal blood was used to prime the circuit. Blood flowing from a cannula, inserted into the posterior vena cava anterior to the liver, passed into a reservoir and was carried to an oxygenator by a rotary pump. Oxygenation was originally achieved with a bubble oxygenator, by gassing with moist 95% oxygen and 5% carbon dioxide. The oxygenated blood overflowed into a collecting chamber and passed into a constant-level reservoir; excess blood was returned to the reservoir. Blood from the constant-level reservoir flowed through a heating coil to a cannula

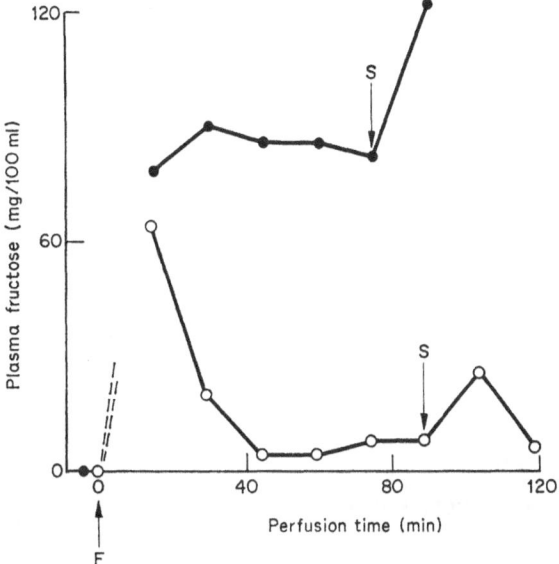

Fig. 4.13. Effect of adding fructose and sorbitol to the perfusate of isolated guinea-pig livers. Fetal age 64 d ●—● and neonatal age 8 d o–o. At F and S 20 mg of fructose and sorbitol, respectively, were added to the recirculating perfusate

inserted in the umbilical vein. Perfusion was started on ligation of the vena cava at a site posterior to the liver.

In more recent experiments a small vertically held spinning-disc oxygenator has been employed instead of the bubble oxygenator.

The perfusion may be carried out with the liver still within the abdominal cavity of the fetus. Alternatively the organ may be removed and placed on a muslin sling suspended across the mouth of a funnel, leading into the reservoir.

With this procedure Andrews *et al.* (1960) showed that the fetal liver was capable, at least from the 122nd day of gestation, of converting sorbitol to fructose, but that the organ was unable to remove fructose. Ability to remove fructose from the perfusate was in fact only seen in livers obtained from lambs three or more days after birth. Premature delivery of the lambs did not appear to influence appreciably the maturation of this enzyme system (Andrews *et al.*, 1961).

The technique has also been used to investigate bile secretion and bilirubin conjugation in fetal sheep (Alexander, Andrews, Britton & Nixon, 1970), removal of insulin (Alexander, Andrews, Britton, Cohen & Nixon, in preparation) and the removal of glucagon (Alexander, Assan & Nixon, in preparation).

The isolated fetal guinea-pig liver has also been studied for removal from the perfusate of sorbitol and fructose. Results (Fig. 4.13) suggest that, as in the sheep, appearance of fructokinase is postnatal.

Human fetal livers obtained at therapeutic abortion have been perfused by a recirculating technique through the umbilical vein by Kekomäki *et al.* (1970). These authors were able to show that labelled leucine was incorporated into the plasma proteins.

Perfusion of fetal heart

It is surprising that so few observations have been made upon this organ which lends itself so well to perfusion studies.

A method for studying the isolated heart of fetal sheep was introduced by Born *et al.* (1956). By using an artificial oxygenator they were able to demonstrate that increased oxygen tensions caused constriction of the ductus arteriosus, as did also adrenaline and nor-adrenaline.

Fetal mouse hearts have been maintained in organ culture by Wildenthal (1971); a considerable prolongation of survival time was achieved by the addition of insulin to the supporting medium.

A few investigations on human fetal hearts obtained at therapeutic abortion have been made. Lloyd (1929) demonstrated that calcium improved the strength of contraction, while Baker (1953) showed that adrenaline, nor-adrenaline and acetylcholine caused a constriction of the coronary vessels.

Perfusion of fetal lungs

The lungs of fetal sheep have been perfused by Dawes, Mott & Widdicombe (1955) and Born *et al.* (1955). In this technique the lungs were ventilated with air while the tissue was perfused with blood via the pulmonary trunk, after the ductus arteriosus and the great veins had been tied. Blood from the lungs returned to the left atrium and from there to a reservoir. The distensibility of the lungs was found to increase considerably from about the 110th day of gestation, and from this age, when the lungs are ventilated, there is an increase of pulmonary blood flow.

Isolated placenta

The ease with which the full-term human placenta may be obtained must account in some measure for the numerous investigations involving perfusion techniques; see, for example, the reports by Thomas & Varangot (1949); Goerke *et al.* (1961); Krantz, Panos & Evans (1962); Nesbitt *et al.* (1970); Hamrin *et al.* (1971). It should be borne in mind that the tissue obtained at this time has fulfilled its biological role and as a consequence may have a different physiology and biochemistry to that of a younger placenta. Furthermore it will have been subjected to mechanical and anoxic stresses during vaginal delivery, and in the subsequent clinical examination for completeness of delivery.

In the method of Krantz *et al.* (1962) the perfusate was introduced directly into the intervillous space through numerous tubes. The venous outflow passed into an oxygenator and from there to a reservoir upon which the pump drew. A recirculating perfusion was also established in the placenta through cannulated umbilical vessels. With this technique Krantz and his colleagues were able to demonstrate the transplacental passage of urea, creatinine, glucose, fructose and glycine. Absence of a physical continuity was indicated by the lack of passage of a bacteriophage across the placenta, when this material was added to the perfusate of one compartment. With this method the transplacental passage of palmitate (Szabo, Grimaldi & Jung, 1969), removal of glucose (Howard & Krantz, 1967) and aspects of glucose metabolism (Krantz *et al.*, 1971) have also been studied.

Isolated umbilical blood vessels

The behaviour of umbilical blood vessels has received some attention. To the fetal physiologist they are of great interest since variations in their calibre may affect blood flow and pressure in the feto-placental vascular system, and influence the rate of transfer of substances in both directions across the placenta. To the pharmacologist the vessels offer the opportunity of investigating the action of drugs upon the smooth muscle of a vascular system that is usually considered to be devoid of nerves.

Several methods have been used in the study of the response of the vessels to various agents. Von Euler (1938) and Aström & Samelius

(1957) took human placentae at term, cannulated the two umbilical arteries and perfused the vessels, with the placentae attached, by means of a pump drawing upon Ringer-Locke or Tyrode solutions. The frequency and stroke volume of the pump was adjusted to give a suitable perfusion pressure, which was monitored by a mercury manometer or by a pressure transducer. The effect of injected drugs upon the vessels could be judged from changes in the perfusion pressure and in the rate of outflow from the umbilical vein. Segments of umbilical arteries can also be perfused by gravity, and their response to added drugs assessed by a flow recorder attached to the other end of the vessel (Gokhale *et al.*, 1966). Another approach has been to examine the response of a spiral strip cut from the umbilical vessel (Somlyo, Woo & Somlyo, 1965; Dyer, 1970). One end was attached to the bottom of an isolated organ bath, filled with a suitable solution such as that of Krebs, and the other end connected to either an isotonic or isometric recording system. The response of the vessel was tested by the addition of drugs to the solution. A useful feature of this type of approach is that several preparations from the same umbilical vessel may be investigated at the same time.

Lewis (1968) examined the reactivity of umbilical arteries and their intra-abdominal portions in the sheep. In addition to observing the effect of drugs, he noted that increased oxygen tensions and lowered temperatures both caused vasoconstriction. Similar responses were found in segments of the intra- and extra-abdominal umbilical arteries.

Studies on the umbilical vessels by these techniques have shown that they constrict to many naturally occurring substances such as adrenaline, nor-adrenaline, 5-hydroxytryptamine and angiotensin. Currently there is considerable interest in the role of prostaglandins in reproductive physiology, and it is worth recalling that von Euler (1938) showed that these substances, which had recently been described by him, exerted a powerful constrictor action upon the umbilical vessels.

Isolated fetal bladder

Differences between the composition of bladder urine and of samples collected without a prolonged stay within the fetal bladder have suggested that the bladder wall possesses certain permeability characteristics (see Alexander & Nixon, 1961). Recently the transport of sodium and the influence of the antidiuretic hormone upon this movement have been investigated by France, Saunders & Stanier (1972) in the fetal sheep bladder. The technique has been to cut open the bladder so as to form a sheet, which is then clamped between two permeability cells.

Permeability of the isolated fetal membranes

Several studies have been made on the passage of substances across the isolated fetal membranes. A schematic diagram of a simple apparatus for use in such investigations is shown in Fig. 4.14. It consists of two perspex cells with a circular area for exchange of 15 cm^2 and a capacity of 15 ml.

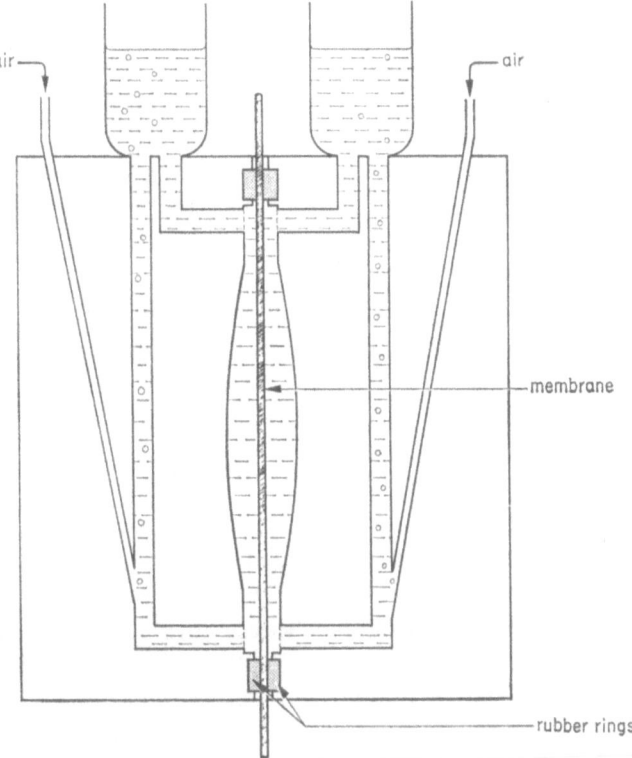

Fig. 4.14. Schematic diagram of a simple apparatus for investigating the passage of substances across fetal membranes. The membrane is placed between the rubber rings and the whole bolted together

Movement of fluid across the membrane surface is ensured by aeration. A flap of membrane *in situ* is cut and, while it is held vertically, the two cells are brought into apposition across the membrane and bolted together. The remaining side of the flap is cut away and the apparatus transferred to a water bath at 37°C. The material to be examined is dissolved in 0·9% NaCl and a 10-ml volume added to one cell; into the other cell is placed a similar volume of fluid without the test material. At intervals similar volumes are removed from the two compartments for analysis. Nixon (1952) using this apparatus examined the perme-

131

ability, in both directions, of the amniotic and allantoic membranes of the sheep towards myo-inositol. Similar studies on the sheep amniotic membrane have also been made with glucose and fructose (Alexander & Nixon, 1953). Impermeability was noted for all three substances; indeed a transfer could only be demonstrated after killing the membrane.

The isolated chorio-allantoic membrane of the pig was shown to transfer sodium by Crawford & McCance (1960). The permeability of the human amniotic membrane at term to many substances including albumen was demonstrated by Garby (1957). Water was shown to pass by Barton & Baker (1967) and by Seeds (1967). The human chorion was shown to be permeable to glucose, but utilization of this substance by the membrane was also found to occur (Battaglia *et al.*, 1962).

Isolated fetal skin

The permeability of isolated portions of human fetal skin towards sodium have been studied in diffusion chambers by Lind & Hytten (1972).

Concluding comments

Interest in the intrauterine life is gaining momentum, for the fetus is no longer so inaccessible to experimentation. The methods now available, the product in most cases of research during the last 25 years, permit a wide range of physiological and biochemical investigations to be carried out. In fetal physiology, no less than in other sciences, advancement in the state of our knowledge depends upon the introduction of techniques.

An analytical approach is now possible in metabolic investigations in which different preparations may be brought into use to answer specific problems. For example, it may be possible in the exteriorized fetus to form some assessment of the rate of utilization of a substance from an examination of the umbilical vein–umbilical artery concentration differences between simultaneously obtained blood samples, and from a knowledge of the umbilical blood flow. Such determination may, however, be difficult to carry out with precision, owing to small concentration differences across the placenta, insensitivity of analytical methods, and errors in the measurement of umbilical blood flow; furthermore, the value obtained applies only for that moment in time. If the same substance is added to the perfusate of an isolated fetus, then a progressive diminution in the concentration of the substance in the blood will be observed. The average rate of removal may be ascertained from a consideration of the differences in concentration with time and an assessment of blood volume, or by observing the time taken for the concentration of an injected substance to attain its pre-injection concentration. Another approach is to infuse the substance at a known rate, so as to maintain a constant plasma concentration. Thus inferences based upon a steady state are eliminated. In the same way investigation on the transfer of substances across the placenta may be facilitated, particularly if the rate is slow or if there is rapid fetal uptake, by establishing an *in-situ* recirculating perfusion. For substances passing across the placenta from fetus to mother the combined uterine and placental perfusion preparation may be the method of choice. The potentially large maternal equilibrating space or a rapid maternal disposal mechanism are eliminated, and the passage of a slowly transferred substance magnified, by its accumulation in the recirculating blood. A cautionary note must be added that, while a clearer insight into the functioning of a tissue may be obtained by perfusion studies, the technique may introduce its own complications through lack, for example, of an adequate tissue for eliminating metabolic end products and pharmacologically active substances, through lack of placental or placentally transferred hormones, or because of the absence of a precursor or co-factor. Caution is necessary, too, in the interpretation of results obtained in anaesthetized preparations. However, by the insertion of indwelling catheters and sensory probes, many investigations are now possible on the fetus *in utero* under what may be considered as normal physiological conditions.

A core of general physiological and biochemical knowledge concerning fetal life is now available, albeit largely that of the sheep, which will serve as a scaffold for investigation of species variations, including those of man.

The human fetus and its environment is also becoming more accessible to those concerned in ascertaining its well-being. Traditionally the guide to the status of the fetus has been the auscultation of its heart rate and later through the electrocardiogram. It is possible now to detect fetal stress by the presence of meconium seen in the amniotic fluid with the aid of the amnioscope, to sample the scalp capillary blood and to determine its acid–base parameters, and the concentration of electrolytes and glucose (Saling, 1968). Samples of amniotic fluid may be removed at amniocentesis and examined; a high bilirubin concentration is associated with haemolytic disease whilst low concentration of lecithin indicates the probability of breathing difficulties at birth of the respiratory-distress-syndrome type (Gluck *et al.*, 1971). Studies on the cells present in the amniotic fluid may reveal not only the sex of the fetus but also chromosomal abnormalities. The alimentary tract may be visualized under X-ray when the fetus swallows iodophendylate presented to it by an intra-amniotic fluid injection; this substance is also adsorbed by the vernix and so outlines the soft parts of the fetus revealing perhaps structural abnormalities (Lennon, 1967). In the future it may be possible to raise the nutritional status of fetuses by administering material to the amniotic fluid. Successful blood transfusions into the peritoneal cavity of the fetus have been carried out by Liley (1963) in cases of haemolytic disease.

Jackson (1969) has peered into the future of experimental fetal physiology. He saw the extensive use of biomedical telemetry in the fetal sheep, where various physiological and biochemical parameters could be measured from sensors inserted at a prior operation, according to a series of precoded instructions, over the remainder of gestation.

References

ALEXANDER, D. P., ANDREWS, W. H. H., BRITTON, H. G. & NIXON, D. A. (1970) Bilirubin in the foetal sheep. *Biology of the Neonate*, **15**, 103–111.

ALEXANDER, D. P., ANDREWS, R. D., HUGGETT, A. ST. G., NIXON, D. A. & WIDDAS, W. F. (1955) The placental transfer of sugars in the sheep: studies with radioactive sugar. *Journal of Physiology*, **129**, 352–366.

ALEXANDER, D. P., ASSAN, R., BRITTON, H. G. & NIXON, D. A. (1971) Glucagon permeability of the sheep placenta. *Journal of Physiology*, **216**, 63–64P.

ALEXANDER, D. P., BRITTON, H. G., COHEN, N. M. & NIXON, D. A. (1969a) Plasma concentrations of insulin, glucose, free fatty acids and ketone bodies in the foetal and newborn sheep and the response to a glucose load before and after birth. *Biology of the Neonate*, **14**, 178–193.

ALEXANDER, D. P., BRITTON, H. G., COHEN, N. M. & NIXON, D. A. (1969b) Foetal metabolism. In *Foetal Autonomy*, Ciba Foundation Symposium, ed. Wolstenholme, G. E. W. & O'Connor, Maeve. pp. 95–113. London: Churchill.

ALEXANDER, D. P., BRITTON, H. G., COHEN, N. M. & NIXON, D. A. (1970) Response of the sheep foetus and young lamb to anti-insulin serum. *Biology of the Neonate*, **15**, 142–155.

ALEXANDER, D. P., BRITTON, H. G., MASHITER, K., NIXON, D. A. & SMITH, F. G. (1970) The response of the foetal sheep *in-utero* to intravenous glucose. *Biology of the Neonate*, **15**, 361–367.

ALEXANDER, D. P., BRITTON, H. G. & NIXON, D. A. (1964a) Glucose and fructose metabolism in the isolated perfused sheep foetus. *Journal of Physiology*, **171**, 45–46P.

ALEXANDER, D. P., BRITTON, H. G. & NIXON, D. A. (1964b) Survival of the foetal sheep at term following short periods of perfusion through the umbilical vessels. *Journal of Physiology*, **175**, 113–124.

ALEXANDER, D. P., BRITTON, H. G. & NIXON, D. A. (1966a) Observations on the isolated foetal sheep with particular reference to the metabolism of glucose and fructose. *Journal of Physiology*, **185**, 382–399.

ALEXANDER, D. P., BRITTON, H. G. & NIXON, D. A. (1966b) Metabolism of ketone bodies by the sheep foetus. *Journal of Physiology*, **186**, 100–101P.

ALEXANDER, D. P., BRITTON, H. G. & NIXON, D. A. (1967) Acetate metabolism in the isolated sheep foetus. *Journal of Physiology*, **190**, 295–307.

ALEXANDER, D. P., BRITTON, H. G. & NIXON, D. A. (1968) Maintenance of sheep fetuses by an extracorporeal circuit for periods up to 24 hours. *American Journal of Obstetrics and Gynecology*, **102**, 969–975.

ALEXANDER, D. P., BRITTON, H. G. & NIXON, D. A. (1969) Comparison of the metabolism of the human fetus and fetal sheep: problems of measurement: effects of glucose administration. In *Perinatal Medicine*, ed. Huntingford, P. J., Huter, K. A. & Saling, E. pp. 183–187. Stuttgart: Thieme.

ALEXANDER, D. P., BRITTON, H. G. & NIXON, D. A. (1970) Metabolism of fructose and glucose by the sheep foetus: studies on the isolated perfused preparation with radioactively labelled sugars. *Quarterly Journal of Experimental Physiology*, **55**, 346–362.

ALEXANDER, D. P., BRITTON, H. G., NIXON, D. A. & COX, B. D. (1970) Amino acid metabolism in the sheep foetus. *Biology of the Neonate*, **15**, 304–308.

ALEXANDER, D. P., HUGGETT, A. ST. G., NIXON, D. A. & WIDDAS, W. F. (1955) The placental transfer of sugars in the sheep: the influence of concentration gradient upon the rate of hexose formation as shown in umbilical perfusion of the placenta. *Journal of Physiology*, **129**, 367–383.

ALEXANDER, D. P. & NIXON, D. A. (1953) *In-vitro* studies on the permeability of the amniotic membrane of the sheep to fructose, glucose and inositol. *Journal of Physiology*, **120**, 26P.

ALEXANDER, D. P. & NIXON, D. A. (1961) The foetal kidney. *British Medical Bulletin*, **17**, 112–117.

ALEXANDER, D. P. & NIXON, D. A. (1963) Reabsorption of glucose, fructose and meso-inositol by the foetal and post-natal sheep kidney. *Journal of Physiology*, **167**, 480–486.

ANDREWS, W. H. H., BRITTON, H. G., HUGGETT, A. ST. G. & NIXON, D. A. (1960) Fructose metabolism in the isolated perfused liver of the foetal and new-born sheep. *Journal of Physiology*, **153**, 199–208.

ANDREWS, W. H. H., BRITTON, H. G. & NIXON, D. A. (1961) Fructose and lactic acid metabolism in the perfused liver of premature lambs. *Nature*, **191**, 1307–1308.

ARTHUR, G. H. (1969) The fetal fluids of domestic animals. *Journal of Reproduction and Fertility*, **Supp. 9**, 45–52.

ASDELL, S. A. (1946) *Patterns of Mammalian Reproduction*. Ithaca: Comstock.

ASTRÖM, A. & SAMELIUS, U. (1957) The action of 5-hydroxytryptamine and some of its antagonists on the umbilical vessels of the human placenta. *British Journal of Pharmacology and Chemotherapy*, 12, 410–414.

BAKER, E. & MORGAN, E. H. (1970) Iron transfer across the perfused rabbit placenta. *Life Sciences*, 9, 765–772.

BAKER, J. B. E. (1953) Some observations upon the isolated perfused human foetal heart. *Journal of Physiology*, 120, 122–128.

BANGHAM, D. R., HOBBS, K. R. & TEE, D. E. H. (1960) Transmission of serum protein from foetus to mother in the rhesus monkey. *Lancet*, ii, 1173–1174.

BARCLAY, A. E., FRANKLIN, K. J. & PRICHARD, M. M. L. (1944) *The Foetal Circulation*. Oxford: Blackwell.

BARCROFT, J. (1946) *Researches on Pre-natal Life*. Oxford: Blackwell.

BARCROFT, J. & BARRON, D. H. (1937a) Experimental 'chronic' lesions in the central nervous system of the sheep's foetus. *Journal of Physiology*, 89, 55–56P.

BARCROFT, J. & BARRON, D. H. (1937b) Movements in midfoetal life in the sheep embryo. *Journal of Physiology*, 91, 329–351.

BARNARD, C. N. (1957) A method of operating on fetal dogs *in-utero*. *Surgery*, 41, 805–807.

BARRON, D. H. & BARCROFT, J. (1938) A case of amputation of leg, 90 days before birth. *Journal of Physiology*, 93, 29–30P.

BARTON, T. C. & BAKER, C. (1967) Permeability of human amnion and chorion membrane. *American Journal of Obstetrics and Gynecology*, 98, 562–567.

BASMAJIAN, J. V. & RANNEY, D. A. (1961) Chemomyelotomy: substitute for general anaesthesia in experimental surgery. *Journal of Applied Physiology*, 16, 386.

BASSETT, E. G. (1965) The anatomy of the pelvic and perineal regions of the ewe. *Australian Journal of Zoology*, 13, 201–241.

BATTAGLIA, F. C., HELLEGERS, A. E., MESCHIA, G. & BARRON, D. H. (1962) *In-vitro* investigations of the human chorion as a membrane system. *Nature*, 196, 1061–1063.

BECK, A. D. (1970) Intrauterine renal surgery: technique for exposing the fetal kidney during the last two thirds of gestation. *Investigative Urology*, 8, 182–187.

BENZI, G., BERTÉ, F., CREMA, A. & ARRIGONI, E. (1968) Uterine-placental-fetal preparation *in-situ* on the dog. Investigation of metabolizing activity and tissue distribution. *Journal of Pharmaceutical Science*, 57, 1031–1032.

BERTON, J. P. (1970) Effets de la néphrectomie bilatérale chez le foetus de Lapin (survie et metabolisme). *Compte rendu hebdomadaire des seances de l'Academie des sciences*, 271, 219–222.

BISSONNETTE, J. M. & GURTNER, G. H. (1970) Transfer of inert gases and tritiated water across the sheep placenta. *The Physiologist*, 13, 150.

BORN, G. V. R., DAWES, G. S. & MOTT, J. C. (1955) The viability of premature lambs. *Journal of Physiology*, 130, 191–212.

BORN, G. V. R., DAWES, G. S., MOTT, J. C. & RENNICK, B. R. (1956) The constriction of the ductus arteriosus caused by oxygen and by asphyxia in newborn lambs. *Journal of Physiology*, 132, 304–342.

BRITTON, H. G., HUGGETT, A. ST. G. & NIXON, D. A. (1967) Carbohydrate metabolism in the sheep placenta. *Biochimica et biophysica acta*, **136**, 426–440.

BUDDINGH, F., PARKER, H. R. & ISHIZAKI, G. (1969) Technique for long-term study of kidney in fetal sheep. *American Journal of Veterinary Research*, **30**, 663–667.

CALLAGHAN, J. C. & DE LOS ANGELES, J. (1961) Long term extracorporeal circulation in the development of an artificial placenta for respiratory distress of the newborn. *Surgical Forum*, **12**, 215–217.

CALLAGHAN, J. C., DE LOS ANGELES, J., BORACCHIA, B., FISK, R. L. & HALLGREN, R. (1963) Studies in the development of an artificial placenta. Possible use of long-term extracorporeal circulation for respiratory distress of the newborn. *Circulation*, **27**, 686–690.

CALLAGHAN, J. C., MAYNES, E. A. & HUG, H. R. (1965) Studies in lambs of the development of an artificial placenta. Review of nine long-term survivors of extracorporeal circulation maintained in a fluid medium. *Canadian Journal of Surgery*, **8**, 208–213.

CAMPBELL, A. G. M., DAWES, G. S., FISHMAN, A. P., HYMAN, A. I. & JAMES, G. B. (1966) The oxygen consumption of the placenta and foetal membranes in the sheep. *Journal of Physiology*, **182**, 439–464.

CAMPLING, J. D. & NIXON, D. A. (1954) The inositol content of foetal blood and foetal fluids. *Journal of Physiology*, **126**, 71–80.

CHAMBERLAIN, G. (1968) Artificial placenta. *American Journal of Obstetrics and Gynecology*, **100**, 615–626.

CHEZ, R. A. & HUTCHINSON, D. L. (1969) The use of experimental surgical techniques in the pregnant Macaca mulatta. *Annals of the New York Academy of Sciences*, **162**, 249–253.

CHINARD, F. P., DANESINO, V., HARTMANN, W. L., HUGGETT, A. ST. G., PAUL, W. & REYNOLDS, S. R. M. (1956) The transmission of hexoses across the placenta in the human and the rhesus monkey (Macaca mulatta). *Journal of Physiology*, **132**, 289–303.

COHNSTEIN, J. & ZUNTY, N. (1884) Untersuchungen über das Blut, den Kreislauf und die Athmung beim Säugethier-Fötus. *Pflügers Archiv für die gesamte Physiologie des Menschen und der Tiere*, **34**, 173–233.

COLE, G. J. & MORRIS, B. (1971a) The growth and development of lambs thymectomized *in utero*. *Australian Journal of Experimental Biology and Medical Science*, **49**, 33–53.

COLE, G. J. & MORRIS, B. (1971b) The cellular and humoral response to antigens in lambs thymectomized *in utero*. *Australian Journal of Experimental Biology and Medical Science*, **49**, 55–73.

COMLINE, R. S. & SILVER, M. (1970) Daily changes in foetal and maternal blood of conscious pregnant ewes with catheters in umbilical and uterine vessels. *Journal of Physiology*, **209**, 567–586.

COMLINE, R. S. & SILVER, M. (1972) The composition of foetal and maternal blood during parturition in the ewe. *Journal of Physiology*, **222**, 233–256.

COWEN, R. H. & LAURENSON, R. D. (1959) A technique of operating upon the fetus of the rabbit. *Surgery*, **45**, 321–323.

CRAWFORD, J. D. & McCANCE, R. A. (1960) Sodium transport by the chorioallantoic membrane of the pig. *Journal of Physiology*, **151**, 458–471.

DANCIS, J. & MONEY, W. L. (1960) Transfer of sodium and iodo-antipyrine across the guinea pig placenta with an *in situ* perfusion technique. *American Journal of Obstetrics and Gynecology*, **80**, 215–220.

DANCIS, J., MONEY, W. L., SPRINGER, D. & LEVITZ, M. (1968) Transport of amino acids by the placenta. *American Journal of Obstetrics and Gynecology*, **101**, 820–829.

DAVIES, H. C. & DAVIES, R. E. (1965) Biochemical aspects of oxygen poisoning. In *Handbook of Physiology—Respiration II*, pp. 1047–1058. Washington: American Physiological Society.

DAWES, G. S. (1968) *Foetal and Neonatal Physiology*. Chicago: Year Book Medical Publishers.

DAWES, G. S., FOX, H. E., LEDUC, B. M., LIGGINS, G. C. & RICHARDS, R. T. (1972) Respiratory movements and rapid eye movement sleep in the foetal lamb. *Journal of Physiology*, **220**, 119–143.

DAWES, G. S., FOX, H. E. & RICHARDS, R. T. (1972) Variations in asphyxial gasping with fetal age in lambs and guinea-pigs. *Quarterly Journal of Experimental Physiology*, **57**, 131–138.

DAWES, G. S., JACOBSON, H. N., MOTT, J. C. & SHELLEY, H. J. (1960) Some observations on foetal and newborn rhesus monkeys. *Journal of Physiology*, **152**, 271–298.

DAWES, G. S., MOTT, J. C. & WIDDICOMBE, J. G. (1955) The patency of the ductus arteriosus in newborn lambs and its physiological consequences. *Journal of Physiology*, **128**, 361–383.

DAWES, G. S. & PARRY, H. B. (1965) Premature delivery and survival in lambs. *Nature*, **207**, 330.

DAWSON, R. F. & ROBSON, J. M. (1940) A technique for the perfusion of the foetal placental circulation. *Proceedings of the Society for Experimental Biology and Medicine*, **43**, 758–761.

DICZFALUSY, E. (1969) Steroid metabolism in the foeto-placental unit. *Excerpta Medica International Congress Series*, **183**, 65–109.

DICZFALUSY, E. (1970) Steroid synthesis and catabolism in the human feto-placental unit. In *Fetal Growth and Development*, ed. Waisman, H. A. & Kerr, G. R. pp. 111–137. New York: McGraw-Hill.

DIXIT, S. P. & COPPOLA, E. D. (1970) Intrauterine thymectomy in the canine fetus. *Canadian Journal of Surgery*, **13**, 170–176.

DROST, M. (1968) Bilateral adrenalectomy in the fetal lamb. *Experimental Medicine and Surgery*, **26**, 61–65.

DROST, M. & HOLM, L. W. (1968) Prolonged gestation in ewes after foetal adrenalectomy. *Journal of Endocrinology*, **40**, 293–296.

DYER, D. C. (1970) The pharmacology of isolated sheep umbilical cord blood vessels. *Journal of Pharmacology and Experimental Therapeutics*, **175**, 365–370.

EMMANOUILIDES, G. C., HOBEL, C. J., YASHIRO, K. & KLYMAN, G. (1972) Fetal responses to maternal exercise in the sheep. *American Journal of Obstetrics and Gynecology*, **112**, 130–137.

EMMANOUILIDES, G. C., TOWNSEND, D. E. & BAUER, R. A. (1968) Effects of single umbilical artery ligation in the lamb fetus. *Pediatrics*, **42**, 919–927.

ELY, P. A. (1966) The placental transfer of hexoses and polyols in the guinea pig as shown by umbilical perfusion of the placenta. *Journal of Physiology*, **184**, 255–271.

FABER, J. J. & HART, F. M. (1966) The rabbit placenta as an organ of diffusional exchange. *Circulation Research*, **19**, 816–833.

FISCHER, W. M. (1967) *In situ* perfusion on both sides of the guinea-pig placenta. In *Intra-uterine Dangers to the Foetus*, ed. Horsky, J. & Stembera, Z. K. pp. 155–159. Amsterdam: Excerpta Medica Foundation.

FISHER, J. H., DE ALMEIDA, M., NETTELBLAD, S. A. C. & DELUCA, F. G. (1964) Technique of thymectomy in newborn puppies and intrauterine thymectomy in dog fetuses. *Transactions. American Society for Artificial Internal Organs*, **10**, 244–246.

FOLKART, G. R., DANCIS, J. & MONEY, W. L. (1960) Transfer of carbohydrate across the guinea-pig placenta. *American Journal of Obstetrics and Gynecology*, **80**, 221–223.

FRANCE, V. M., SAUNDERS, N. R. & STANIER, M. W. (1972) Sodium transport in foetal sheep urinary bladder. *Journal of Physiology*, **224**, 23P.

GARBY, L. (1957) Studies on transfer of matter across membranes with special reference to the isolated human amniotic membrane and the exchange of amniotic fluid. *Acta Physiologica Scandinavica*, **40**, suppl. 137.

GLUCK, L., KULOVICH, M. V., BORER, R. C., BRENNER, P. H., ANDERSON, G. G. & SPELLACY, W. N. (1971) Diagnosis of the respiratory distress syndrome by amniocentesis. *American Journal of Obstetrics and Gynecology*, **109**, 440–445.

GOERKE, R. J., MCKEAN, C. M., MARGOLIS, A. J., GLENDENING, M. B. & PAGE, E. W. (1961) Studies of the isolated perfused human placenta. I. Methods and organ responses. *American Journal of Obstetrics and Gynecology*, **81**, 1132–1136.

GOKHALE, S. D., GULATI, O. D., KELKAR, L. V. & KELKAR, V. V. (1966) Effects of some drugs on human umbilical artery *in vitro*. *British Journal of Pharmacology and Chemotherapy*, **27**, 332–346.

GOLDSMITH, E. I. & MOOR-JANKOWSKI, I. (1969) ed. *Experimental medicine and surgery in primates*. *Annals of the New York Academy of Sciences*, **162**, Art. 1, 1–704.

GOODLIN, R. C. (1962) Foetal incubator. *Lancet*, **i**, 1356–1357.

GOODLIN, R. C. (1963) An improved fetal incubator. *Transactions American Society for Artificial Internal Organs*, **9**, 348–350.

GOODLIN, R. C. & PERRY, D. (1964) Hyperbaric oxygen in resuscitation of asphyxiated newborn rabbits. *Lancet*, **ii**, 1124.

GOODLIN, R. C. & RUDOLPH, A. M. (1970) Tracheal fluid flow and function in fetuses *in utero*. *American Journal of Obstetrics and Gynecology*, **106**, 597–606.

GREENBERG, E. M. (1951) A plan for an artificial uterus. 145th Annual Convention of the Medical Society of the State of New York. Paper deposited in National Library of Medicine, Bethesda, Maryland, U.S.A.

HAMRIN, C. E., CONGER, W. L., LINDSTROM, R. N., SHIER, R. W. & DILTS, P. V. (1971) Placental perfusion device. *American Journal of Obstetrics and Gynecology*, 110, 422–423.

HARNED, H. S., TANDYSH, M. A., McGARRY, M., KEEVE, J. & KUSSEROW, B. (1957) Use of the pump oxygenator to sustain life during neonatal asphyxia of lambs. *American Journal of Diseases in Children*, 94, 530–531.

HESS, A. (1957) The experimental embryology of the foetal nervous system. *Biological Reviews*, 32, 231–260.

HEYMANN, M. A. & RUDOLPH, A. M. (1967) Effect of exteriorization of the sheep fetus on its cardiovascular function. *Circulation research*, 21, 741–745.

HODARI, A. A. & THOMAS, L. (1969) Experimental surgical procedures upon the fetus in obstetric research. *Obstetrics and Gynecology*, 34, 204–211.

HOPKINS, L., McFADYEN, I. R. & YOUNG, M. (1971) Placental transfer and foetal uptake of free amino acids in the pregnant ewe. *Journal of Physiology*, 215, 11–12P.

HOWARD, J. M. & KRANTZ, K. E. (1967) Transfer and use of glucose in the human placenta during *in-vitro* perfusion and the associated effects of oxytocin and papaverine. *American Journal of Obstetrics and Gynecology*, 98, 445–458.

HUGGETT, A. ST. G. (1927) Foetal blood-gas tension and gas transfusion through the placenta of the goat. *Journal of Physiology*, 62, 373–384.

HUGGETT, A. ST. G. & NIXON, D. A. (1961) Fructose as a component of the foetal blood in several mammalian species. *Nature*, 190, 1209.

HUGGETT, A. ST. G., WARREN, F. L. & WARREN, V. N. (1951) Origin of the blood fructose in the foetal sheep. *Journal of Physiology*, 113, 258–273.

HUGGETT, A. ST. G. & WIDDAS, W. F. (1951) The relationship between mammalian foetal weight and conception age. *Journal of Physiology*, 114, 306–317.

HUTCHINSON, D. L., BASHORE, R. A. & WILL, D. W. (1962) Creatinine equilibrium between mother and nephrectomized primate fetus. *Proceedings of the Society for Experimental Biology and Medicine*, 110, 395–396.

HUTCHINSON, D. L., WESTOVER, J. L. & WILL, D. W. (1962) The destruction of the maternal and fetal pituitary glands in subhuman primates. Techniques and preliminary observations. *American Journal of Obstetrics and Gynecology*, 83, 857–865.

JACKSON, B. T. (1969) Approach to fetal research—present and future. *American Journal of Disease of Children*, 118, 812–816.

JACKSON, B. T. & EGDAHL, R. H. (1960) The performance of complex fetal operations *in-utero* without amniotic fluid loss or other disturbances of fetal–maternal relationships. *Surgery*, 48, 564–570.

JAFFE, R. B. & LEDGER, W. J. (1966) *In-vivo* steroid biogenesis and metabolism in the human term placenta. I. *In-situ* placental perfusion with isotopic pregnenolone. *Steroids*, 8, 61–78.

JOST, A. (1947) Expériences de décapitation de l'embryon de Lapin. *Compte rendu hebdomadaire des seances de l'Academie des sciences*, 225, 322–324.

KARVONEN, M. J. & RÄIHÄ, N. (1954) Permeability of the guinea-pig placenta. *Acta physiologica scandinavica*, 31, 194–202.

KAYDEN, H. J., DANCIS, J. & MONEY, W. L. (1969) Transport of lipids across the guinea pig placenta. *American Journal of Obstetrics and Gynecology*, **104**, 564–572.

KEKOMÄKI, M., SEPPÄLÄ, M., SCHWARTZ, A., RAIVIO, K., EHNHOLM, C. & SEPPÄLÄ, I. J. T. (1970) Plasma protein synthesis by the isolated perfused human fetal liver. *Scandinavian Journal of Clinical and Laboratory Investigation*, **25**, suppl. 113, 100.

KISKEN, W. A. & SWENSON, N. A. (1968) A technique of intrauterine thymectomy in the rabbit. *Surgery*, **63**, 546–548.

KRANTZ, K. E., BLAKEY, J., YOSHIDA, K. & ROMINTO, J. A. (1971) Demonstration of viability of perfused human term placenta. *Obstetrics and Gynecology*, **37**, 183–191.

KRANTZ, K. E., PANOS, T. C. & EVANS, J. (1962) Physiology of maternal-fetal relationship through the extracorporeal circulation of the human placenta. *American Journal of Obstetrics and Gynecology*, **83**, 1214–1228.

LANMAN, J. T. & SCHAFFER, A. (1968) Gestational effects of fetal decapitation in sheep. *Fertility and Sterility*, **19**, 598–605.

LAWN, L. & McCANCE, R. A. (1961) An artificial placenta. *Journal of Physiology*, **158**, 2–3P.

LAWN, L. & McCANCE, R. A. (1962) Ventures with an artificial placenta. Principles and preliminary results. *Proceedings of the Royal Society B*, **155**, 500–509.

LAWN, L. & McCANCE, R. A. (1964) Artificial placentae. A progress report. *Acta paediatrica*, **53**, 317–325.

LAWN, L. & McCANCE, R. A. (1967) Artificial placentae: comparative results with two gas exchangers. *Quarterly Journal of Experimental Physiology*, **52**, 157–167.

LENNON, G. G. (1967) Intrauterine foetal visualization. *Journal of Obstetrics and Gynaecology of the British Commonwealth*, **74**, 227–229.

LERNER, U. & DICZFALUSY, E. (1968) A new method for the *in-vitro* perfusion of the human foeto-placental unit. *Excerpta Medica International Congress Series*, **170**, 19.

LEVITZ, M., CONDON, G. P., MONEY, W. L. & DANCIS, J. (1960) The relative transfer of estrogens and their sulfates across the guinea-pig placenta: sulfurylation of estrogens by the placenta. *Journal of Biological Chemistry*, **235**, 973–977.

LEWIS, B. V. (1968) The response of isolated sheep and human umbilical arteries to oxygen and drugs. *Journal of Obstetrics and Gynaecology of the British Commonwealth*, **75**, 87–81.

LIGGINS, G. C. (1968) Premature parturition after infusion of corticotrophin or cortisol into foetal lambs. *Journal of Endocrinology*, **42**, 323–329.

LIGGINS, G. C. (1969a) Premature delivery of foetal lambs infused with glucocorticoids. *Journal of Endocrinology*, **45**, 515–523.

LIGGINS, G. C. (1969b) The foetal role in the initiation of parturition in the ewe. In *Foetal Autonomy*. Ciba Foundation Symposium, ed. Wolstenholme, G. E. W. & O'Connor, Maeve. pp. 218–244. London: Churchill.

LIGGINS, G. C., KENNEDY, P. C. & HOLM, L. W. (1967) Failure of initiation of parturition after electrocoagulation of the pituitary of the foetal lamb. *American Journal of Obstetrics and Gynecology*, **98**, 1080–1086.

LILEY, A. W. (1963) Intrauterine transfusion of the foetus in haemolytic disease. *British Medical Journal*, **ii**, 1107–1109.

LIND, T. & HYTTEN, F. E. (1972) Fetal control of fetal fluids. In *Physiological Biochemistry of the Fetus*, ed. Hodari, A. A. & Mariona, F. G. pp. 54–65. Springfield: Thomas.

LITTLE, W. A., NASSER, D. & SPELLACY, W. N. (1971) Carbohydrate metabolism in the primate fetus. *American Journal of Obstetrics and Gynecology*, **109**, 732–743.

LLOYD, W. D. M. (1929) The action of calcium on the isolated foetal heart. *Journal of Pharmacology and Experimental Therapeutics*, **36**, 185–193.

LONDON, W. T., MONEY, W. L. & RAWSON, R. W. (1964) Placental transfer of ^{131}I-labelled iodide in the guinea-pig. *Journal of Endocrinology*, **28**, 247–252.

LUCKEY, T. D. (1963) *Germfree Life and Gnotobiology*. London: Academic Press.

MAY, N. D. S. (1964) *The Anatomy of the Sheep, with Instructions for its Dissection*. 2nd ed. Brisbane: University of Queensland Press.

MELLOR, D. J. (1970) A technique for chronic catheterization of the amniotic and allantoic sacs of sheep foetuses. *Research in Veterinary Science*, **11**, 93–95.

MELLOR, D. J., WILLIAMS, J. T. & MATHESON, I. C. (1972) A technique for chronic catheterization of the bladder of the foetal sheep. *Research in Veterinary Science*, **13**, 87–88.

MERLET, C., HOERTER, J., DEVILLENEUVE, C. & TCHOBROUTSKY, C. (1970) Mise en évidence de mouvements respiratoires chez le foetus d'agneau *in utero* au cours du dernier mois de la gestation. *Compte rendu hebdomadaire des seances de l'Academie des sciences*, **270**, 2462–2464.

MESCHIA, G., COTTER, J. R., BREATHNACH, C. S. & BARRON, D. J. (1965a) The diffusibility of oxygen across the placenta. *Quarterly Journal of Experimental Physiology*, **50**, 466–480.

MESCHIA, G., COTTER, J. R., BREATHNACH, C. S. & BARRON, D. H. (1965b) The hemoglobin, oxygen, carbon dioxide and hydrogen ion concentrations in the umbilical bloods of sheep and goats as sampled via indwelling plastic catheters. *Quarterly Journal of Experimental Physiology*, **50**, 185–195.

MESCHIA, G., MAKOWSKI, E. L. & BATTAGLIA, F. C. (1970) The use of indwelling catheters in the uterine and umbilical veins of sheep for a description of fetal acid-base balance and oxygenation. *Yale Journal of Biology and Medicine*, **42**, 154–165.

MESCHIA, G., WOLKOFF, A. S. & BARRON, D. H. (1959) The oxygen, carbon dioxide, hydrogen-ion concentration in the arterial and uterine venous bloods of pregnant sheep. *Quarterly Journal of Experimental Physiology*, **44**, 333–342.

METCALFE, J., MOLL, W., BARTELS, H., HILPERT, P. & PARER, J. T. (1965) Transfer of carbon monoxide and nitrous oxide in the artificial perfused sheep placenta. *Circulation Research*, **16**, 95–101.

MIKHAIL, G., WIQVIST, N. & DICZFALUSY, E. (1963) Oestriol metabolism in the human foeto-placental unit. *Acta endocrinologica*, **43**, 213–219.

MONEY, W. L. & DANCIS, J. (1960) Technique for the *in-situ* study of placental transport in the pregnant guinea pig. *American Journal of Obstetrics and Gynecology*, **80**, 209–214.

MYERS, R. E., HILL, D. E., HOLT, A. B., SCOTT, R. E., MELLITS, E. D. & CHEEK, D. B. (1971) Fetal growth retardation produced by experimental placental insufficiency in the rhesus monkey. *Biology of the Neonate*, **18**, 379–394.

NEIL, M. V., WALKER, D. G. & WARREN, F. L. (1961) The mechanism of fructose formation in goat placenta with special reference to the possible involvement of sorbitol or of phosphoric acid esters. *Biochemical Journal*, **80**, 181–187.

NESBITT, R. E. L., RICE, P. A., ROURKE, J. E., TORRESI, V. F. & SOUCHAY, A. M. (1970) *In-vitro* perfusion studies of the human placenta. A newly designed apparatus for extracorporeal perfusion achieving dual closed circulation. *Gynecologic Investigations*, **1**, 185–203.

NICHOLAS, J. S. (1925) Notes on the application of experimental methods upon mammalian embryos. *Anatomical Record*, **31**, 385–394.

NICHOLAS, J. S. (1934) Experiments on developing rats. I. Limits of foetal regeneration; behavior of embryonic material in abnormal environments. *Anatomical Record*, **58**, 387–413.

NIEDERHUBER, J. E., SHERMETA, D., TURCOTTE, J. G. & GIKAS, P. W. (1971) Kidney transplantation in the fetal lamb. Technique and histopathology of rejection. *Transplantation*, **12**, 161–166.

NIXON, D. A. (1952) *Inositol metabolism in reproduction*. M.Sc. Thesis. University of London.

NIXON, D. A. (1963) The transplacental passage of fructose, urea and meso-inositol in the direction from foetus to mother, as demonstrated by perfusion studies in the sheep. *Journal of Physiology*, **166**, 351–362.

NIXON, D. A. (1964) Perfusion of the sheep foetus. *Journal of Physiology*, **171**, 1P.

NIXON, D. A. (1968) Concentration of free meso-inositol in the plasma of perfused sheep foetuses. *Biology of the Neonate*, **12**, 113–120.

NIXON, D. A. (1972a) Carbohydrate metabolism in fetal life. In *Physiological Biochemistry of the Fetus*, ed. Hodari, A. A. & Mariona, F. G. pp. 96–116. Springfield: Thomas.

NIXON, D. A. (1972b) Extracorporeal maintenance of fetuses and neonates. *Bibliography of Reproduction*, **19**, 593–598 & 747–750.

NIXON, D. A., ALEXANDER, D. P. & HUGGETT, A. ST. G. (1966) Influence of umbilical blood fructose on the production of fructose by the perfused placenta of sheep. *Nature*, **209**, 918–919.

NIXON, D. A., BRITTON, H. G. & ALEXANDER, D. P. (1963) Perfusion of the viable sheep foetus. *Nature*, **199**, 183–185.

NIXON, D. A., HUGGETT, A. ST. G. & AMOROSO, E. C. (1966) Fructose as a component of the foetal blood and foetal fluids of the Bush Baby (Galago senegalensis senegalensis). *Nature*, **209**, 300–301.

NIXON, D. A. & WRIGHT, G. H. (1964) The intestinal absorption of fructose in the foetal sheep. *Biology of the Neonate*, **7**, 167–171.

OLSEN, D. B. & ALLRED, L. G. (1968) A surgical approach to bovine fetal catheterization and hemodynamics. *American Journal of Surgery*, **116**, 715–719.

OSBURN, B. I., ESPANA, C., JACKSON, T. & KENNEDY, P. C. (1971) Infection with rubella vaccine virus in the fetal lamb. *Federation Proceedings*, **30**, 252.

OSTERGARD, D. R. & CONTOPOULOS, A. N. (1970) A technique for selective fetal hypophysectomy *in-utero*. *American Journal of Obstetrics and Gynecology*, **108**, 322–323.

PARKES, A. S. ed. *Marshall's Physiology of Reproduction*, 3rd edn. Vol. 1, part 1 (1956), part 2 (1960), Vol. 2 (1952), Vol. 3 (1966). London: Longman.

PARSHALL, C. J. & SILVERSTEIN, A. M. (1969) Surgical approaches to the study of fetal immunology in primate animals. *Annals of the New York Academy of Sciences*, **162**, 254–266.

PEIRCE, E. C., FULLER, E. O., PATON, W. W., SWARTWOUT, J. R., BALLENTINE, M. B., WRIGHT, B. G., JOHNSON, N. J. & JOHNSON, R. L. (1970) Isolated perfusion of the pregnant sheep uterus. *Transactions. American Society for Artificial Internal Organs*, **16**, 318–324.

PIERREPOINT, C. G., ANDERSON, A. B. M., GRIFFITHS, K. & TURNBULL, A. C. (1971) Short-term maintenance of the isolated sheep foetus in the last fifth of pregnancy. *Acta endocrinologica*, **66**, 35–49.

PLENTL, A. A. & FRIEDMAN, E. A. (1962) Isotope tracer studies on the carbon dioxide exchange in pregnant primates. *American Journal of Obstetrics and Gynecology*, **84**, 1242–1252.

QUILLIGAN, E. J., HON, E. H., ANDERSON, G. G. & YEH, S. Y. (1968) Fetal cephalic metabolism in sheep. *American Journal of Obstetrics and Gynecology*, **102**, 716–724.

REYNOLDS, M. L. & YOUNG, M. (1971) The transfer of free α amino nitrogen across the placental membrane in the guinea-pig. *Journal of Physiology*, **214**, 583–597.

REYNOLDS, S. R. M. & PAUL, W. M. (1955) Circulatory responses of the fetal lamb *in-utero* to increase of intrauterine pressure. *Bulletin of the Johns Hopkins Hospital*, **97**, 383–394.

REYNOLDS, S. R. M., PAUL, W. M. & HUGGETT, A. ST. G. (1954) Physiological study of the monkey fetus *in-utero*: a procedure for blood pressure recording, blood sampling and injection of the fetus under normal conditions. *Bulletin of the Johns Hopkins Hospital*, **95**, 256–268.

ROSENKRANTZ, J. G., SIMON, R. C. & CARLISLE, J. H. (1968) Fetal surgery in the pig with a review of other mammalian fetal techniques. *Journal of Pediatric Surgery*, **3**, 392–397.

RUDOLPH, A. M. & HEYMANN, M. A. (1967) Validation of the antipyrine method for measuring fetal umbilical blood flow. *Circulation Research*, **21**, 185–190.

SALING, E. (Translated by Loeffler, F. E. 1968) *Foetal and Neonatal Hypoxia in Relation to Clinical Obstetric Practice*. London: Arnold.

FETAL AND PLACENTAL PHYSIOLOGY

SCHINCKEL, P. G. & FERGUSON, K. A. (1953) Skin transplantation in the foetal lamb. *Australian Journal of Biological Sciences*, 6, 533–546.

SEEDS, A. E. (1967) Water transfer across the human amnion in response to osmotic gradients. *American Journal of Obstetrics and Gynecology*, 98, 568–571.

SILVERSTEIN, A. M., UHR, J. W., KRANER, K. L. & LUKES, R. J. (1963) Fetal response to antigenic stimulus. II. Antibody production by the fetal lamb. *Journal of Experimental Medicine*, 117, 799–812.

SMEATON, T. C., COLE, G. J., SIMPSON-MORGAN, M. W. & MORRIS, B. (1969) Techniques for the long-term collection of lymph from the unanaesthetized foetal lamb *in utero*. *Australian Journal of Experimental Biology and Medical Science*, 47, 565–572.

SOMA, L. R., WHITE, R. J. & KANE, P. B. (1971) Surgical preparation of a chronic maternal–fetal model in pregnant sheep: a technique for the measurement of middle uterine blood flow, umbilical blood flow and fetal sampling in the awake sheep. *Journal of Surgical Research*, 11, 85–94.

SOMASUNDARAM, K. & PRATHAP, K. (1970) Intra-uterine healing of skin wounds in rabbit foetuses. *Journal of Pathology and Bacteriology*, 100, 81–86.

SOMLYO, A. V., WOO, C. Y. & SOMLYO, A. P. (1965) Responses of nerve-free vessels to vasoactive amines and polypeptides. *American Journal of Physiology*, 208, 748–753.

SUZUKI, K. & PLENTL, A. A. (1969) Chronic implantation of instruments in the neck of the primate fetus for physiologic studies and production of hydramnios. *American Journal of Obstetrics and Gynecology*, 103, 272–281.

SWENSON, E. A. (1925) The use of cerebral anemia in experimental embryological studies upon mammals. *Anatomical Record*, 30, 147–151.

SZABO, A. J., GRIMALDI, R. D. & JUNG, W. F. (1969) Palmitate transport across perfused human placenta. *Metabolism*, 18, 406–415.

TCHOBROUTSKY, C., CLAUVEL, M. & LAURENT, D. N. (1966) Extracorporeal oxygenation in puppies and in newborn and fetal lambs. *American Journal of Obstetrics and Gynecology*, 96, 367–381.

THOMAS, J. A. (1948) Nouveaux procédés de perfusion physiologique et aseptique permettant la survie prolongée d'organes ou d'organismes pesant plusieurs kilogrammes. *Journal de physiologie*, 40, 123–146.

THOMAS, J. A., SALOMON, L. & SALOMON, L. (1948a) Recherches préliminaires sur la culture en masse du virus de la fièvre aphteuse inoculé au foetus de vache, entretenu en survie par perfusion aseptique. *Compte rendu hebdomadaire des seances de l'Academie des sciences*, 227, 310–311.

THOMAS, J. A., SALOMON, L. & SALOMON, L. (1948b) La survie des grands foetus de mammifères perfusés aseptiquement. *Journal de physiologie*, 40, 233–250.

THOMAS, J. A., SALOMON, L., SALOMON, L. & LAMY, F. (1948) La survie expérimentale aseptique des grands foetus de mammifères. *Compte rendu hebdomadaire des seances de l'Academie des sciences*, 226, 966–968.

THOMAS, J. A. & VARANGOT, J. (1949) La survie du placenta humain perfusé aseptiquement. *Compte rendu hebdomadaire des seances de l'Academie des sciences*, 228, 132–133.

THOMASSON, B. H. & RAVITCH, M. M. (1969) Fetal surgery in the rabbit. *Surgery*, **66**, 1092–1102.

TSOULOS, N. G., COLWILL, J. R., BATTAGLIA, F. C., MAKOWSKI, E. L. & MESCHIA, G. (1971) Comparison of glucose, fructose and oxygen uptake by fetuses of fed and starved ewes. *American Journal of Physiology*, **221**, 234–237.

TWARDOCK, A. R. & AUSTIN, M. K. (1970) Calcium transfer in perfused guinea-pig placenta. *American Journal of Physiology*, **219**, 540–545.

VAN DUYNE, C. M., PARKER, H. R., HAVEL, R. & HOLM, L. W. (1960) Free fatty acid metabolism in fetal and newborn sheep. *American Journal of Physiology*, **199**, 987–990.

VON EULER, U. S. (1938) Action of adrenaline, acetylcholine and other substances on nerve-free vessels (human placenta). *Journal of Physiology*, **93**, 129–143.

WALKER, C. H. M., ROSE, R. L. & SIMON, S. L. (1962) Closed partial perfusion in puppies. *American Journal of Diseases of Children*, **104**, 296–301.

WELLS, L. J. (1950) Subjection of fetal rats to surgery and repeated subcutaneous injections: methods and survival. *Anatomical Record*, **108**, 309–332.

WESTIN, B., NYBERG, R. & ENHÖRNING, G. (1958) Technique for perfusion of the previable human fetus. *Acta paediatrica*, **47**, 339–349.

WIDDAS, W. F. (1952) The inability of diffusion to account for placental glucose transfer in the sheep and consideration of the kinetics of a possible carrier. *Journal of Physiology*, **118**, 23–29.

WILDENTHAL, K. (1971) Long-term maintenance of spontaneously beating mouse hearts in organ culture. *Journal of Applied Physiology*, **30**, 153–157.

WILSON, R., BIRD, C. E., WIQVIST, N., SOLOMONS, S. & DICZFALUSY, E. (1966) Metabolism of progesterone by the perfused adrenalectomized human fetus. *Journal of Clinical Endocrinology*, **26**, 1155–1159.

ZAPOL, W. M., KOLOBOW, T., PIERCE, J. E., VUREK, G. G. & BOWMAN, R. L. (1969) Artificial placenta: two days of total extrauterine support of the isolated premature lamb fetus. *Science*, **166**, 617–618.

5 Isolated Extracorporeal Placentation of the Fetal Lamb

WARREN M. ZAPOL and
THEODOR KOLOBOW

Until recently, the isolated premature lamb fetus was an unstable subject for physiological, biochemical and pharmacological study (Alexander *et al.*, 1966, 1968). Total artificial placentation with a membrane lung was performed in fetal lambs via the umbilical vessels for 19 h by Callaghan *et al.* (1965), and for up to 24 h by Alexander *et al.* (1968). Tchobroutsky (1966) perfused fetal and newborn lambs for approximately 3 h with the aid of an artificial lung constructed with tetrafluoroethylene membranes. Extracorporeally perfused fetuses succumbed from haemorrhage, oedema and haemolysis. Safer blood-handling methods, and advances in polymer membranes and biocompatible materials, have allowed us to devise a total support system for prolonged isolated fetal perfusion. This system, combined with parenteral nutritional support, has maintained the sheep fetus in a stable state for several days (Zapol *et al.*, 1969).

Improved extracorporeal perfusion techniques are useful, not only because they enable investigators to study fetal life more effectively, but also because they hold promise for the development of methods of treating newborn respiratory distress. Despite recent advances in prenatal diagnosis and postnatal treatment, respiratory distress of the newborn remains a major cause of death in the premature infant weighing less than one kilogram (Avery, 1968). Extracorporeal perfusion with membrane oxygenators can now support infants in severe acute pulmonary failure for over one week. By preventing hypoxia and hypercapnoea the membrane lung can maintain extrauterine life, while otherwise intolerable pulmonary damage heals (White *et al.*, 1971). Repair may be possible because extracorporeal support can relieve the infant's lungs of their primary burden of respiratory-gas exchange and of the handicaps of conventional therapy: high ventilator pressures and high inspired-oxygen concentrations (Zapol & Kitz, 1972; Hill *et al.*, 1972).

In this chapter we shall detail our techniques of fetal support, present some physiological observations made during total fetal and partial newborn-lamb perfusion, and review recent partial perfusions of infants with severe respiratory-distress syndrome.

General aspects

Membrane lungs

Disposable silicone-membrane oxygenators allow safe prolonged exchange of respiratory gases without causing the adverse effects of a raw blood–gas interface. Membrane oxygenators have the following advantages: (*a*) they minimize mechanical trauma to cells (Hill *et al.*, 1968); (*b*) they minimize interface denaturation of large globular proteins (Zapol *et al.*, 1969); (*c*) they reduce formation of cell–fibrin emboli (Allerdyce *et al.*, 1966); (*d*) they completely eliminate the hazard of gas emboli which can occlude blood vessels (Gullino, 1966); and (*e*) they can contain a fixed prime volume, thus, extracorporeal blood volume does not fluctuate.

Gas-exchange theory

Silicone membrane lungs have been carefully designed to balance the forces governing gas transfer (Fig. 5.1).

A fully oxygenated unstirred blood film adjacent to the synthetic membrane in a membrane lung resists further oxygen transport into blood (Marx, 1960; Buckles, 1966). This phenomenon is due mainly to the low solubility of oxygen in plasma. No such diffusion resistance

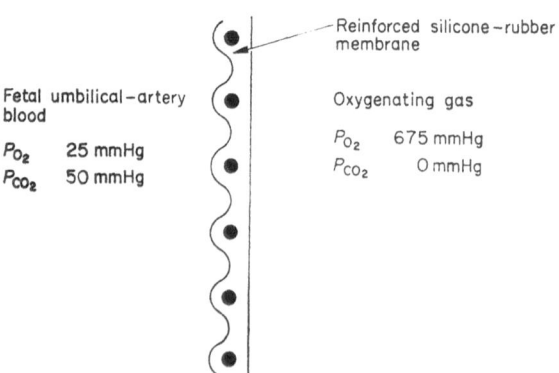

Fetal umbilical-artery blood

P_{O_2} 25 mmHg
P_{CO_2} 50 mmHg

Reinforced silicone–rubber membrane

Oxygenating gas

P_{O_2} 675 mmHg
P_{CO_2} 0 mmHg

Fig. 5.1. Diagram to illustrate the function of the silicone membrane lung in balancing the forces governing gas transfer

exists in the case of carbon dioxide, as carbon dioxide is about 25 times more soluble in plasma than is oxygen. A stagnant saturated blood layer next to the membrane causes an apparent increase in membrane diffusion resistance, and thus an apparent increase in membrane thickness (Galletti & Brecher, 1962). However, the physiological partial-pressure gradient across the membrane favours oxygen diffusion into blood 13 times more than carbon dioxide diffusion in the reverse direction. The net result is oxygen and carbon dioxide exchange are nearly equal in a membrane lung.

To construct a practical membrane lung, blood-film and membrane thickness must be minimized, blood-path length and membrane surface area must be adequate, and oxygen transfer across the boundary layer must be increased by stirring. Differences in oxygen-transfer rates of various membrane lungs depend primarily on the way blood mixing is accomplished at the stagnant blood–membrane interface (Bartlett et al., 1969). Carbon dioxide elimination remains primarily a function of the material, thickness and surface area of the membrane.

Design of the membrane lungs

The first clinically useful membrane oxygenator had polyethylene membranes (Clowes et al., 1956). Tetrafluoroethylene and silicone rubber membranes have since been evaluated. Silicone rubber, with its high gas permeability and relatively good blood compatibility, has proven superior (Galletti et al., 1966). However, silicone rubber membranes (0·05–0·13 mm) are inherently weak, and tear easily. Therefore, they are often re-inforced with a fabric support of fibreglass, nylon or dacron mesh (Kolobow & Zapol, 1970). The stronger membranes of the GE–Pierce lung and the Dow capillary lung contain copolymers of silicone rubber and polycarbonate (Pierce, 1970; Dutton et al., 1971). However, these membranes have lower gas-transfer permeabilities and poorer blood compatibility than pure silicone rubber (Bruck, 1972).

Limiting blood-film thickness is desirable but technically difficult. It is not possible to attain a uniform blood film equivalent to one erythro-cyte diameter (7 μm), the thickness encountered in the capillaries of the human lung, because very short blood-flow paths would be necessary to avoid large pressure drops.

If the blood is stirred adequately to reduce blood-boundary-layer diffusion resistance, membrane lungs can be constructed with thicker blood films (250 μm). Presently available membrane lungs attain safe perfusion pressures by sandwiching thick blood films between many sheets of membrane; capillary membrane lungs of only 1 litre/min blood-flow capacity require thousands of parallel 10- to 18-cm-long tubes ($1 \cdot 5 \times 10^{-2}$ to $3 \cdot 0 \times 10^{-2}$ cm internal diameter; $2 \cdot 5 \times 10^{-3}$ cm wall) for safe blood-perfusion pressures (Dutton *et al.*, 1971). Table 5.1 presents performance data and cost of several commercially available membrane lungs.

Membrane-lung requirements for use in fetal perfusion

(1) All blood-contacting surfaces should be disposable after a single use.

(2) The membrane-lung-perfusion circuit must have a small and constant blood-priming volume.

(3) A membrane lung with a low perfusion pressure is preferred.

(4) Manifolds of the membrane lung should be transparent to make any adherent thrombi visible.

(5) Surfaces contacting blood must not be conducive to protein denaturation.

(6) Surfaces contacting blood must be non-toxic and hypothrombogenic.

(7) The membrane lung must have a reliable and steady gas-transfer capacity of 40 cm^3 oxygen and carbon dioxide per m^2 of membrane area per minute for periods lasting several days. Total fetal respiratory-gas transfer must be achieved at extracorporeal blood flows of 400–500 ml/min.

Table 5.1. Commercially available membrane lungs for fetal perfusion

Lung	Membrane material	Surface area (m²)	Unit cost ($US)	Prime volume* (ml)	O₂ transfer (cm³/m² min) (flow 0·5-1 l/min)	CO₂ transfer (cm³/m² min) (flow 0·5-1 l/min)
Kolobow spiral coil	DMPS†	1	135	100	40–60	40–60
Bramson IMS	DMPS	1·6	180	330	30±5	–‡
Landé Edwards	DMPS	1	95	180	35	18·4±2·9
GE–Pierce	DMPS & Polycarbonate	1	150	200	50±13·4	43·1±5·6
Travenol	DMPS	1·5	150	320	50	–‡

* Oxygenator alone (does not include circuitry).

† Dimethylpolysiloxane.

‡ Data not available.

The spiral-coil membrane oxygenator

The spiral-coil membrane lung designed by Kolobow is shown in Fig. 5.2 (Kolobow & Bowman, 1963). It consists of a single length of saran screen spacer enveloped by a continuous layer of fabric-reinforced silicone-

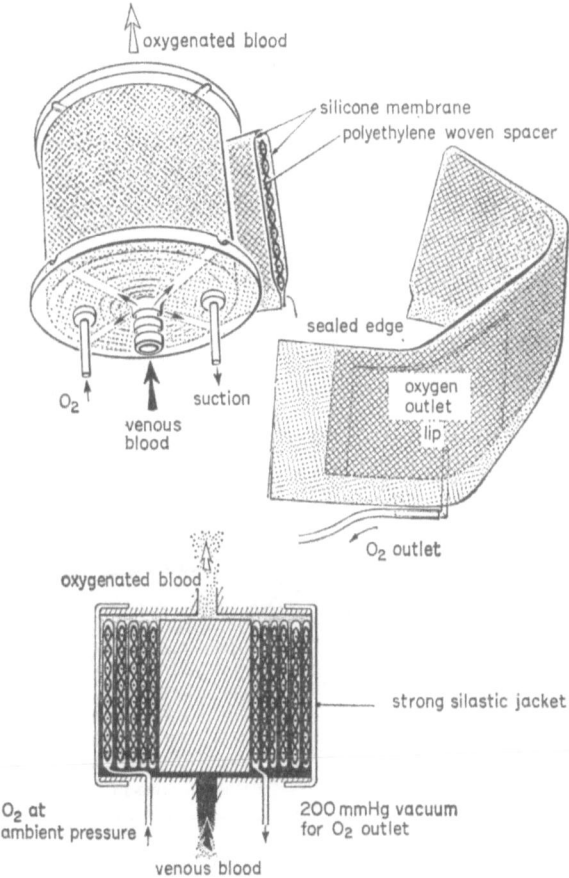

Fig. 5.2. Schematic diagram of the disposable, spiral-coil, silicone-rubber-membrane oxygenator

rubber membrane. Oxygen-inlet and outlet ports are incorporated into the membrane-lung envelope. The membrane envelope, while under a tension of several hundred grams, is wrapped around a spool. A strong silicone-rubber jacket provides a leakproof seal for the enclosed blood-oxygenating compartment. Blood enters the lung through an inflow channel and passes between individual layers of the membrane, where

151

gas exchange occurs. Oxygenated blood exits from the outflow channel of the spool.

The modular construction of this design permits fabrication of larger units while still utilizing the basic membrane-envelope design: a larger capacity unit requires either a wider or longer piece of silicone-membrane envelope (Plate 5.1, p. 164). By serially adding membrane lungs of 0·4-m² surface area, we are able to tailor our gas-transfer capabilities to the weight of the perfused fetus. However, this modular technique increases

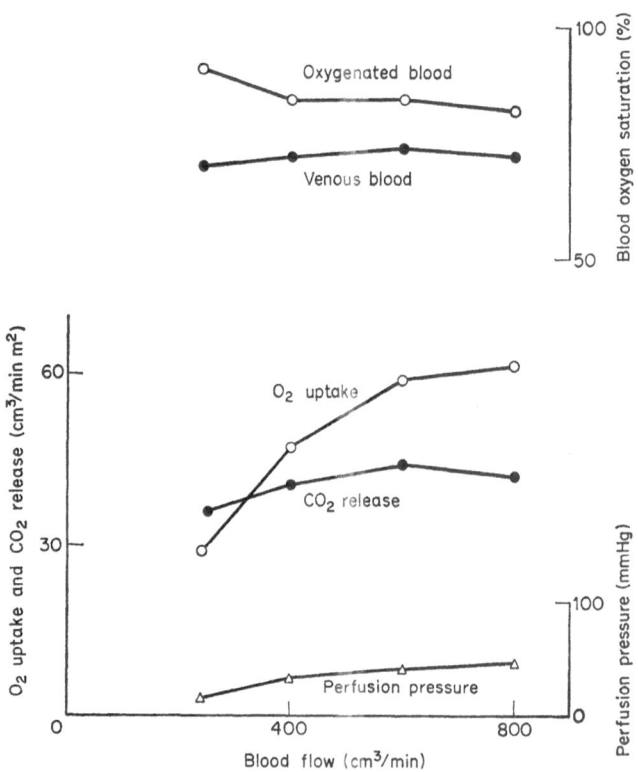

Fig. 5.3. *In-vivo* venovenous perfusion with a 10-cm-wide membrane envelope, 0·44 m² Silastic 0·1 mm; pH 7·4; P_{CO_2} 40–43 Torr; haematocrit reading 46 %

the extracorporeal prime volume in excess of the added surface area since the manifolds of the lung contain considerable blood volume.

A significant safety feature of this membrane lung is its use of oxygen at slightly sub-atmospheric pressure (−50 Torr). Gas emboli cannot be released into the bloodstream through an inadvertent membrane leak. Should a membrane leak occur, a small amount of blood will leak into the gas phase, and in the process often seal the pinhole. Minimally oscillating suction between −200 Torr and ambient pressure at 8 cycles per minute reduces the oxygen-partial-pressure gradient, but mixing produced in the blood compartment enhances oxygen exchange. Spiral-coil membrane-lung respiratory-gas transfers and blood-pressure change during *in-vivo* perfusion are shown in Fig. 5.3.

The spiral-coil membrane lung can be readily sterilized with humidified ethylene oxide at 25°C. Great care must be taken to adequately exhaust the toxic ethylene oxide gas.

Blood pumping and controls

The cardinal principle of safe pumping is to minimize suction on blood (Blackshear *et al.*, 1965). Blood is continuously gravity-drained into a closed silicone-rubber reservoir through 6·35-mm internal diameter

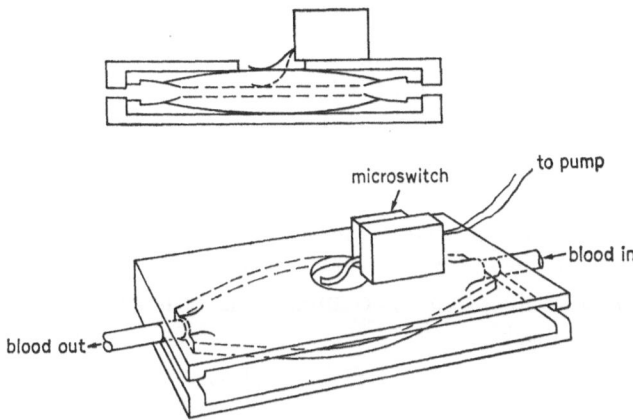

Fig. 5.4. Volume-sensing blood-reservoir bag and holder

tubing. The reservoir volume is monitored by a volume-sensing micro-switch (Fig. 5.4). If the reservoir empties, the pump is shut off and an alarm sounds. The pump restarts automatically when the reservoir bag is refilled.

In fetal perfusion we use an occlusive roller pump (Sarns model 6202). A battery power source makes the pump portable. A 9·53-mm internal diameter, 12·7-mm external diameter silicone pump chamber with 34-ml-stroke volume has been used for over 300 h without rupture.

For partial large-blood-flow perfusions in lambs and human beings, we use a precision-made pump chamber of segmented polyurethane reinforced with radial-ply fiberglass filament (Kolobow & Zapol, 1971).

153

The wall thickness of the pump chamber is 0·9 mm, its internal diameter 1·5–2·0 cm. The pump rollers are permanently fixed to provide a pump-chamber wall clearance of 0·05–0·10 mm. We have not observed rupture of pump chambers after over two months of pumping.

Blood catheters

Extracorporeal respiratory support with a pump oxygenator requires large blood flows. Flexible, thin-walled, large-bore cannulae can better achieve large flows than can other commonly used catheters for the following reasons: (*a*) metal and many plastic (polytetrafluoroethylene, polyethylene) cannulae are stiff and may cause vessel perforation; (*b*) other plastics (polyvinyl, silicone rubber) are flexible but thick-walled, thereby limiting blood flow; (*c*) non-reinforced cannulae may kink, and flow may become obstructed when the fetus is active.

It is difficult to cannulate the aorta of fetal lambs through the umbilical arteries, a procedure necessary for total extracorporeal respiratory support with an artificial placenta. The first 4 or 5 cm of the umbilical arteries can easily be cannulated with conventional large-bore catheters, but these catheters cannot be readily advanced beyond the junction of the umbilical and iliac arteries.

If the cannula does not enter the internal iliac artery, the proximal umbilical artery spontaneously constricts after a few hours of perfusion. This diminishes extracorporeal blood flow, causing hypoxia and fetal death. For successful perfusion, cannulae must be large bore, thin-walled and extremely flexible.

A segmented polyurethane elastomer, Lycra, has clot-inducing properties similar to those of silicone rubber (by the Lee-White clotting test and the inferior-vena-cava ring tests). Lycra is also highly resistant to biologic degradation (Boretos & Pierce, 1968). By embedding a stainless-steel spring wire between coats of Lycra, Kolobow was able to produce a non-kinking, thin-walled, steel-spring-reinforced cannula having the elastic properties of an elastomer with the strength of a coiled steel spring (Kolobow & Zapol, 1970) (Fig. 5.5).

Our cannulae consist of a short piece of silicone-rubber tubing bonded to polyurethane tubing with room-temperature-vulcanizing silicone-rubber adhesive. The catheter can be permanently moulded to any desired angle by a 30-min exposure to dry heat at 160°C, or by auto-claving at 30 lbf/in² (Fig. 5.6).

Catheters are sterilized by ethylene oxide gas or 70% ethanol. Neither vessel erosion nor perforation occur. The reinforced portion of the

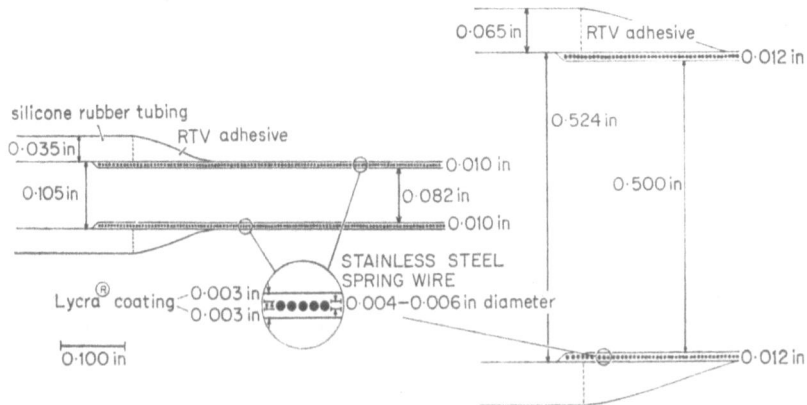

Fig. 5.5. Small- and large-bore catheters with the same wall thickness (0·25 to 0·27 mm) are shown. The catheter is bonded to Silastic with a room-temperature vulcanizing adhesive

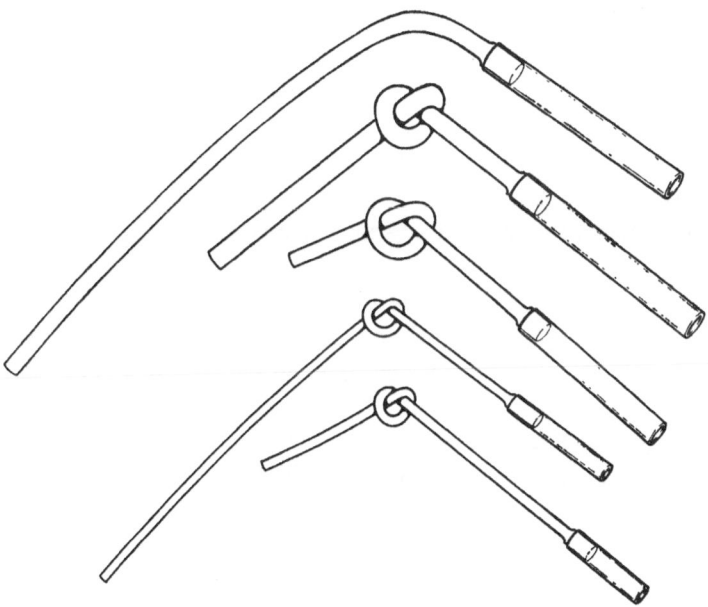

Fig. 5.6. Catheters of 2·54 to 5·28-mm external diameter, with a wall thickness of between 0·25 mm and 0·27 mm

catheter must not be clamped lest the steel spring reinforcement becomes irreversibly deformed. The silicone-rubber connector can be clamped safely.

Perfusion circuitry

The only materials that come into contact with the fetal bloodstream are silicone-rubber tubing and membrane, the polycarbonate oxygenator spool coated with silicone fluid, tetrafluoroethylene plastic connectors and segmented polyurethane. We do not use any polyvinyl tubing since organotin compounds and plasticizers are used in their manufacture. These are known cytotoxic agents which can leach into the bloodstream (Atkins *et al.*, 1968).

Initial experiments: partial bypass in newborn lambs

Our studies enabled us to assess the biological response of healthy animals to long-term extracorporeal blood oxygenation with the spiral-coil membrane lung. A preliminary series of studies, performed without blood pumping, analysed only the effects of membrane oxygenation (Kolobow *et al.*, 1968). Newborn lambs were used as the experimental animal because their term weight (2·5–3·5 kg) closely approximates that of the human infant.

We did not use continuous general anaesthesia or immobilization since prolonged anaesthesia for several days causes progressive atelectasis as well as red-blood-cell sequestration and anaemia (Indeglia *et al.*, 1967).

AV shunting with membrane oxygenation

One- to eight-day-old Hampshire and Corriedale lambs were anaesthetized briefly with Halothane via an endotracheal tube. One internal carotid and one external jugular vessel were cannulated with Teflon cannulae (internal diam. 2·54 mm; external diam. 3·18 mm) by clean but not sterile techniques. Lambs were initially heparinized at 3 mg/kg, the cannulae were clamped, and the lambs allowed to awaken and feed whole cow's milk from a bottle. Animals were given intramuscular penicillin and streptomycin every 12 h. They were transferred to cages and connected to the inflow line of a circuit consisting of a 0·2 m² spiral-membrane oxygenator, bubble trap and ultrasonic flow cell, all primed with heparinized lactated Ringer's solution (Fig. 5.7). Heparin was subsequently infused at a constant rate of from 1·0 to 6·0 mg/kg h. Oxygen inflow was humidified at 40°C to prevent water-vapour diffusion across the membrane. The lambs were permitted to move around while tethered on a 0·5-m chain to the enclosure wall, and kept warm with an infra-red lamp. Blood pressure, pulse and perfusion pressure were continuously monitored with transducers connected to blood inflow and outflow lines. Blood flow was monitored continuously by a dual-frequency ultrasonic phase flow-meter (Noble, 1968). The bubble trap was removed from the circuitry approximately 15 min after perfusion began. Blood leaks, which could compromise gas transfer or terminate an experiment, did not develop. Membrane oxygenators were replaced

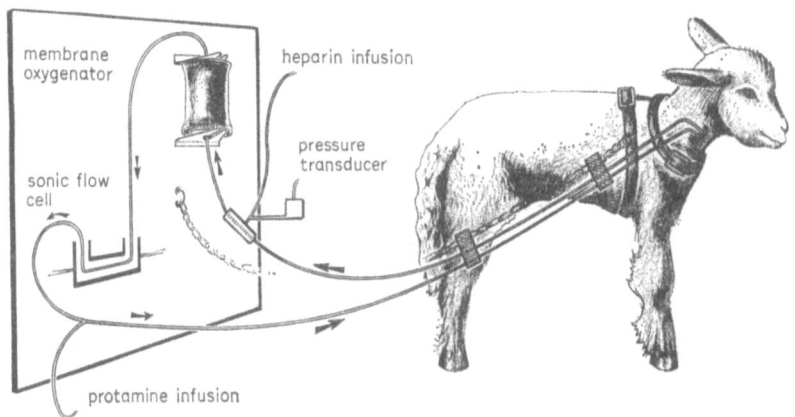

Fig. 5.7. Experimental set-up within the animal's cage

if it became apparent, by external inspection or by noting a change in gas transfer, that thrombus formation had occurred within the membrane lung.

Daily tests of membrane-oxygenator efficiency were performed after lambs were made hypoxic by a brief exposure to a 7% oxygen and 93% nitrogen gas mixture. Blood-gas tensions were determined by the Radiometer system, oxygen saturations by a reflectance oximeter. Haemoglobin was measured by the cyanmethaemoglobin method. Samples for plasma-free haemoglobin determinations were drawn at the start and finish of each perfusion, and analysed by the method of Hunter *et al.* (1950). Whole fresh adult-sheep blood, after compatible cross-match, was transfused when warranted.

Following cannulation many of the lambs appeared fatigued, but recovered strength in a few hours, bottle-fed eagerly and seemed surprisingly healthy. We consider clinical status vital in any description of long-term perfusion.

These perfusions provided evidence that spiral-membrane lungs can transfer oxygen reliably for several days. While it was possible that perfusion produced traumatic effects on red cells, white cells, platelets and plasma proteins, these effects, whatever their nature, did not limit survival of normal newborn lambs.

Continuous partial extracorporeal bypass was carried out in eight lambs at 40–85 ml/kg min (Kolobow *et al.*, 1968). Six lambs were long-term survivors of up to 96 h of perfusion. These studies revealed a progressive

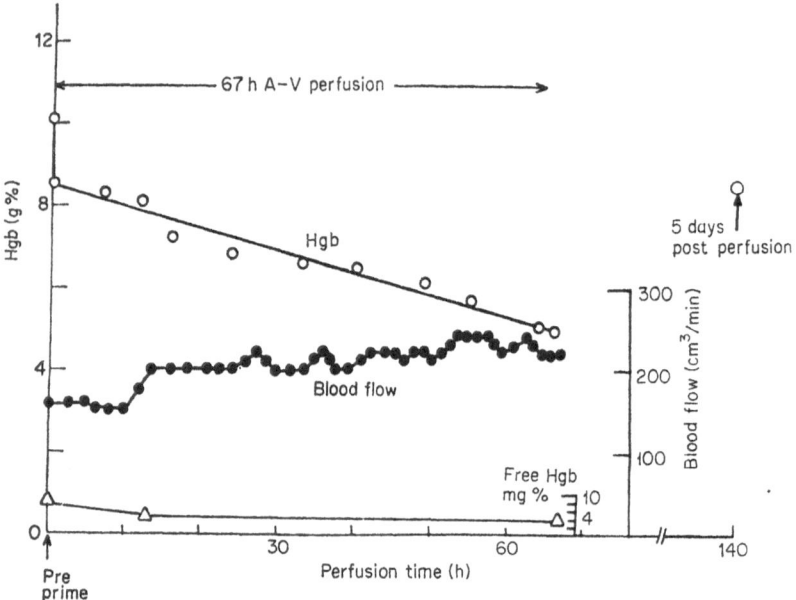

Fig. 5.8. A-V shunting with membrane blood oxygenation in a three-day-old 3·6-kg lamb

anaemia without elevation of plasma-free haemoglobin (Fig. 5.8). Acidosis and oedema did not occur. Autopsy specimens after prolonged perfusion showed grossly and microscopically normal lungs. This is significant because the lung is the sensitive target organ to receive thrombi and would show deleterious changes if they were produced by long-term AV shunting with membrane-lung blood oxygenation. These studies were the first to suggest that long-term extracorporeal gas exchange was feasible.

Venovenous pumping with membrane oxygenation
Adverse haematological changes due to extracorporeal pumping of blood were considered a barrier to prolonged circulatory assistance

(Shea *et al.*, 1968). No significant progress in artificial placentation could be expected until we had learned how to safely pump and oxygenate large quantities of blood outside the body.

We undertook to design an extracorporeal venovenous (pumping) perfusion system that would cause minimal blood trauma and animal morbidity during and after prolonged perfusion. Drawing on the experience of others with unsuccessful *in-vivo* extracorporeal perfusions and successful intrathoracic pumping (Pierce *et al.*, 1967; LaFarge *et al.*,

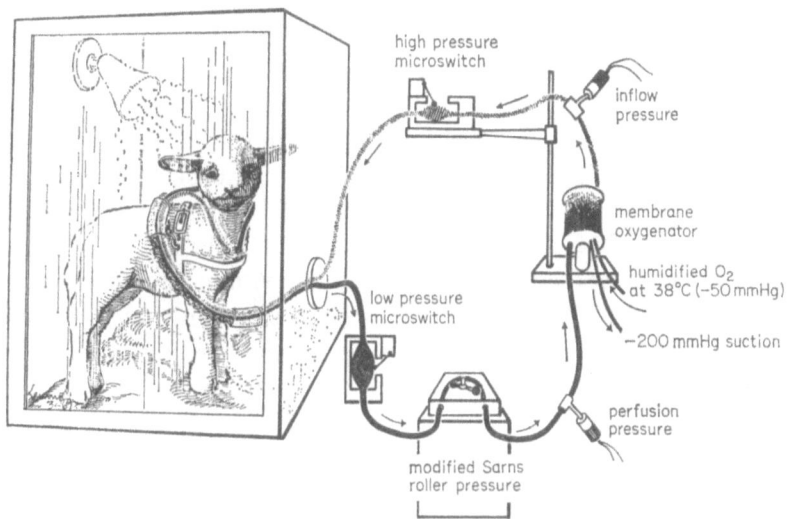

high pressure
microswitch

inflow
pressure

membrane
oxygenator

humidified O_2
at 38°C (–50 mmHg)

low pressure
microswitch

–200 mmHg suction

perfusion
pressure

modified Sarns
roller pressure

Fig. 5.9. The perfusion circuit

1968), we felt it necessary to limit blood suction in blood conduits, to use a precisely non-occlusive pump, and to employ only a silicone-rubber, tetrafluoroethylene or polyurethane blood interface. We excluded all blood-gas interfaces, avoided sudden pressure changes within the perfusion circuit, and used the spiral-coil membrane oxygenator (Kolobow *et al.*, 1969) (Fig. 5.9).

A model 3500 Sarns roller pump was modified for smooth operation at low speed. A thin-walled, large-bore (1·53 cm internal diameter) segmented polyurethane pumping chamber was fabricated with 0·886 ± 0·025-mm walls and the rollers centred and adjusted to give an exact clearance from 0·025 to 0·040 mm. This pump produced a flow of 500 ml/min at 7 rev/min against a hydrostatic head of 60 Torr. Retrograde flow with the roller engaged and stopped was 20 ml/min.

No blood-heat exchanger was used. Instead, convective heat loss was minimized by enclosing the blood tubing, pump and membrane-lung gas manifold within a polyvinyl hood. With the animal chamber enclosed, a thermostatically-controlled heat lamp maintained a constant environmental temperature of 36°C.

A 0·4-m^2 disposable spiral-coil-membrane blood oxygenator having a priming volume of 70 ml was incorporated in some of our perfusion studies. Used membrane lungs were disassembled, cleaned, rechecked for gas leaks and rewrapped. Some units were re-used up to six times and did not deteriorate in gas-exchange capacity nor cause any increased thrombogenesis. Silicone-rubber tubing within the animal cage was supported by a stainless steel spring wound around the tubing to prevent accidental kinking.

Under clean but not sterile conditions three- to eight-weeks-old Hampshire and Corriedale lambs were placed in a supine position and briefly anaesthetized with Halothane. The animals were initially heparinized with 4–5 mg/kg heparin, and the right and left external jugular veins were both cannulated with a 4·56 mm internal diameter, 5·08 mm external diameter reinforced segmented-polyurethane cannulae. One of the cannulae was passed into the inferior vena cava to the level of the diaphragm (verified either by X-ray or pressure tracings through the catheter). The other cannula returned blood to the superior vena cava. Meticulous attention was paid to haemostasis by limiting the area of dissection, and employing electrocautery. These catheters were shaped to permit easy positioning within the vena cava and were sufficiently flexible to allow motion within the vena cava during perfusion of an alert animal.

Once awake, the lambs were transferred into the heated animal enclosure and connected to the perfusion circuit, which had been primed with 350 ml of lactated Ringer's solution. The priming solution contained 8 mg/100 ml heparin and one million units of Penicillin G. Lambs were given continuous heparin infusion at a rate of 1·5 mg/kg h directly into the blood inflow line. Intramuscular procaine penicillin and streptomycin were given every 12 h. The animals were tethered to the cage wall by a chain and allowed limited freedom of activity. They fed *ad libitum* on standard forage and also drank water without restriction.

Figure 5.10 illustrates the data recorded during a 7-d venovenous perfusion with a spiral-membrane lung at 45 ml/kg min. This sheep was a long-term survivor. When sacrificed, these animals did not show any gross or microscopic pathological findings.

Anaemia appearing in these older animals was minimal, and could be accounted for entirely by blood-sampling loss (80–350 ml/lamb). Control plasma free haemoglobin was less than 2 mg%. Bleeding from cutdown sites was insignificant or absent following careful initial haemostasis with electrocautery. Post-perfusion haematocrit readings continued to rise in all animals, and reached their normal values within two to three weeks. Of special interest was the absence of significant thrombocytopaenia during perfusion. Platelet counts always ranged above 200,000/mm³. Venous-blood pH remained within the normal range throughout all long-term perfusion experiments, suggesting normal acid–base balance. Animals either had stable weights or lost 0·5–1·5 kg during perfusion. Fluid retention was never observed. The introduction of a membrane lung into the pumping circuit did not appear to have any detectable deleterious effect on the parameters monitored during similar pumping experiments without membrane oxygenation, nor on the clinical status of the animal. We are puzzled by the lack of red blood cell destruction in these studies, in contrast to the significant anaemia we produced in 1- to 8-d-old lambs (AV shunting studies). It is possible that fetal red cells are more susceptible to extracorporeal blood handling than adult-type red cells.

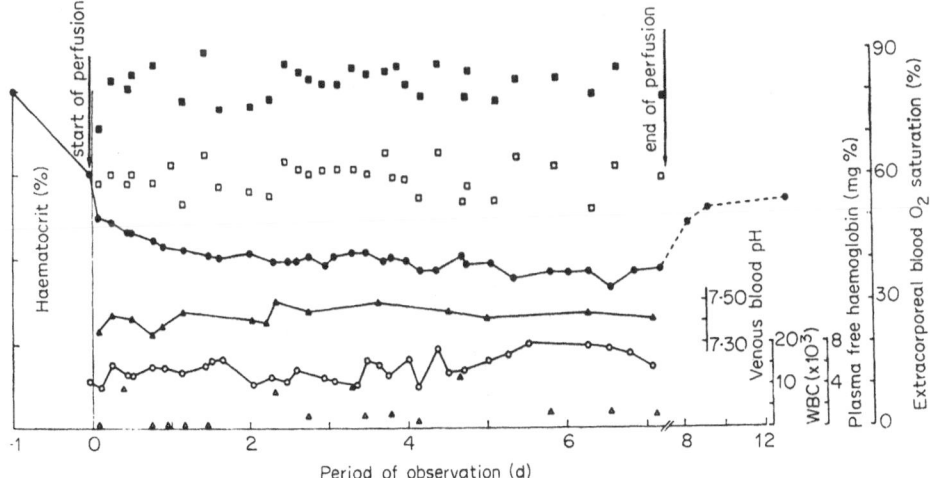

Fig. 5.10. Seven days of venovenous blood pumping with membrane blood oxygenation in a 10·8-kg lamb. V-V pumping with membrane lung; flow 500 ml/min = 45 ml/kg min. ● Haematocrit, ▲ plasma free haemoglobin, □ venous O₂ saturation, ■ oxygenated-blood O₂ saturation, △ venous-blood pH, o WBC

A week of constant use did not cause membrane-lung clotting. Membrane lungs were not replaced unless definite impairment of oxygen transfer across the unit was documented. Throughout the perfusions lambs ate and drank with their characteristic voracious appetites.

We have extended the duration of long-term venovenous perfusion to 16 d (Fig. 5.11) by reducing heparin infusion after the first 1–2 d to allow a Lee-White clotting time (25°C) of 3–6 h (control 10 min) (Kolobow *et al.*, 1971). When the membrane lung developed thrombus (manifested by increased perfusion pressure or reduced gas-transfer efficiency) it was safely changed within a few minutes. In this lamb no blood was

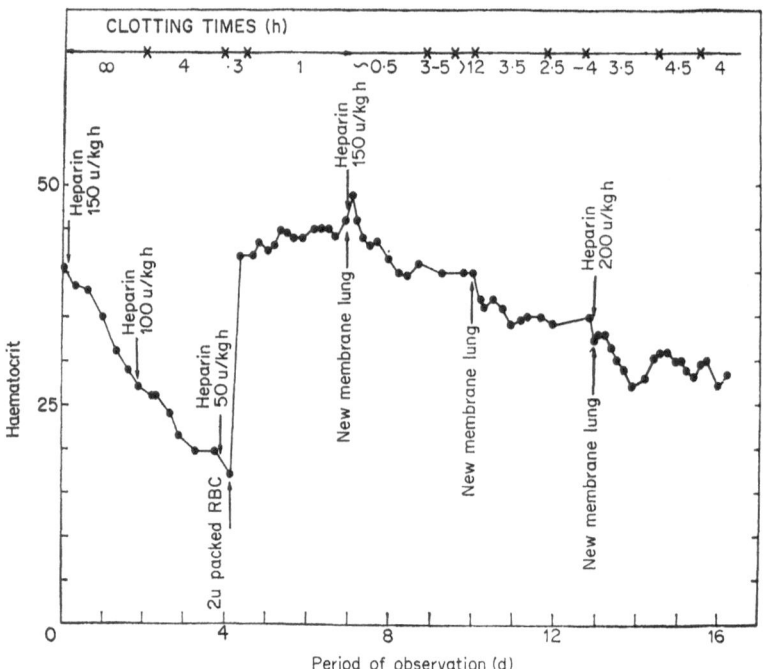

Fig. 5.11. Sixteen days of venovenous perfusion with the spiral membrane lung. 12·9-kg lamb. Extra-corporeal blood flow, 450 ml/min (day 7–9, 100 ml/min)

transfused after day 4 of perfusion. Platelet counts on day 14 ranged from 300,000 to 400,000/mm³. This animal exsanguinated from an accidentally disconnected blood tube. Gross and microscopic examination at autopsy revealed no abnormalities.

Venoarterial pumping

In these studies extracorporeal venoarterial (VA) membrane-lung blood oxygenation (partial heart–lung bypass) was performed in alert un-anaesthetized 4- to 6-week-old lambs with a perfusion circuit similar to that previously described for venovenous pumping (Kolobow et al., 1970) (Fig. 5.9). Venous-drainage polyurethane catheters, modified with side holes, permitted large blood flows. Oxygenated blood was returned to the carotid artery through a smaller wire-reinforced polyurethane cannula.

Prolonged VA blood pumping with spiral-membrane blood oxygenation at 30–65 cm^3/kg min was performed in eight alert lambs for periods from 2 to 7 d. Six lambs were long-term survivors.

Bypass flow was kept arbitrarily at approximately 50 ml/kg min, an easily-maintained flow rate with large-bore venous catheters. In contrast to the previously described venovenous perfusions, we used whole-blood prime to increase haematocrit readings and allow assessment of changes in haemodynamic parameters. The life-span of donor red-blood cells is relatively short, and in part accounted for a slow fall in lamb haematocrit readings of 3–4 points per day. Plasma haemoglobin levels in these studies ranged from 2 to 10 mg/100 ml, or slightly more than we found in earlier perfusion studies. This minimal elevation can also be explained on the basis of donor blood. Platelet counts always remained high and approximated control values (300,000–500,000/mm^3).

At elective sacrifice, we did not find myocardial or subendocardial ecchymoses. Thorough gross and microscopic examination did not reveal other organ damage. We therefore concluded that non-synchronous arterial return of membrane-lung-oxygenated blood by a roller pump is not detrimental to myocardial or other internal organ integrity, at least not with the pulse-wave forms generated by our blood pump. We further concluded that the performance of the membrane lung is safe and reliable during prolonged venoarterial perfusion.

We observed occasional spontaneous bleeding in several long-term perfusions; we experienced two cases of haematoma of the thigh and one case of intraperitoneal bleeding. These haemorrhages occurred despite elevated platelet counts. None of the spontaneous bleeding episodes occurred before the fourth day of perfusion. Heparin levels used in this series of experiments were high (1·5 mg/kg h) but were believed necessary in order to prevent clotting of sheep blood within the extra-corporeal circuit.

Total perfusion for isolated support of the fetus

Maternal care

Hampshire and Corriedale ewes with dated gestation (130–145 d) were used for Caesarian section (Dawes, 1968). Initially, size and number of fetuses was confirmed by radiographic examination. Difficulty in obtaining a standard radiographic magnification in the pregnant ewe, however, led us to abandon these efforts. Inability to determine whether a 2- or 4-kg-sheep fetus would be delivered made it mandatory to provide gas exchange for the largest possible fetus. Frequently our ewes produced twins. This allowed a second attempt of fetal perfusion if the first twin was not successfully cannulated or if it succumbed within 6 h of perfusion.

Ewes near term can develop toxaemia of pregnancy. This disease can be precipitated by their move from the pasture, change in diet and abnormal handling, or excitement. To avoid these problems, we brought our ewes in from the farm about 24 h before Caesarian section.

Caesarian section and fetal cannulation

Ewes were anaesthetized in the standing position by performing a lumbar subarachnoid puncture with a No. 18-gauge spinal needle in the lumbosacral space at the level of the iliac crest. Clear colourless cerebrospinal fluid was aspirated. A dose of 5–7 ml of 2% xylocaine was injected slowly (1 ml/s). Two hours of operating time was easily provided by this dose. The ewe was immediately lifted and placed supine on a level table and the fore and rear hocks immobilized. An intravenous No. 18-gauge-catheter infusion was begun in the jugular vein and 1 l of lactated Ringer's solution was administered. Anaesthetic level was tested with pin pricks. We initially monitored maternal arterial pressure. This parameter remained consistently stable and we do not believe routine blood-pressure monitoring is necessary during Caesarian section. Ewes were allowed to breathe room air spontaneously. If the fetal cannulation was difficult, we allowed the ewe to breathe at an FIO_2 of 1·0.

Sheep became bloated if given green food or hay within 24 h of induction of anaesthesia. Under a spinal subarachnoid block, excitement may cause swallowing of air, compounding endogenous generation of gas. Under general anaesthesia an oro-gastric tube may be passed to help decompress the stomach; this therapy is not possible with spinal anaesthesia. Visual excitatory stimuli following the induction of spinal anaesthesia can be greatly reduced by blindfolding the sheep. Occasion-

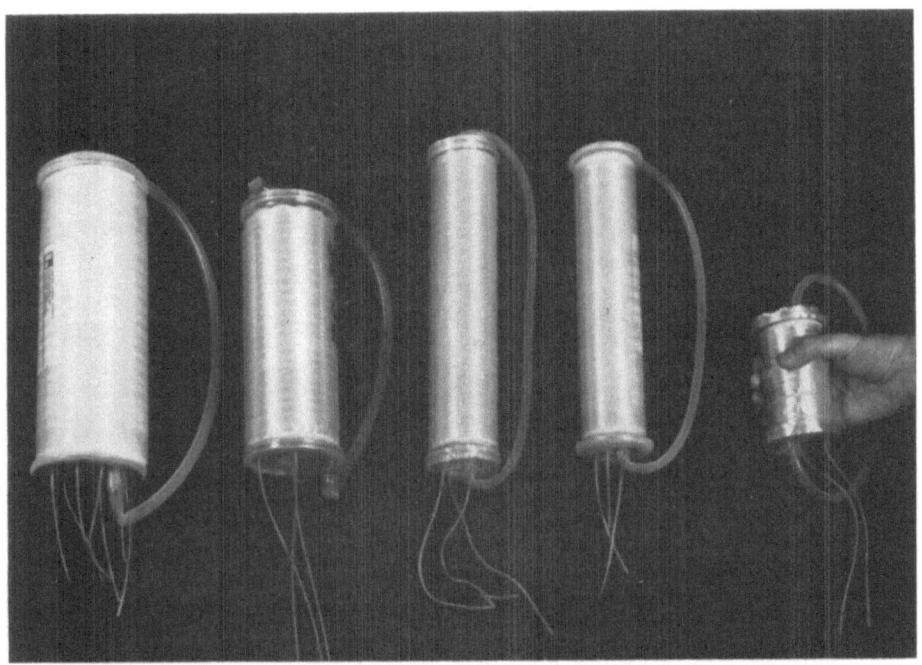

Plate 5.1. Kolobow spiral-coil membrane lungs: 5 m², 2·5 m², 1 m², 0·8 m², 0·4 m²

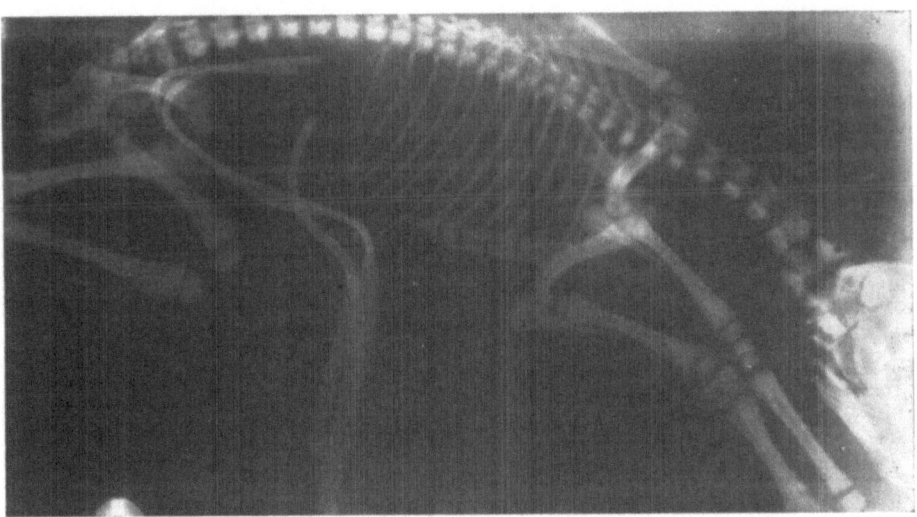

Plate 5.2. Radiograph demonstrating placement of umbilical cannulae in the fetal lamb

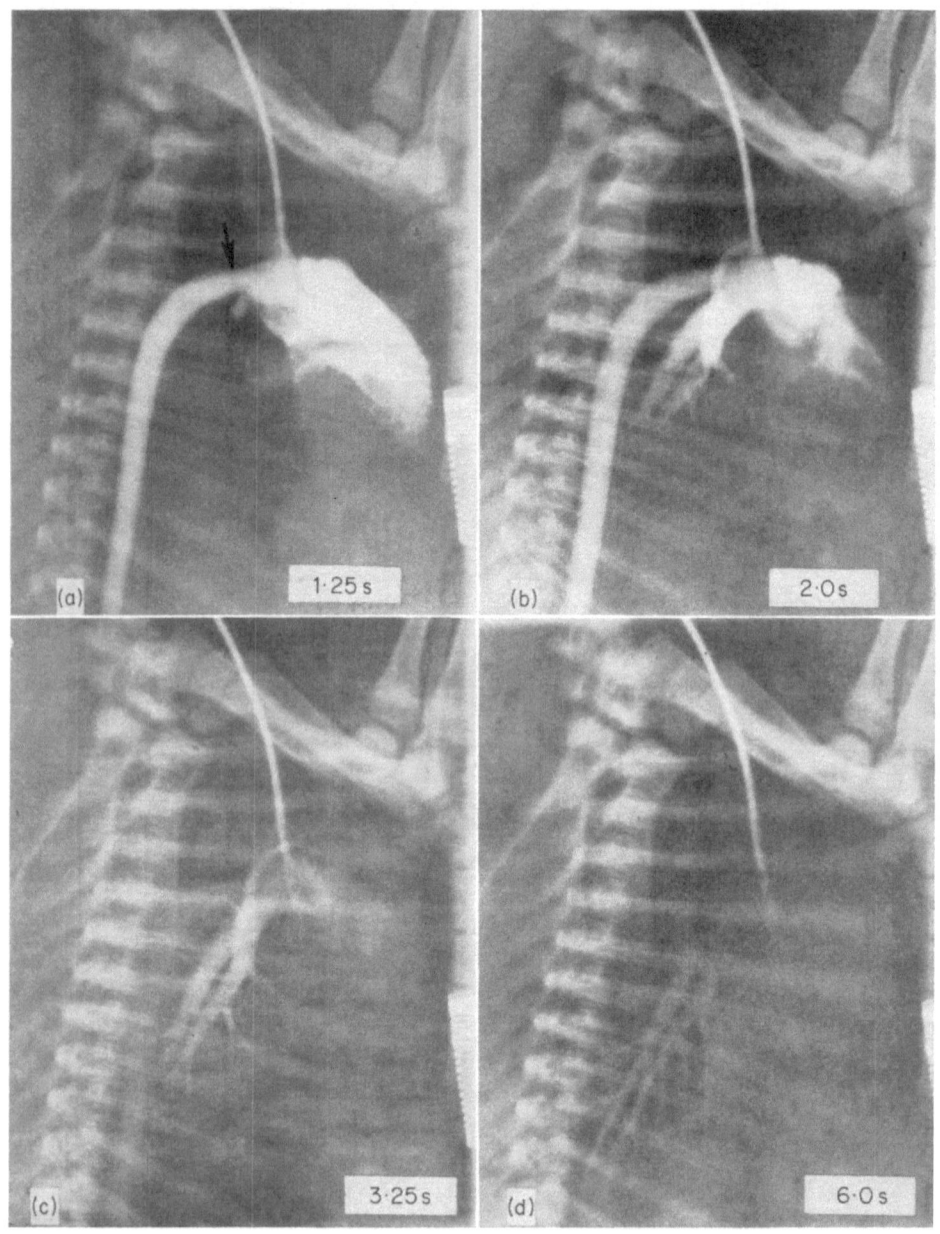

Plate 5.3. Serial radiographs of fetus No. 5 at an umbilical arterial-oxygen tension of 19 Torr

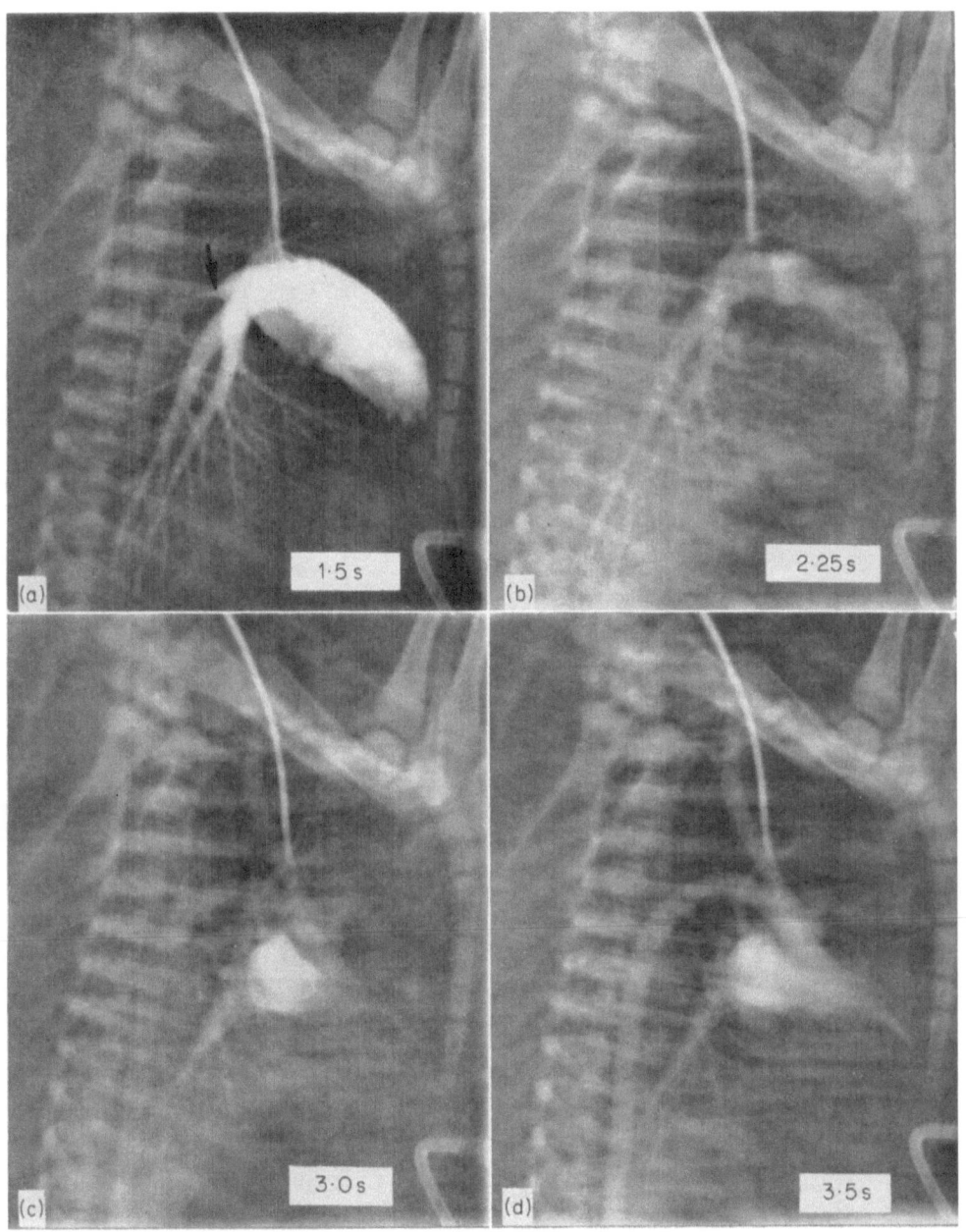

Plate 5.4. Serial radiographs of fetus No. 5 at an umbilical arterial-oxygen tension of 56 Torr

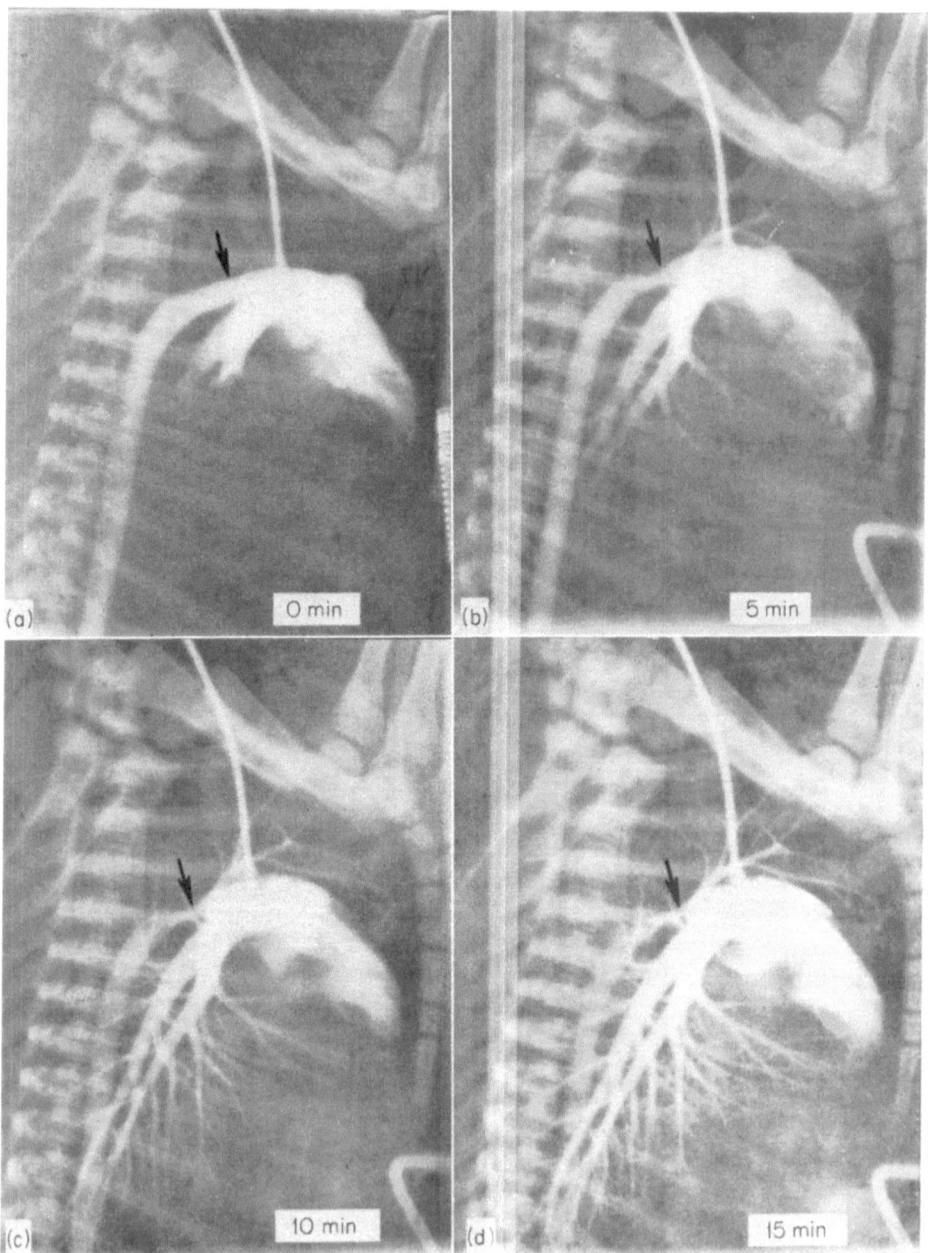

Plate 5.5. Angiographic study of fetus No. 5. The control angiogram (0 min) was made at an umbilical-arterial Po_2 of 19 Torr. The other three angiograms were made at 5, 10 and 15 min after raising the umbilical-artery Po_2 to 56 Torr

ally we have used intermittent intravenous doses of thorazine to further sedate the sheep.

The abdomen was thoroughly scrubbed with soap, prepared with Betadine and alcohol, and draped. A lower midline abdominal incision was made, avoiding the engorged abdominal veins. The peritoneum was opened with care, so that the bowel would not eviscerate. The uterus was delivered anteriorly and palpated. An area of uterus relatively clear of placental cotyledons was incised and the fetal rear hocks and lower abdomen were delivered. We kept this uterine incision small; the fetus could therefore occlude the incision. Amniotic fluid was retained and submerged the fetal head.

Heat loss of the exposed wet fetus was minimized by warming the operating room and using infra-red heating lamps, although this procedure made conditions uncomfortable for the surgical team. Warm saline packs or padding on the fetus were an unsatisfactory solution; they hindered operative procedures.

We found that successful cannulation required a minimum of three operative personnel. One assistant immobilized the fetal limbs and provided proper fetal positioning. The surgeon and an assistant prepared and cannulated the umbilical vessels. Since the umbilical cord exterior to the fetus was necessary only for cannulation, we found that infiltration of a 10% aqueous solution of formaldehyde into the intervascular space of the cord from 2 to 10 cm distal to the fetus prevented spasm of the fetal vessels and provided good vessel fixation for easy dissection. Extreme care was taken not to inject this noxious solution intravascularly. The vessels were thoroughly dissected for 10 cm from the fetus; care was taken not to occlude the vessels. With a 26-gauge needle, 500 u/kg of heparin (weight of fetus was estimated) was injected into the umbilical vein and 2 min were allowed for mixing. One umbilical artery was tied off 8 cm from the fetus with a silk suture. A second suture was passed around the vessel 2 cm from the umbilicus and a $\frac{1}{4}$- to $\frac{1}{2}$-cm transverse incision was made 4 cm from the umbilicus. A 2·6-mm external diameter, 2·1-mm internal diameter, 20-cm-long wire-reinforced segmented-polyurethane cannula was rapidly introduced with a stainless-steel stylet. As soon as the cannula was introduced 2 cm, the stylet was withdrawn, and the silicone-rubber portion of the cannula was occluded with a haemostat or a tetrafluoroethylene plug. Under steady pressure, the cannula was rapidly advanced into the aorta, to a position below the renal arteries. If the cannula did not enter the iliac artery, the intra-abdominal umbilical artery would slowly constrict, causing progressive reduction of blood flow. The cannula was anchored to the umbilical artery with several silk sutures. After aortic cannulation blood was easily sampled to determine fetal blood-gas tensions, pH and other biochemical parameters.

The two umbilical veins of the fetal sheep merge in the subcutaneous and abdominal muscular tissue and enter the peritoneal cavity as a single vessel running cephalad a short distance to the liver. By technique similar to that described previously, one umbilical vein was tied off and cannulated with a larger wire-reinforced cannula (3·8 mm external diameter, 3·3 mm internal diameter, 8 cm long). The cannula was advanced up to

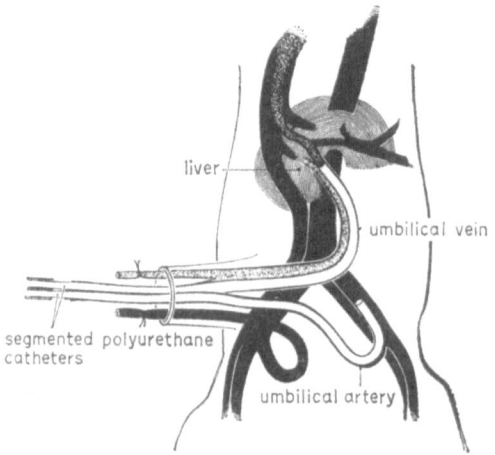

Fig. 5.12. Umbilical cannulation in the fetal lamb

the liver. This catheter was also anchored to the vessel with several silk sutures (Fig. 5.12, Plate 5.2).

With a mean arterial pressure of 50 Torr we achieved a maximum blood flow of 200 ml/min with a single well-placed reinforced cannula. We found this size of cannula suitable for fetuses weighing over 750 g. In fetuses weighing more than 2 kg, we placed a second umbilical arterial cannula immediately after instituting partial bypass. In fetuses weighing less than 2 kg we placed a smaller polyurethane cannula in the second umbilical artery for aortic pressure monitoring and blood sampling.

For angiographic studies we cannulated the fetal external jugular vein with a French 5 or 8 vinyl catheter. The catheter was advanced into the superior vena cava, the wound electrocauterized, and then closed with methyl-2-cyanoacrylate adhesive (Eastman Monomer 910).

Fetal perfusion circuit

Umbilical arterial blood flowed through a 0·95-cm internal diameter silicone-rubber tube into a volume-sensing silicone reservoir (fully expanded volume 10 ml) (Fig. 5.13). The blood was then pumped by a Sarns Model 3500 occlusive roller pump (0·95-cm internal diameter tubing) through a single 0·4-m² spiral-coil membrane oxygenator, a silicone-rubber bubble trap (15-ml volume), and returned via 0·635-cm

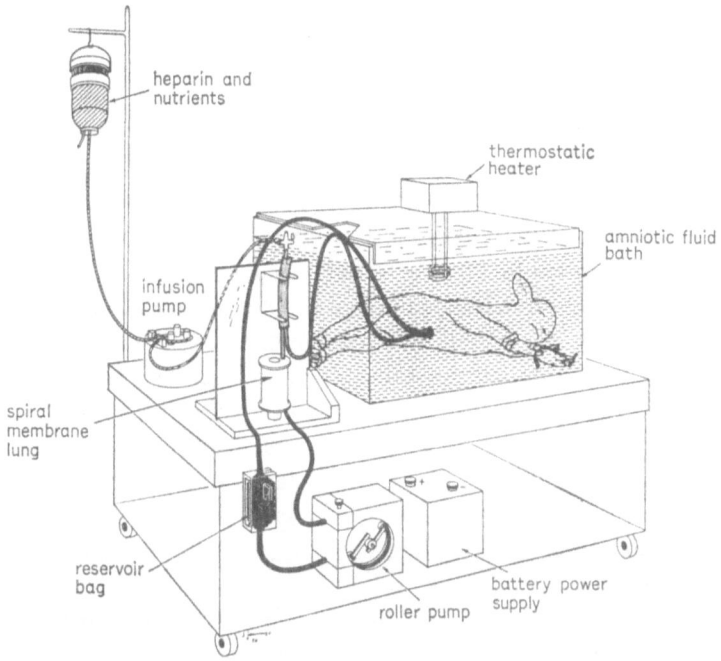

Fig. 5.13. Diagram of perfusion apparatus for artificial placentation

internal diameter tubing to the umbilical vein (Fig. 5.13). Two oxygenators in series were used (0·8 m² in total surface area) if the estimated fetal weight exceeded 3 kg. All blood conduits were constructed of silicone rubber with tetrafluorethylene connectors.

The extracorporeal circuit was primed with 240 ml of adult sheep cells (310 ml if two oxygenators were used). Blood donors were from the same flock as the fetus. The blood was drawn in heparinized, siliconized glass bottles and administered within 12 h of collection. The blood was centrifuged, the plasma decanted and the cells suspended in lactated Ringer's solution. This suspension was centrifuged, the supernatant decanted and the cells administered as prime after filtration through dacron mesh. When the prime and fetal blood were combined, the resulting increase in haematocrit value (approximately 50%) provided a large oxygen-carrying capacity. Blood samples were replaced with equal volumes of Ringer's lactate.

Fetal environment

A portable fetal support cart was constructed (Fig. 5.13) to permit transport of the isolated fetus to distant radiographic facilities. Storage batteries provided power for blood pumping and gas suction while in transit.

Umbilical-arterial-blood gas tensions and pH were measured intermittently with a capillary pH electrode, Clark oxygen electrode, Severinghaus carbon dioxide electrode and a Radiometer PHM27/PHA927 gas analyser. All electrodes were in a water bath at 39°C, and measurements were corrected for fetal rectal temperature. Fetal blood lactate and pyruvate were measured by the techniques of Olson (1962) and Segal *et al.* (1956). Plasma-free haemoglobin was determined by the technique of Hunter *et al.* (1950). Umbilical arterial outflow was clamped for 3 s, and the values of the initial pressure pulse were recorded with a Statham P23Db pressure transducer and Sanborn recorder Model 150. The measurement of umbilical-artery instantaneous-clamp pressure obviates the need for an additional arterial cannulation in the heparinized fetus. The pressure so recorded is approximately 20 Torr higher than average aortic blood pressure simultaneously measured with an indwelling aortic catheter.

Spirometry of the membrane-lung gas phase was used during isolated fetal perfusion (Fig. 5.14). This technique is more reliable and accurate than conventional intermittent blood-gas analyses that are subject to multiple sources of error. In addition, spirogram tracings provided continuous 'on line' readout of oxygen consumption. An initial one hour period for nitrogen washout was allowed when the membrane lung was on 100% oxygen.

Fetal carbon dioxide production was calculated from the gas flow through the membrane envelope (measured with a rotameter) and the carbon dioxide concentration of gas leaving the membrane envelope (measured by a Beckman IR carbon dioxide Analyser model 315A). All calculations were corrected to s.t.p.

The fetus was suspended in a covered acrylic tank (61 cm × 48 cm × 30·5 cm). The fetal fore and rear hocks were loosely tied and anchored to projections on the bottom of the tank. The tank contained 30 l of

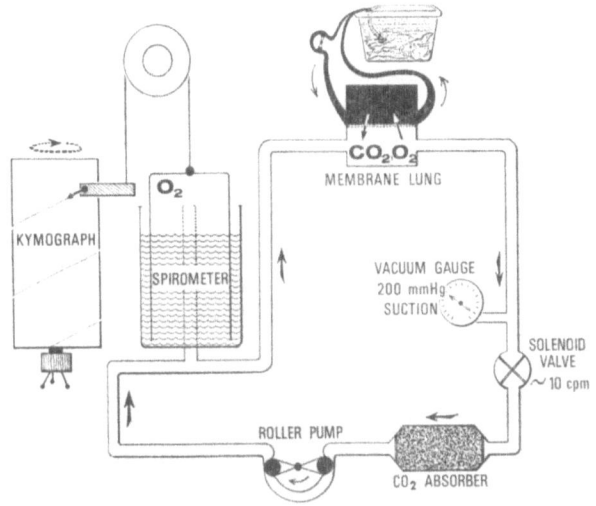

Fig. 5.14. Continuous monitoring system for respiratory-gas analysis

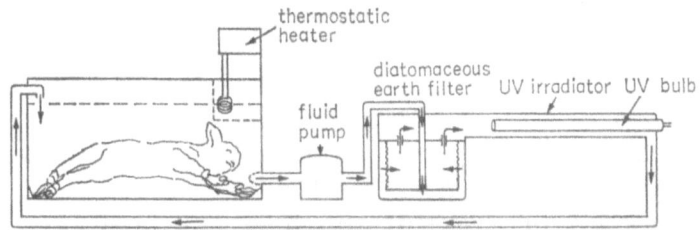

Fig. 5.15. The synthetic-amniotic-fluid conditioner

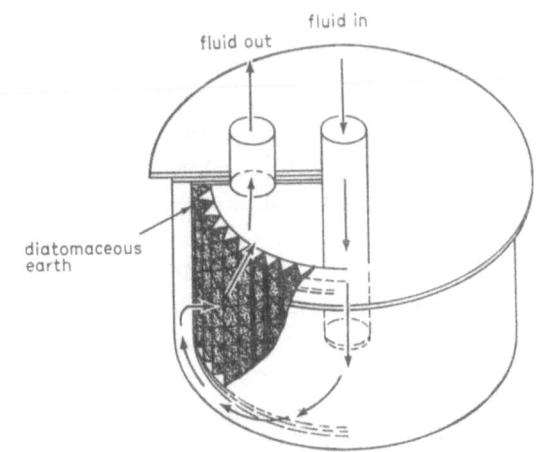

Fig. 5.16. The amniotic-fluid filter

169

synthetic amniotic fluid composed of ingredients in concentrations found by sampling natural sheep amniotic fluid: sodium (130 mEq/l), potassium (6 mEq/l), chloride (100 mEq/l), calcium (7 mg/100 ml) and phosphorus (3 mg/100 ml). We did not add glucose or amino acids since these would foster bacterial growth. Despite addition of antibiotics to the bath we frequently cultured resistant bacteria. To remedy this we constructed an amniotic-fluid conditioner (Fig. 5.15) comprised of a fluid pump, di-atomaceous-earth filter (Fig. 5.16), and ultra-violet-radiation source. Five litres of synthetic amniotic fluid were contained in the amniotic-fluid conditioner. The filter removed bath turbidity caused by fetal desquamation and occasional passage of meconium, which otherwise would reduce the sterilizing capacity of transmitted ultraviolet light.

Monitoring during perfusion

Fetal umbilical arterial blood passed through an in-line cuvette placed in the silicone-rubber tubing before the reservoir bag; a reflectance oximeter thus allowed continuous monitoring of fetal blood oxygen saturation (Vurek *et al.*, 1970).

Metabolic observations during perfusion

Biochemical and physiological parameters monitored during a 55-h perfusion of a 3·05-kg male Hampshire fetus of 125-d gestation age are illustrated in Fig. 5.17. Blood pH was stable in the range of 7·40 ± 0·05, with P_{CO_2} between 30 and 50 Torr. No buffers were necessary. Fetal rectal temperature varied between 39·2°C ± 0·5°C. When blood flow through the artificial placenta was 70–100 ml/kg min oxygen saturation in the umbilical artery was between 55 and 75%. To compensate for early metabolic acidosis we increased initial blood flows through the membrane lung up to 150 ml/kg min. In addition we removed excess carbon dioxide by increasing gas flow through the membrane lung.

Fetal blood volume was initially adjusted with packed, washed, adult-sheep red cells to maintain an umbilical arterial systolic pressure of 85 Torr. No further infusions of red cells or plasma were necessary. Haemoglobin fell from 16 g/100 ml to 8 g/100 ml in 55 h, an amount consistent with sampling losses of 265 ml. This volume was replaced by Ringer's lactate solution. Free haemoglobin in the plasma decreased from an initial 100 mg/100 ml to 13 mg/100 ml at 15 h of perfusion, and persisted below 10 mg/100 ml for the remaining 40 h of the experiment. Platelets measured at 24 h of perfusion were 199,000/mm³. Total bilirubin remained below 1 mg/100 ml. Plasma glucose remained at

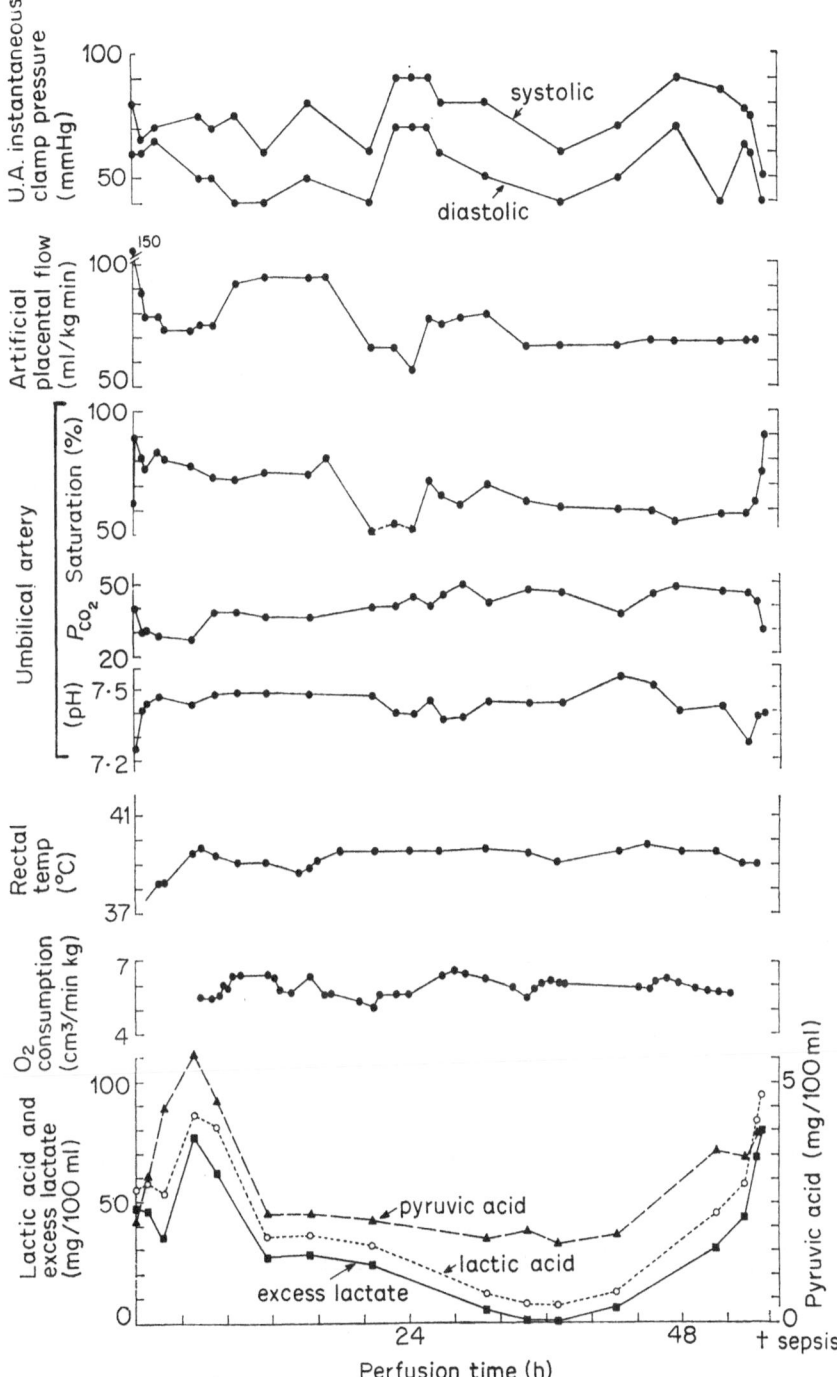

Fig. 5.17. Physiological and biochemical parameters monitored during a 55-h perfusion of a 3·05-kg male Hampshire fetus at 125 days of gestation. Excess lactate was computed by a modification of the method of Huckabee (1968)

approximately 100 mg/100 ml. Lactic acid rose to a high of 84 mg/100 ml at 5 h of perfusion, then fell sharply to about 10 mg/100 ml and remained so (Fig. 5.17). Potassium in the plasma was about 6 mEq/l. Creatinine fell from an initial value of 2·5 mg/100 ml to less than 1 mg/100 ml.

Fifteen sheep fetuses perfused over an 18-month period achieved a stable state similar to that illustrated in Fig. 5.17. The primary factors in achieving a successful perfusion were (*a*) a rapid and successful cannulation allowing large blood flows (> 75 cc/kg min); (*b*) adequate con-

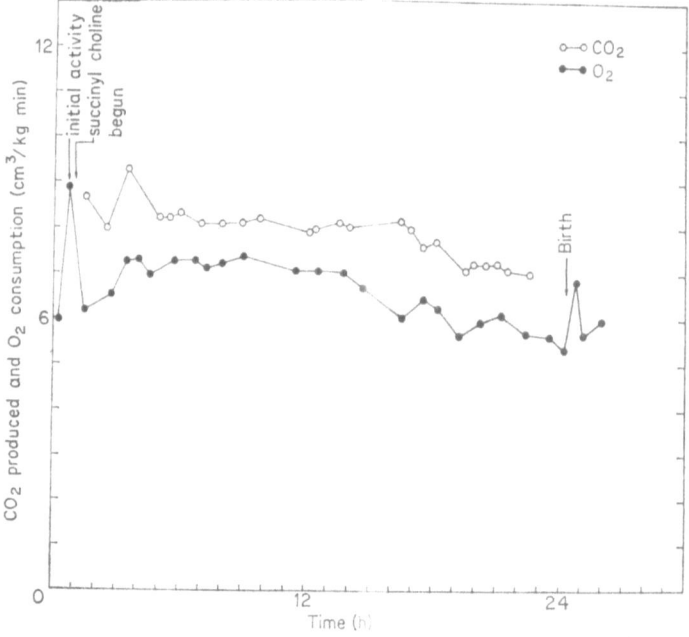

Fig. 5.18. Oxygen and carbon dioxide exchange data during continuous *in-vivo* monitoring. 3·2-kg lamb, 127 days' gestation

tinuous levels of systemic heparinization; (*c*) adequate functional membrane-lung surface area for sufficient oxygen and carbon dioxide exchange. All fetuses who were successfully placed on long-term perfusion and perfused more than 24 h eventually succumbed. The reasons for failure were at times mechanical (e.g. failure of solenoid vacuum switch, pump failure, etc.); they were occasionally related to cannulation (a sheep fetus can kick out a cannula if his extremities are unrestrained), haemorrhage, and septicaemia. At times a fetus would undergo cardiac arrest without evident metabolic cause. The fetus in Fig. 5.17 abruptly underwent cardiac arrest after 55 h of perfusion and stopped extracting oxygen from the arterial blood. Cultures of the fetal blood and the amniotic bath yielded *Klebsiella* and *Aerobacter* species resistant to the infused antibiotics.

We tried to achieve an anabolic state with respect to fetal nutrition. We took the pioneering work of Dudrick *et al.* (1968) in prolonged parenteral alimentation of puppies and that of human infants as a model of anabolism without oral feeding. To 250 ml of commercially available modified fibrin hydrolyzate (containing 5 g of fibrin per 100 ml, Aminosol, Abbott), we added 150 ml of dextrose solution (50 g/100 ml) and 10 ml of vitamin supplement (Multi-Vitamin Infusion, MVI, U.S.V. Pharmaceutical Corp.). This made 410 ml of nutritional infusate. Heparin 1·5 mg/kg h and antibiotics (penicillin and sodium colistimethate) were added to the infusion. The solution was continuously infused by a roller pump at a rate of 8·4 ml/h.

Respiratory gas-exchange rates obtained during a 24-h perfusion of a 3·2-kg Hampshire fetus at 127 d of gestation are shown in Fig. 5.18. Initial levels of oxygen consumption following cannulation rose to 9·5 cm^3/kg min with activity, and decreased to 7 cm^3/kg min following succinylcholine administration and paralysis. Because carbon dioxide release was continuously greater than oxygen uptake, the respiratory quotient was greater than one, apparently signifying net metabolic fat production. Blood-glucose levels were 128, 76 and 168 mg/100 ml at 3, 14 and 18 h of perfusion, respectively.

Behavioural observations during perfusions

During perfusion fetuses rested quietly on the bottom of the artificial amniotic bath. Spontaneous movement of the head or extremities was observed every few minutes and occasionally a fetus became hyperactive. Marked reduction in arterial oxygen tensions was an obvious drive for respiratory movements. General hyperactivity and the passage of meconium were additional responses to respiratory stimuli. We did not observe hypercapnoea without hypoxia to be a powerful respiratory stimulant.

Fetuses exhibited a strong sucking and swallowing reflex to oral tactile stimulation, as well as a diffuse withdrawal reflex to noxious stimuli (e.g. electrocautery or pinching). They exhibited a startle reflex if the bath was sharply tapped.

All fetuses perfused beyond 24 h passed meconium into the synthetic amniotic bath. Frequently a fetus passed meconium several times. Bath filtration became necessary to allow visual observation of the fetus.

Cardiovascular observations during artificial placentation

Physiological mechanisms that trigger the pronounced changes occurring in fetal circulation at birth are only partially understood. Current knowledge is mainly limited to studies of the exteriorized lamb fetus, which exhibits a progressively decreasing placental flow and dies within a few hours (Heymann & Rudolph, 1967). With the previously described system for isolated maintenance of the sheep fetus, we were able to observe the anatomic and haemodynamic changes in fetal circulation—such as closure of the ductus arteriosus and increase in pulmonary blood flow—produced by alterations of blood-oxygen tensions under carefully controlled conditions of amniotic life. We have chosen the technique of angiocardiography (Zapol *et al.*, 1971) for these observations, as any extensive operative invasion of the fetus may itself induce marked changes in fetal circulation. Following such studies, the subjects have been 'delivered' from the artificial placenta and have completed the birth process.

Several different respiratory-gas mixtures were passed through the gas compartment of the membrane lung. We could vary fetal arterial respiratory-blood-gas tensions independently of placental blood flow by changing the diffusion gradients across the silicone-rubber membrane and returning blood with the desired respiratory gas tensions into the umbilical vein.

The fetus was transferred into a shallow (12-cm) thermoregulated amniotic-fluid bath for radiographic study. A standard dose of 4 ml of meglumine iothlamate was hand-injected as rapidly as possible into the external-jugular-vein catheter. Angiograms were recorded either with a 16- or 35-mm cine camera at 60 frames per s (simultaneously recorded on videotape for immediate review), or with a Schonander cut-film changer at a speed of four films per s for 4 s, and then three films per 2 s for 8 s.

Cineangiograms were analysed on a Vanguard Motion Analyzer (Model 16-CD, Vanguard Instrument Corporation, New York, N.Y.). Pulmonary circulation time, computed by the method of Kaplan and Rudolph (Kaplan & Rudolph, 1969), was defined as the interval between appearance of contrast medium in the main branch pulmonary arteries and the opacification of the left atrium. The time required for the contrast medium to disappear from the peripheral pulmonary arteries is referred to as 'pulmonary-artery clearance time'. Determination of this time made it possible to quantify pulmonary circulation in states of very low pulmonary blood flow when the left atrium did not opacify. The diameter of the ductus arteriosus was measured in the lateral projection just beyond its origin from the pulmonary artery. Simultaneous measurement of the injection-catheter diameter made it possible to convert the radiological ductal measurement to actual dimensions. Estimates of pulmonary circulation time by two observers agreed within ten cine frames. In most cases agreement on ductus arteriosus size was within 10%.

After completion of angiocardiographic study, fetuses Nos. 2, 3, 4 and 6 were delivered from the tank, dried and warmed. They were then intubated with a No. 12 French endotracheal tube, placed on a volume-controlled respirator and ventilated with air. The artificial placental circulation was terminated, and the umbilical and neck cannulae were removed. When adequate physical strength had returned, the lambs were extubated, fed and placed in warm cages.

Table 5.2 presents data from six perfusions. Lambs Nos. 2 and 3 were delivered after approximately 12 h of perfusion and are long-term (over

Table 5.2. *Data from six perfusion experiments**

Cine	Perfusion time (min)	Placental flow (ml/kg min)	pH	P_{CO_2} (Torr)	P_{O_2} (Torr)	Ductus-arteriosus size (mm i.d.†)	Pulmonary-circulation time (s)	Pulmonary-artery clearance (s)
				Umbilical artery				
Fetus No. 1: 3·6 kg								
1	403	95	7·44	36	25	4·4	>9	2·0
	404: 21% oxygen							
2	423	95	7·45	36	15	4·1	>10·5	4·1
	424: 100% oxygen							
3	438	95	7·41	40	32	2·8	2·1	1·7
4	454	95	7·41	40	32	2·8	2·2	1·4
5	470	95	7·42	38	36	2·3	2·4	1·8
6	475	95	7·42	38	36	2·5	2·4	1·8
Fetus No. 2: 4 kg								
1	192	115	7·32	49	30	<0·9	1·6	1·5
	202: 70% oxygen							
2	218	115	7·30	45	19	4·4	2·4	2·1
	219: 40% oxygen							
3	223	115	7·31	44	13	4·6	4·6	3·4
4	234	115	7·30	47	18	4·8	4·1	3·0
	238: 100% oxygen							
5	252	110	7·32	42	34	<0·9	1·6	1·1
	256: 5% carbon dioxide, 95% oxygen							
6	268	110	7·19	59	34	<0·9	1·5	1·5
	272: 100% oxygen							
	283	55						
7	308	55	7·29	40	16	4·3	>12	3·8
	310	100						
8	334	100	7·30	42	36	<0·9	1·4	1·4

175

Table 5.2 (*continued*)

Cine	Perfusion time (min)	Placental flow (ml/kg min)	pH	P_{CO_2} (Torr)	P_{O_2} (Torr)	Ductus-arteriosus size (mm i.d.†)	Pulmonary-circulation time (s)	Pulmonary-artery clearance (s)
				Umbilical artery				
Fetus No. 3: 2·6 kg								
1	291	80	7·49	32	37	< 1·0	1·8	–
	321: 21% oxygen							
2	396	80	7·48	35	21	4·1	> 18·4	11·4
	406: 100% oxygen							
3	411	80	7·45	32	41	2·1	6·5	4·4
4	416	80	7·45	32	41	< 1·0	2·4	–
5	439	80	7·47	31	41	1·0	2·4	–
	440: 5% carbon dioxide, 95% oxygen							
6	475	80	7·25	60	43	< 1·0	2·2	–
Fetus No. 4: 3·8 kg								
	0: 21% oxygen							
1	385	83	7·34	48	18	6·5	> 10	–
	395: 100% oxygen							
2	415	83	7·36	53	29	2·2	2·0	–
	448: 5% carbon dioxide, 95% oxygen							
3	483	83	7·21	75	28	2·4	2·5	–
Fetus No. 5: 4 kg								
	0: 21% oxygen							
	225: 15% oxygen							
1	252	130	7·10	23	19	5·9	> 13	–
	253: 100% oxygen							
2	258	130	7·10	23	56	3·2	3·1	–
3	263	130	7·11	21	56	2·0	2·0	–
4	268	130	7·12	21	54	1·8	1·8	–
5	283	130	7·12	21	52	1·4	1·5	–
Fetus No. 6: 2·8 kg								
1	371	130	7·48	30	63	2·8	2·5	–
	372: 10% oxygen							
2	404	120	7·48	31	22	5·6	> 15	–
	424: 100% oxygen							
3	429	120	7·48	31	64	4·0	3	–
4	435	120	7·49	30	65	2·8	2	–
5	444	120	7·49	30	66	1·4	1·7	–
	460: 5% carbon dioxide, 95% oxygen							
6	493	106	7·25	61	71	2·1	2·7	–

* The membrane-lung gas phase is 100% oxygen unless otherwise noted.
† Internal diameter.

one year) apparently normal survivors. Lamb No. 6 died 18 h post-perfusion of an iatrogenic pneumothorax; lambs Nos. 4 and 5 died of pneumonia 6 weeks after perfusion. Fetus No. 1 was perfused 42 h but died when inadvertently the membrane-lung gas flow was interrupted. The fetus was normal at autopsy, without haemorrhages, swelling or oedema. Oxygen consumption, measured for 12 h by spirometry of the membrane-lung gas supply, was 6.5 ± 0.5 cm^3/kg min. Plasma-free haemoglobin fell from an initial level of 74 mg/100 ml at 2 h of perfusion to 40 mg/100 ml at 42 h. Lactic and pyruvic acids were 70 and 4.6 mg/100 ml, respectively, at 11 h of perfusion, stabilizing at about 16 and 2 mg/100 ml, respectively, during the last 24 h of perfusion.

Figure 5.19 illustrates physiological criteria monitored during prolonged perfusion of fetus No. 3. Cineangiogram No. 1 was taken 5 h after perfusion began; pH was about 7.5; P_{O_2} was 37 Torr; and umbilical blood flow was 80 cm^3/kg min (Table 5.2). The inside diameter of the ductus arteriosus measured less than 1 mm; the pulmonary circulation time was only 1.8 s. Gas-phase oxygen concentration of the membrane lung was then decreased from 100 to 21%. Umbilical arterial-blood P_O fell to 21 Torr; pH, P_{CO_2}, and placental flow remained constant. Cineangiogram No. 2 showed that after 75 min the ductus arteriosus was dilated to 4.1 mm and the pulmonary circulation time was lengthened to more than 18.4 s. We then raised umbilical arterial P_{O_2} by increasing the oxygen concentration in the membrane lung to 100%. Cineangiograms No. 3 (5 min), No. 4 (10 min) and No. 5 (33 min after the sudden rise in blood-oxygen tension) showed progressive closure of the ductus arteriosus and increasing pulmonary blood flow. Next, a mixture of 95% oxygen and 5% carbon dioxide was placed in the gas compartment of the membrane lung. Thirty-five minutes were allowed for the induction of respiratory acidosis, with P_{CO_2} rising to 60 Torr and pH falling to 7.25. A final cineangiogram (No. 6) did not show significant changes in either ductus arteriosus diameter or pulmonary circulation time. After 11 h of perfusion, the fetus was delivered, intubated and ventilated with a respirator. Cannulae were removed, and 4 h later the fetus was extubated. It is now three years old.

Results of superior-vena-cava injection in fetus No. 5 (P_{O_2} 19 Torr) are shown in Plate 5.3. The ductus arteriosus appeared maximally dilated (5.9 mm), and pulmonary circulation time was greater than 13 s. Most of the contrast material bypassed the lung through the ductus arteriosus and opacified the descending aorta. Pulmonary artery opacification was prolonged; the left atrium never opacified.

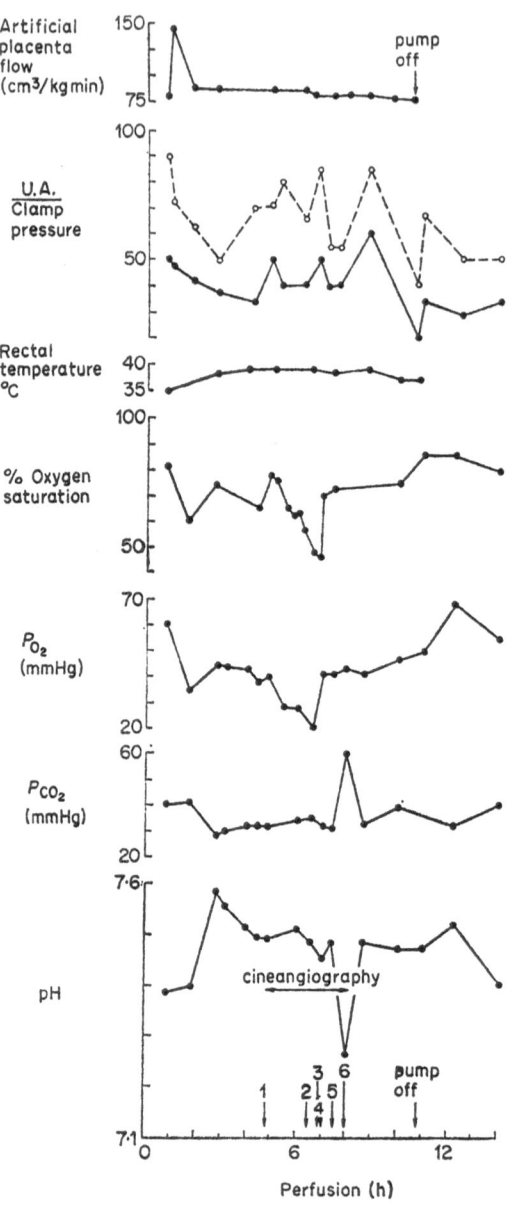

Fig. 5.19. Physiological criteria monitored during perfusion of fetus No. 3

Radiographs of the same fetus, perfused for 30 min at constant placental blood flow, pH and P_{CO_2}, with an umbilical arterial oxygen tension raised to 56 Torr, are shown in Plate 5.4. The ductus arteriosus stump was identified, but little contrast material passed into the descending aorta. Pulmonary blood flow was rapid, with dense opacification of the left atrium and a 1½-s pulmonary circulation time. The left ventricle, ascending aorta and brachiocephalic trunk were clearly visible.

The effect of raising blood P_{O_2} on ductus-arteriosus diameter, pulmonary-artery clearance time and pulmonary-circulation time is shown in Figs. 5.20, 5.21 and 5.22 (fetus No. 2). The ductal diameter exhibited

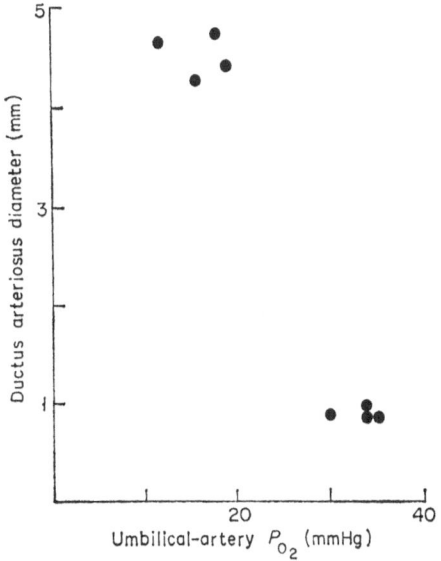

Fig. 5.20. Ductus-arteriosus diameter as a function of umbilical artery P_{O_2} for fetus No. 2

an inverse relationship with P_{O_2}: large ductal diameters of 4·3–4·8 mm were found at oxygen tensions from 12 to 18 Torr, while marked ductal constriction occurred at tensions between 30 and 35 Torr. Both pulmonary artery-clearance time and pulmonary-circulation time were long at low oxygen tensions, but they shortened at elevated oxygen tensions (Table 5.2).

We have also been able to lower umbilical artery P_{O_2} by decreasing blood flow through the membrane lung. Once again, pulmonary-circulation time increases and the ductus dilates widely (Table 5.2, fetus No. 2, cine 7).

Ductal closure following metabolic acidosis (pH 7·11) caused by prolonged cannulation was observed in fetus No. 5. Following a rapid increase of oxygen concentration within the membrane-lung gas compartment, blood-oxygen tensions rose in 2–3 min to within 95% of the

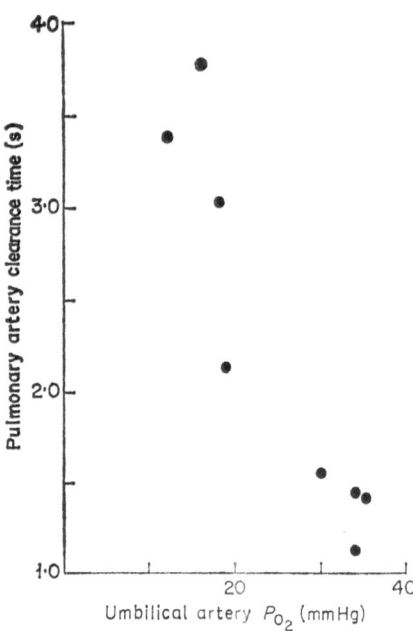

Fig. 5.21. Pulmonary-artery clearance time as a function of umbilical-artery P_{O_2} for fetus No. 2

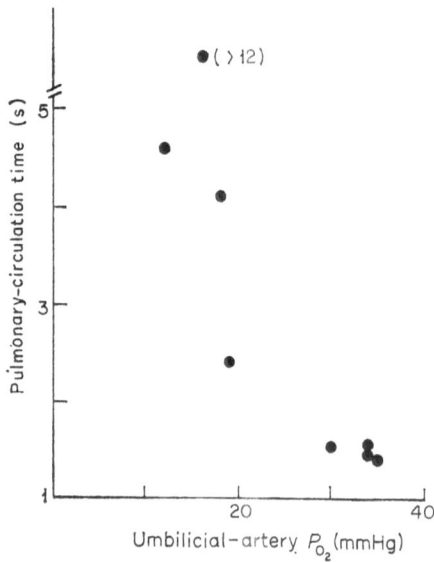

Fig. 5.22. Pulmonary-circulation time as a function of umbilical-artery P_{O_2} in fetus No. 2

final value of 56 Torr. The results of serial angiocardiographic studies at 0, 5, 10 and 15 min, Plate 5.5 (fetus No. 5), show rapid closure of the ductus arteriosus and a marked increase in pulmonary blood flow.

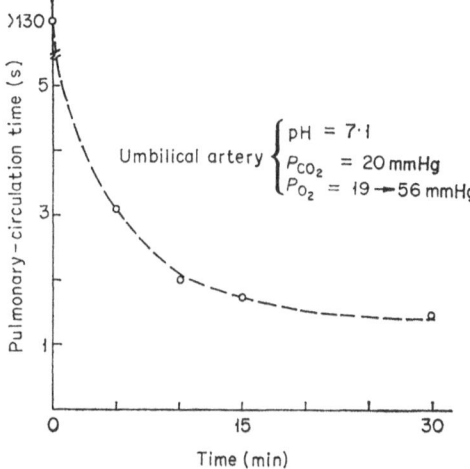

Fig. 5.23. Pulmonary-circulation time as a function of time following an increase in blood-oxygen tension (fetus No. 5)

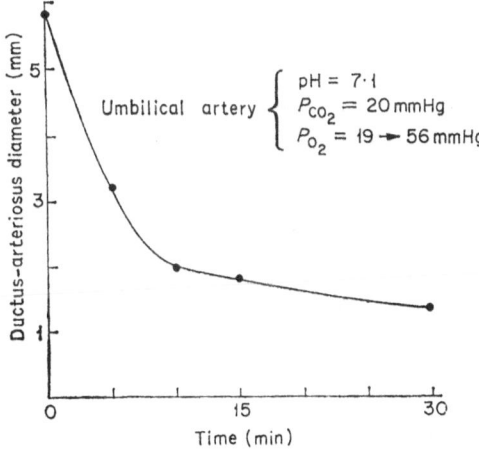

Fig. 5.24. Ductus-arteriosus diameter as a function of time following an increase in blood-oxygen tension (fetus No. 5)

Changes in pulmonary-circulation time and in ductus-arteriosus diameter with time are given in Figs. 5.23 and 5.24. The ductus arteriosus constricted to one-third the control diameter 10 min after blood-oxygen tension was suddenly raised above fetal physiologic levels (13–20 Torr).

In fetus No. 6 (pH 7·48) rapid constriction of the ductus arteriosus and increased pulmonary flow followed rapid elevation of the blood-oxygen tension from 22 to 64 Torr (Figs. 5.25 and 5.26). The ductus arteriosus

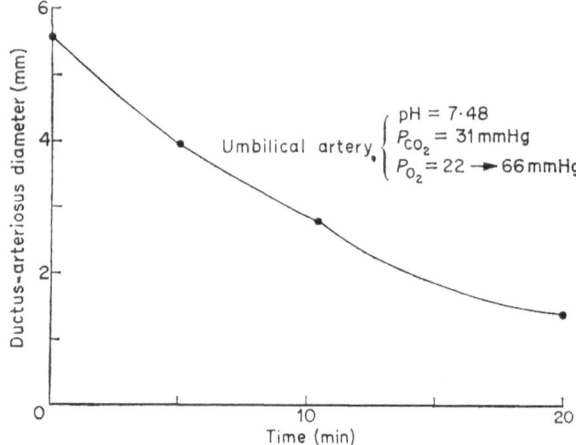

Fig. 5.25. Ductus-arteriosus diameter as a function of time following an increase in blood-oxygen tension (fetus No. 6)

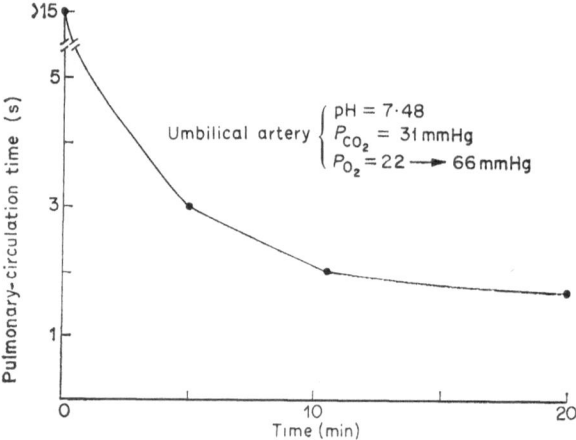

Fig. 5.26. Pulmonary-circulation time as a function of time following an increase in blood-oxygen tension (fetus No. 6)

constricted to half the control diameter within 11 min after the rise in blood-oxygen tension.

A thorough review of the findings of previous investigators on the response of the ductus arteriosus to oxygen tensions is given in our previously published article (Zapol *et al*, 1971).

We have used angiocardiography to study the fetal circulation as this yields a maximum amount of anatomic and haemodynamic data, without injury to the fetus. The technique is limited in that it does not provide quantitative measurements of flow.

Siassi *et al.* (1972) recently reported initial investigations of fetal circulation performed with similar artificial-placentation techniques. Using dye dilution he determined in seven fetal lambs at placental flows of 100–130 ml/kg min the data shown in Table 5.3.

Table 5.3. *Data obtained by Siassi (1972) on seven fetal lambs*

Umbilical-artery P_{O_2} (Torr)	Cardiac output (ml/kg min)	Left ventricular output (ml/kg min)	Pulmonary-circulation time (s)
20–30	305 ± 28	153 ± 25	$5 \cdot 8 \pm 0 \cdot 8$
35–45	–	269 ± 26	$2 \cdot 1 \pm 0 \cdot 3$

During our angiographic experiments we achieved relatively stable physiological conditions which allowed us to record fetal circulatory responses to change of a single variable. Rectal temperature, pH, P_{CO_2} and P_{O_2} of fetal blood, as well as placental-blood flow were controlled by the artificial placenta. The placental flows studied (80–130 ml/kg min) were less than those reported for late-gestation fetal lambs (150–200 ml/kg min) (Meschia *et al.*, 1965). Extracorporeal blood flow was limited by catheter size and arterial pressure. All blood that drained into the silicone-rubber reservoir was pumped through the extracorporeal circuit. It would be impossible to control these variables in an exteriorized fetus connected to a biological placenta.

In the exteriorized fetal preparation the natural placenta removes contrast material. In our studies, it was removed by the fetal kidney, with marked and rapid opacification of the bladder (Doppman *et al.*, 1970). Hypernatraemia, acidosis and hypotension, the major complications of the injection of large amounts of meglumine iothalamate, did not occur. In three animals, four injections of meglumine iothalamate at 10- to 14-min intervals showed no significant change of ductus diameter or pulmonary circulation time.

Our data reveal the reversibility of the newborn fetal cardiovascular conversion. It also shows that a near-term sheep can be maintained with a newborn circulation for many hours and then be born in good condition. The fetal studies presented here show that raising blood-oxygen tension from 13–20 up to 30–60 Torr will constrict the ductus arteriosus and produce a large increase in pulmonary-blood flow. Most of this response occurs within 5 min.

In the sheep fetus with a newborn circulatory pattern, neither metabolic nor respiratory acidosis significantly dilates the constricted ductus arteriosus or alters pulmonary-blood flow. Blood-oxygen tensions appear to be the primary factor determining the circulatory pattern of the fetus.

In the majority of our long-term perfusions, umbilical artery P_{O_2} was maintained above 35 Torr, and the 'hyperoxic' fetus had large amounts of pulmonary-blood flow and a constricted ductus arteriosus. We plan future long-term lamb perfusions at lower umbilical-artery oxygen tensions more consistent with intrauterine fetal life.

'Birth' following cessation of extracorporeal circulation

Several of our lambs over 130 d gestation that were perfused were 'delivered' successfully following artificial placentation for up to 12 h. To insure a safe transition to terrestrial life, we removed the fetus from the bath and dried it thoroughly, maintaining rectal temperature above 38°C. The fetus was then intubated with a French 14 cuffed endotracheal tube and ventilated with a volume-controlled respirator at an FIO_2 of 0·5. If the lungs were well ventilated the fetal arterial blood became fully saturated and in 15 min placental blood flow was terminated. Umbilical-arterial-clamp blood pressure and gas tensions were measured frequently. If these parameters were stable, the cannulae were removed and the vessels individually tied with silk sutures. When the lamb showed adequate muscular strength he was given a trial of spontaneous ventilation on room air. If he was successful in maintaining adequate arterial-blood gas tensions, he was extubated. Lambs were bottle-fed immediately and placed with a mother ewe.

Partial bypass for respiratory support of newborn infants

Mechanical ventilation with positive expiratory pressure is frequently inadequate for the therapy of overwhelming respiratory-distress syndrome of the newborn (Gregory, 1971). High ventilator pressures can cause pneumothorax and rapid deterioration of pulmonary gas exchange. Oxygen toxicity can destroy the lungs or cause severe pulmonary fibrosis. In low-birth-weight premature infants, temporary partial or total extracorporeal gas exchange may be particularly effective in supporting life until pulmonary disease can heal.

Several groups have reported their techniques of extracorporeal respiratory support for respiratory-distress syndrome. Rashkind *et al.* (1967) perfused infants with a bubble oxygenator, using a femoral-artery-to-femoral-vein shunt. Although arterial gas tensions improved, progressive oedema, haemolysis and circulatory failure occurred. Dorson *et al.* (1969) also has reported his efforts with a capillary-membrane lung. To date he has no survivors.

Recently White *et al.* (1971) reported 2- to 10-d venovenous perfusions of three newborn infants with respiratory-distress syndrome. None of the infants survived. Using a multifactorial scoring system to predict infant mortality, he tried to begin perfusion before irreversible pulmonary or systemic hypoxic changes had occurred. For blood drainage, White passed a silicone-rubber catheter (1·98 mm internal diameter, 3·18 mm external diameter) through the right internal jugular vein into the superior vena cava. The cannula returning oxygenated blood (3·35 mm internal diameter, 4·67 mm external diameter) was passed through a cut-down on the abdominal wall a short distance cephalad to the umbilicus, and placed in the umbilical vein.

Table 5.4. *Observations reported by White et al. (1971) on 2- to 10-d venovenous perfusions on three newborn infants*

Patient	Weight (kg)	Bypass flow (ml/min)	Days perfused	Membrane lung O₂ transfer (cm³ O₂/min)	Autopsy data
1	1·0	130	$9\frac{1}{2}$	24	subdural and cerebellar haematomas
2	1·25	70	3	13	intracerebral haemorrhages
3	1·73	130	2	20	subdural and subarachnoid haemorrhages

The perfusion apparatus consisted of a silicone-rubber blood-drainage reservoir, a Holter model RB 161 occlusive roller pump, a 0·4-m² spiral-membrane lung and a bubble trap. Total prime volume was about 200 ml, with 70 ml in the membrane lung. Priming solution was fresh heparinized adult blood. Infants were kept under a radiant heater with additional infra-red heat lamps. Antibiotics (cephalin and gentamycin) were administered. The results of all blood cultures during clinical perfusion were negative. Table 5.4 presents a summary of White's results.

Systemic heparinization remains necessary for extracorporeal perfusion. White found it possible to perfuse infants at a Lee-White clotting time between 30 and 45 min without clotting the membrane lung. Banked blood provides some 4·5 u heparin per ml, and in 200 ml prime this exceeds a complete heparinizing dose for a premature infant. Hypoxic premature infants frequently bleed spontaneously, increasing the risk of heparinizing these infants. Case 1 developed gastrointestinal haemorrhage on the first day of perfusion and severe bleeding on the third day. Apparently caused by overheparinization, haemorrhage was controlled by blood transfusion and protamine administration. Autopsy of Case 3 showed extensive focal haemorrhages of the lungs, thymus and gastrointestinal tract. Incremental blood replacement of 3 ml/h was necessary to counter oozing, bleeding from overheparinization or blood loss from other causes. White noted leukopaenia (2,500/mm^3) and thrombocytopaenia (50,000/mm^3) in all perfusions.

In all three cases extracorporeal blood-gas exchange was adequate for survival; arterial oxygen tensions of 40–60 Torr were easily maintained. Carbon dioxide removal was diminished by adding 5% carbon dioxide to the gas ventilating the membrane lung, and with rare exception, pH was maintained at 7·4–7·5. After bypass was started all three infants ceased spontaneous respiration. It is notable that the chest radiographs of case 1 were initially completely opacified, but radiolucency had returned after 7 d of perfusion. By then, pulmonary function was returning and bypass flow could be reduced to 70 ml/min; spontaneous shallow respirations resumed and the infant maintained adequate arterial-gas tensions. Throughout bypass the infant manifested good neurological function, eye movements, sucking, and voluntary and reflex movements of the limbs were present. However, on the ninth day the baby abruptly became unresponsive and ceased breathing.

The future of isolated extracorporeal perfusion

Current difficulties

(1) *Fetal size.* We can cannulate only sheep fetuses over 100 days' gestation and weighing over 700 g. Perfusion is difficult in these fetuses, and umbilical artery clamp pressure very low. Progressive oedema and acidosis occur early. In the future, it will be necessary to continuously observe their intra-aortic blood pressure and urine output. Renal and hepatic immaturity may limit isolated

perfusion in early gestation unless provision is made for haemo-dialysis.

(2) *Sepsis*. There is a dual problem of bacterial growth in blood and bath. In our experiments diatomaceous-earth filtration, ultra-violet irradiation and antibiotics retarded growth in the bath. In the future fetuses may be maintained in a warm air environment or sealed liquid container. Blood sepsis is a more difficult problem. Instead of using three-way stopcocks for blood sampling (and thus providing sites of bacterial growth) we interpose 3-cm lengths of thick-walled latex rubber tubing in the sampling 'T' connectors. We sterilize the surface with 70% ethanol and sample blood by puncturing the tubing. When partially perfusing lambs, we wash the polypropylene stopcocks with a saline solution containing penicillin and streptomycin to remove residual blood. The stop-cocks are enclosed in polyethylene bags to prevent dust accumula-tion, and they are changed every 5 d. Gas flow to the membrane lungs is filtered through a 0·22-μm Millipore filter to prevent any possibility of bacterial transfer through membrane-lung pinholes.

(3) *Haemorrhage*. This remains the most difficult problem of artificial placentation. Clinical perfusions of human infants are plagued by spontaneous intracerebral, subdural, gastrointestinal, pulmonary and retroperitoneal haemorrhages, even when perfusions are carried on at minimal heparin doses and Lee-White clotting times of 30–45 min. During lamb perfusion, prolonged clotting times occasionally cause spontaneous retroperitoneal bleeding. Recent advances in bonding heparin to membrane lungs and tubing sur-faces has made possible partial venovenous perfusion of dogs for up to 48 h without systemic heparin (Rea *et al.*, 1971). But we have noted in lambs that while the oxygenator continues to function without clotting, multiple small pulmonary emboli are produced. Research into the compatibility of blood with hypothrombogenic membrane surfaces is essential if we are to reduce severe haemor-rhage in patient and animal perfusion.

(4) *Undetermined effects*. At times apparently stable fetuses underwent cardiac arrest. We could not determine a cause for these deaths. It is unclear whether we were inadequately measuring important electrolyte parameters or were not supplying some placental or maternal factor. The rapid demise of these animals suggests an acute metabolic disturbance or cardiac arrhythmia with ventricular arrest or fibrillation.

Possible explorations

The artificial placenta makes possible a number of physiological and pharmacological studies. We are preparing to investigate the circulatory effects of general and regional anaesthetic agents on the term lamb fetus. In particular we shall study the effects of anaesthetic agents on (a) fetal circulation, (b) fetal response to hypoxic stress and (c) conversion from a fetal to a newborn circulatory pattern. Use of the artificial placenta can separate direct effects of anaesthetics on the fetus from effects secondary to maternal cardiovascular changes. Artificial placentation will allow rapid equilibration of fetal blood with anaesthetic gas tensions within the membrane lung; intravenous agents can similarly achieve rapid equilibration.

Using the sheep fetus we must learn how to safely prolong artificial placentation for over one week. The knowledge gained may be directly applicable to extracorporeal support of the newborn dying of hypoxia from respiratory-distress syndrome.

Future modifications

Widespread use of membrane oxygenation to support the newborn with respiratory-distress syndrome is frustrated by the frequent, massive haemorrhaging that accompanies systemic heparinization. A clinical urgency for cardiorespiratory support systems currently motivates the search for improved membrane-lung material. Several materials, still in early experimental stages, show superior thromboresistance. Trifluoro-propylmethylpolysiloxane (Merker *et al.*, 1969) and the alkylpolysulfones (Crawford & Gray, 1971) are two polymers both with gas-transmission rates adequate for membrane lungs and promising blood compatibility. Heparin-coating techniques (Grode *et al.*, 1969), and thin layers of compatible hydrogels (Merrill & Salzman, 1970), or polymers, need to be assayed for *in-vivo* thromboresistance during partial perfusion. If these materials permit blood oxygenation without heparinization, they will virtually eliminate the risks of extracorporeal perfusion.

The human placenta functions as an arteriovenous shunt with the heart as a blood pump. We found initially that fetuses became hypotensive after cannulation and could not successfully maintain an arteriovenous shunt through 0.2-m^2, 5-cm-wide spiral-coil lungs. Inadequate placental blood flow led to hypoxia and death. When we determined (in partial perfusions) that extracorporeal blood pumping could be safely performed, we pumped blood through the membrane lung. In future perfusions, it may be possible to allow a fetus to effect an AV shunt through a membrane lung after his systemic blood pressure has initially equilibrated. A lower-resistance membrane lung will be necessary to allow large blood flows. The fetus will then be able to reflexly control placental blood flow with his arterial blood pressure. With the eventual replacement of the blood-pumping chamber and shortened blood tubing, we will be able to perform total gas transfer for the term fetus with only 100-ml prime volume.

Artificial placentation has just begun clinical application. Requirements for clinical cardiac and respiratory support demand further research into the basic interactions of thrombus formation on polymers in the bloodstream (Salzman, 1971). Major progress in safe artificial placentation is expected as soon as such non-thrombogenic materials are available.

Acknowledgement

We wish to acknowledge Drs. J. Doppman and J. Pierce for their technical assistance and valuable efforts in performing the fetal experiments.

References

ALEXANDER, D. P., BRITTON, H. G. & NIXON, D. A. (1966) Maintenance of the isolated foetus. *British Medical Bulletin*, **22**, 9–12.

ALEXANDER, D. P., BRITTON, H. G. & NIXON, D. A. (1968) Maintenance of sheep fetuses by an extracorporeal circuit for periods up to 24 hours. *American Journal of Obstetrics and Gynecology*, **102**, 969–975.

ALLERDYCE, D. B., YOSHIDA, S. H. & ASHMORE, P. G. (1966) The importance of microembolism in the pathogenesis of organ dysfunction caused by prolonged use of the pump oxygenator. *Journal of Thoracic and Cardiovascular Surgery*, **52**, 706–715.

ATKINS, R., ROBINSON, W., & EISEMAN, B. (1968) Cytotoxins released from plastic perfusion apparatus. *The Lancet*, 1014–1015.

AVERY, M. E. (1968) *The Lung and its Disorders in the Newborn Infant*, 2nd ed. Vol. 1. Philadelphia: W. B. Saunders Co.

BARTLETT, R. H., KITTREDGE, D., NOYES, B. S., Jr., WILLARE, R. H. III & DRINKER, P. A. (1969) Development of a membrane oxygenator: overcoming blood diffusion limitation. *Journal of Thoracic and Cardiovascular Surgery*, **58**, 795–800.

BLACKSHEAR, P. L., DORMAN, F. D. & STEINBACH, J. H. (1965) Some mechanical effects that influence hemolysis. *Transactions, American Society for Artificial Internal Organs*, **11**, 112–117.

BORETOS, V. W. & PIERCE, W. S. (1968) Segmented polyurethane: a polyether polymer. *Journal of Biomedical Materials Research*, **2**, 121–130.

BRUCK, S. (1972) Biomaterials in medical devices. *Transactions. American Society for Artificial Internal Organs*, **18**, 1–8.

BUCKLES, R. G. (1966) An analysis of gas exchange in a membrane oxygenator. Ph.D. Thesis, Massachusetts Institute of Technology, Cambridge, Massachusetts.

CALLAGHAN, J. C., MAYNES, E. A. & HUG, H. R. (1965) Studies on lambs of the development of an artificial placenta. *Canadian Journal of Surgery*, **8**, 208–213.

CLOWES, G. H. A., HOPKINS, A. L. & NEVILLE, W. E. (1956) An artificial lung dependent upon diffusion of oxygen and carbon dioxide through plastic membranes. *Journal of Thoracic Surgery*, **32**, 630–637.

CRAWFORD, J. E. & GRAY, D. N. (1971) Preparation and properties of some poly (alpha-olefin sulfones). *Journal of Applied Polymer Science*, **15**, 1881–1888.

DAWES, G. S. (1968) *Foetal and Neonatal Physiology*, Appendix: Some Experimental Methods. Chicago: Year Book Medical Publishers.

DOPPMAN, J. L., ZAPOL, W., KOLOBOW, T. & PIERCE, J. (1970) Angiocardiography of fetal lambs on artificial placenta. *Investigative Radiology*, **5**, 181–186.

DORSON, W., Jr., BAKER, E., COHEN, M. L., MEYER, B., MOLTHAN, M., TRUMP, D. & ELGAS, R. (1969) A perfusion system for infants. *Transactions. American Society for Artificial Internal Organs*, **15**, 155–160.

DUDRICK, S. J., WILMORE, D. W., VARS, H. M. & RHOADS, J. E. (1968) Long-term total parenteral nutrition with growth, development, and positive nitrogen balance. *Surgery*, **64**, 134–142.

DUTTON, R. C., MATHER, F. W. III & WALKER, S. N. (1971) Development and evaluation of a new hollow-fiber membrane oxygenator. *Transactions. American Society for Artificial Internal Organs*, **17**, 331–336.

GALLETTI, P. M. & BRECHER, G. A. (1962) *Heart-Lung Bypass*, p. 50. New York: Grune and Stratton.

GALLETTI, P. M., SNIDER, M. T. & DANIELE, S. A. (1966) Gas permeability of plastic membranes for artificial lungs. *Medical Research Engineering*, **5**, 20–23.

GREGORY, G. A. (1971) Treatment of the idiopathic respiratory-distress syndrome with continuous positive airway pressure. *New England Journal of Medicine*, **284**, 1333–1340.

GRODE, G. A., ANDERSON, S. J., GROTTA, H. M. & FALB, R. D. (1969) Nonthrombogenic materials via a simple coating process. *Transactions. American Society for Artificial Internal Organs*, **15**, 1–6.

GULLINO, P. M. (1966) In vitro perfusion of tumors. In *Organ Perfusion and Preservation*, ed. Folkman, J., Hardison, W. G., Rudolf, L. E. & Veith, F. J. Ch. 65, pp. 877–898. New York: Appleton-Century-Crofts.

HEYMANN, M. A. & RUDOLPH, A. M. (1967) Effect of exteriorization of the sheep fetus on its cardiovascular function. *Circulation Research*, **21**, 741–745.

HILL, J. D., BRAMSON, M. L., KLEINHENZ, R., RAISON, J. C. A., HACKEL,

A., OSBORN, J. J. & GERBODE, F. (1968) Lung morphology following prolonged venovenous perfusion with the Bramson membrane lung. In *Organ Perfusion and Preservation*, ed. Folkman, J., Hardison, W. G., Rudolf, L. E. & Veith, F. J. Ch. 16, pp. 177–188. New York: Appleton-Century-Crofts.

HILL, J. D., O'BRIEN, T. G., MURRAY, J. J., DONTIGNY, L., BRAMSON, M. L., OSBORN, J. J., & GERBODE, F. (1972) Prolonged extracorporeal oxygenation for acute post-traumatic respiratory failure (shock-lung syndrome). *The New England Journal of Medicine*, **286**, 629–634.

HUCKABEE, W. E. (1958) Relationships of pyruvate and lactate during anaerobic metabolism. I. Effects of infusion of pyruvate or glucose and of hyperventilation. *Journal of Clinical Investigations*, **37**, 244–271.

HUNTER, F. T., GROVE-RASMUSSEN, M. & SOUTTER, L. (1950) A spectrophotometric method for quantitating hemoglobin in plasma or serum. *American Journal of Clinical Pathology*, **20**, 429–433.

INDEGLIA, R. A., SHEA, M. A. & BERNSTEIN, E. F. (1967) Studies of the biologic response to prolonged anesthesia and immobilization. *Transactions. American Society for Artificial Internal Organs*, **13**, 151–156.

KAPLAN, N. & RUDOLPH, A. M. (1969) Physiologic studies of pulmonary circulation and ductus arteriosus in sheep and goat fetuses. *Investigative Radiology*, **4**, 68–82.

KOLOBOW, T. & BOWMAN, R. L. (1963) Construction and evaluation of an alveolar membrane artificial heart-lung. *Transactions. American Society for Artificial Internal Organs*, **9**, 238–242.

KOLOBOW, T., ZAPOL, W., PIERCE, J. E., KEELEY, A. F., REPLOGLE, R. L. & HALLER, A. (1968) Partial extracorporeal gas exchange in alert newborn lambs with a membrane artificial lung perfused via an A-V shunt for periods up to 96 hours. *Transactions of the American Society for Artificial Internal Organs*, **14**, 328–334.

KOLOBOW, T., ZAPOL, W. & PIERCE, J. (1969) High survival and minimal blood damage in lambs exposed to long-term (1 week) veno-venous pumping with a polyurethane chamber roller pump with and without a membrane blood oxygenator. *Transactions. American Society for Artificial Internal Organs*, **15**, 172–177.

KOLOBOW, T., ZAPOL, W. M., SIGMAN, R. L. & PIERCE, J. (1970) Partial cardiopulmonary bypass lasting up to seven days in alert lambs with membrane lung blood oxygenation. *The Journal of Thoracic and Cardiovascular Surgery*, **60**, 781–788; 794–795.

KOLOBOW, T. & ZAPOL, W. (1970) A new thin-walled nonkinking catheter for peripheral vascular cannulation. *Surgery*, **68**, 625–629.

KOLOBOW, T., ZAPOL, W. M. & SIGMAN, R. L. (1970) Design considerations and long-term in vivo studies with the disposable spiral membrane lung. In *Blood Oxygenation*, ed. Hershey, E., pp. 306–320. New York: Plenum Press.

KOLOBOW, T. & ZAPOL, W. M. (1971) Partial and total extracorporeal respiratory gas exchange with the spiral membrane lung. In *Mechanical Devices for Cardiopulmonary Assistance*, ed. Bartlett, R. H., Drinker, P. A. & Galletti, P. M. *Advances in Cardiology*, Vol. 6, pp. 112–132. Basel: S. Karger.

KOLOBOW, T., SPRAGG, R. G., PIERCE, J. E. & ZAPOL, W. M. (1971) Extended term (to 16 days) partial extracorporeal blood gas exchange with the spiral membrane lung in unanesthetized lambs. *Transactions. American Society for Artificial Internal Organs*, 17, 350–354.

LA FARGE, C. G., BERNHARD, W. F., ROBINSON, T. C., KITRILAKIS, S., YUN, I. & SHIRAHIGE, K. (1968) Physiological consequences of acutely and chronically implanted left ventricular-aortic assist devices. *Transactions. American Society for Artificial Internal Organs*, 14, 316–322.

MARX, T. I., SNYDER, W. E., ST. JOHN, A. D. & MOELLER, C. E. (1960) Diffusion of oxygen into a film of blood. *Journal of Applied Physiology*, 15, 1123–1129.

MERKER, R. L., EYLASH, L. J., MAYHEW, S. H. & WANG, J. Y. C. (1969) The heparinization of silicone rubber using aminoorganosilane coupling agents. In *Proceedings of the Artificial Heart Program Conference (June 9–13, 1969)*, Ch. 3, pp. 29–39. United States Government Printing Office.

MERRILL, E. W. & SALZMAN, E. W. (1970) Polyvinyl alcohol-heparin hydrogel 'G', *Journal of Applied Physiology*, 29, 723–730.

MESCHIA, G., COTTER, J. R., BREATHNACH, C. S. & BARRON, D. H. (1965) The diffusibility of oxygen across the sheep placenta. *Quarterly Journal of Experimental Physiology*, 50, 466–480.

NOBLE, F. W. (1968) Dual frequency ultrasonic fluid flowmeter. *Review of Scientific Instruments*, 39, 1327–1331.

OLSON, G. F. (1962) Optimal conditions for the enzymatic determination of L-lactic acid. *Clinical Chemistry*, 8, 1–10.

PIERCE, E. C. (1970) A comparison of the Lande-Edwards, the Pierce, and the General Electric–Pierce membrane lungs. *Transactions. American Society for Artificial Internal Organs*, 16, 358–364.

PIERCE, W. S., TURNER, M. C., Jr., BORETOS, J. W., METZ, H. D., NOLAN, S. P. & MORROW, A. G. (1967) Mechanical left ventricular assistance: experimental studies using an implantable roller pump. *Transactions. American Society for Artificial Internal Organs*, 13, 299–305.

RASHKIND, W. J., MILLER, W. W., FALCONE, D. & TOFT, R. (1967) Hemodynamic effects of arteriovenous oxygenation with a small-volume artificial extracorporeal lung. *Journal of Pediatrics*, 70, 425–429.

REA, W. J., EBERLE, J. W. & WATSON, J. T. (1971) Gas transfer in a heparinized membrane oxygenator. *Surgical Forum*, 22, 188–190.

SALZMAN, E. W. (1971) Thrombosis in artificial organs. In *Artificial Organs and Cardiopulmonary Support Systems*, ed. Rapaport, E. T. & Merrill, J. P. pp. 97–102. New York: Grune and Stratton.

SEGAL, S., BLAIR, A. E. & WYNGAARDEN, J. B. (1956) An enzymatic spectrophotometric method for the determination of pyruvic acid in blood. *Journal of Laboratory Clinical Medicine*, 48, 137–143.

SHEA, M. A., INDEGLIA, R. A. & BERNSTEIN, E. F. (1968) Hematologic observations during perfusions in sheep. In *Organ Perfusion and Preservation*, ed. Folkman, J., Hardison, W. G., Rudolf, L. E. & Veith, F. J. Ch. 66, pp. 899–909. New York: Appleton-Century-Crofts.

SIASSI, B., TESSLER, I., WU, P. Y. K., MODANLOU, H. & LI, R. (1972) The effect of variation in arterial oxygen tension on fetal circulation. *The American Journal of Cardiology*, **29**, 292.

TCHOBROUTSKY, C. (1966) Extracorporeal oxygenation in puppies and in newborn and fetal lambs. *American Journal of Obstetrics and Gynecology*, **96**, 367–381.

VUREK, G. G., FRIAUF, W. S., PERRY, K. & BRASLOW, N. (1970) Dual wavelength reflectance oximeter for long-term extracorporeal monitoring. *Annual Conference on Engineering in Medicine and Biology*, **12**, 113.

WHITE, J. J., ANDREWS, H. G., RISEMBERG, H., MAZUR, D. & HALLER, J. A. (1971) Prolonged respiratory support in newborn infants with a membrane oxygenator. *Surgery*, **70**, 288–296.

ZAPOL, W. M., LEVY, R. I., KOLOBOW, T., SPRAGG, R. & BOWMAN, R. L. (1969) *In vitro* denaturation of plasma-lipoproteins by bubble oxygenation in the dog. *Current topics in Surgical Research*, **1**, 449–467.

ZAPOL, W., KOLOBOW, T., PIERCE, J. E., VUREK, G. G. & BOWMAN, R. L. (1969) Artificial placenta: two days of total extrauterine support of the isolated premature lamb fetus. *Science*, **166**, 617–618.

ZAPOL, W. M., KOLOBOW, T., DOPPMAN, J. & PIERCE, J. E. (1971) Response of ductus arteriosus and pulmonary blood flow to blood oxygen tensions in immersed lamb fetuses perfused through an artificial placenta. *The Journal of Thoracic and Cardiovascular Surgery*, **61**, 891–903.

ZAPOL, W. M. & KITZ, R. J. (1972) Buying time with artificial lungs. *The New England Journal of Medicine*, **286**, 258.

6 Extracorporeal Maintenance of Small Human Fetuses

M. C. MACNAUGHTON

This chapter is concerned with the extracorporeal maintenance of the human fetus at mid-gestation. Maintenance of premature babies and animals such as the sheep are referred to in other chapters of this book.

In the human subject much information of a physiological nature has been obtained by experiments with fetal preparations, and our understanding of vital physiological and biochemical processes before birth on which the development of a fetus into a normal child depends has been greatly increased. To quote from the Report of the Advisory Group on The Use of Fetuses and Fetal Material for Research (Peel, 1972): 'The whole pre-viable fetus has offered an important opportunity that cannot be obtained in any other way for making observations of great value on the transfer of substances across the human placenta, the reaction of the immature fetus to drugs and on the endocrinological development of the fetus and the development of the placenta.' This report on the ethical, medical, social and legal implications of using fetuses and fetal material for research should be consulted by all those proposing to do this type of work in the U.K. It gives useful guidelines for work in this field.

One of the most important reasons for trying to develop a system of extracorporeal support of the human fetus is that the technique might be used in the management of the premature or severely hypoxic human infant. The immature child has a high mortality rate, one of the major causes of death being the respiratory distress syndrome. A technique that would support infants with this syndrome until recovery of the lungs had occurred would be an invaluable adjunct to present therapy. Experimental work on this aspect is described in Chapter 7.

Much of the work done to date on pre-viable human fetuses has been in the realm of endocrinology. While this chapter is concerned with the isolated fetus it should be remembered that, as far as endocrinology is

195

concerned, the fetus and the placenta form a functional metabolic unit. This concept, introduced by Diczfalusy (1962), has been most rewarding and has greatly helped the general understanding of steroid metabolic processes during pregnancy.

Gestation stage of experiments

In the future it will be difficult to experiment with fetuses of over 20 weeks' gestation because technical advances are reducing the age of viability (Peel, 1972). The optimum period for perfusion experiments is 16–20 weeks' gestation, when the fetus should weigh between 200 and 400 g. Fetuses up to 980 g have been perfused (Chamberlain, 1968), but a few of this weight might well survive and experimentation would now be unacceptable. The upper limit recommended for experimental work is 500 g. This weight normally occurs just under 20 weeks' gestation (Thomson *et al.*, 1968).

Types of apparatus

This depends on the nature of the experiments contemplated. These are of two main types. In the first, the object is to measure fetal metabolism of a substance such as a steroid compound. Here only a single cycle perfusion is required and, therefore, the apparatus necessary is of a relatively simple nature. Fig. 6.1 illustrates the types of apparatus used for fetal-perfusion experiments in the author's laboratory. It consists of a heated perspex tank filled with Ringer-lactate solution at 37°C. The fetus with an umbilical arterial and an umbilical venous catheter inserted is placed in the tank. The arterial cannula is led through a seal in the tank to a beaker so that perfused blood can be collected for analysis. The venous cannula is connected to a bottle of prepared blood (see later), via a blood-warming coil and a drip chamber. The infusion is commenced and once the flow is established the tank is topped up with Ringer-lactate, the lid put on and the water manometer adjusted. Any rise of pressure in the tank due to the input being greater than the outflow will now be shown by a rise of fluid in the manometer and the required adjustments can be made.

It is practical to carry out metabolic experiments with this apparatus and Westin, Nyberg and Enhorning (1958) have maintained the heart

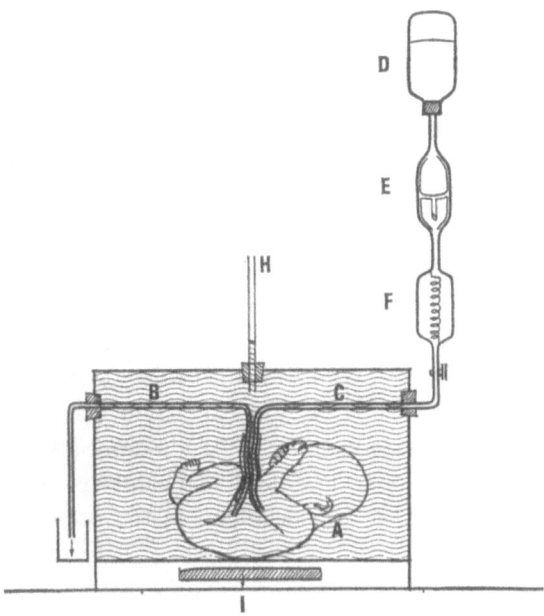

Fig. 6.1. Apparatus for perfusion of the isolated human fetus at mid gestation

A—Fetus immersed in tank containing Ringer-
 lactate solution
B—Catheter from umbilical artery conveying
 perfused blood to beaker
C—Catheter in umbilical vein

D—Oxygenated blood
E—Drip chamber
F—Warming coil
H—Water manometer
I—Heater

rate and fetal movements in pre-viable fetuses for up to 12 h, but most
experiments performed with this type of apparatus are completed in 2 h.

Preparation of the fetus

The fetuses are obtained from women undergoing therapeutic abortion
performed by abdominal hysterotomy. It is important that the uterine
incision at termination should be large enough to allow easy extraction of
the products of conception—otherwise fetuses may be so damaged as to
be unsuitable for experimentation. It is best to remove the amniotic sac
containing the fetus with the placenta intact. In this way some feto-
placental circulation continues while the fetus is being transported to the
adjacent laboratory. This should preferably be sited near to the operating
theatre so that the time from removal from the uterus to initiation of the
perfusion is as short as possible. In the laboratory the amniotic sac is

opened and the cord clamped about 1 cm from the umbilicus. The insertion of catheters into the umbilical vessels is now performed and is facilitated by prior dilation of the vessels with probes of varying sizes. Soft plastic catheters which have previously been filled with heparin solution are inserted into one umbilical artery and also the umbilical vein. A ligature is placed round the cord stump and the cannulae to prevent bleeding from the second umbilical artery, and the fetus is transferred to the perfusion tank. It is not necessary to cannulate the second artery and a good perfusion can be obtained when only one artery is cannulated. The author has found the most suitable cannula to be the Argyle Sterile Premature Infant Feeding Tube size $3\frac{1}{2}$ Fr. This is a soft plastic catheter and does not damage the very fine veins during insertion. The cannulae are connected to the input and output seals in the wall of the tank, and the umbilical vein catheter is connected, via a blood-warming coil and a drip chamber, to the bottle containing the perfusion fluid.

This system is a simple one and relies on the fetal heart to maintain the circulation. No extracorporeal pump is required. The input must be adjusted so that the heart is not overloaded.

Overdistension of the fetus with blood is deleterious to the system. It is usually the liver that first shows signs of distension, the abdomen becoming rather swollen in appearance. If examined closely at this time, veins in the fetal scalp will be seen to be distended, and it is also interesting to note that often the muscles of the arms and legs have a 'Charles Atlas' appearance owing to their distension with blood. If overdistension occurs it can be reduced by slowing the input rate, but sometimes cardiac failure supervenes and the experiment has to be terminated because the perfusion ceases to function. The volume of the fetus can be maintained at a constant state by the use of the manometer inserted into the top of the perfusion chamber. When the level of fluid in the manometer rises it indicates that the input is greater than the outflow, and this can be corrected by reducing the flow rate.

The time from removal of the fetus to the start of perfusion is usually about 20 min, and this time gap does not seem to affect the ability of the fetus to maintain perfusion experiments for 2 h or so thereafter.

The second type of apparatus used is that first described by Lerner & Diczfalusy (1968) and is intended to be used for *in-vitro* perfusion of the intact feto-placental unit.

In the case of steroid production and metabolism in pregnancy, Diczfalusy was the first to suggest that, at least at mid-pregnancy, the human fetus and placenta form a functional unit. They carry out the synthesis of various steroid hormones that the placenta or fetus alone would not be able to complete. It had originally been thought that various hormones or their precursors had been synthesized during pregnancy by endocrine activity of the placenta in a similar way to that which takes place in other endocrine organs such as the adrenals or

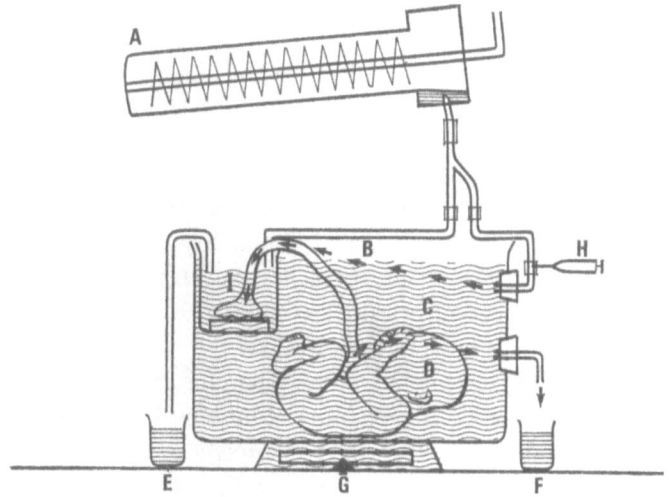

Fig. 6.2. Apparatus for perfusion of the feto-placental unit at mid-gestation (after Lerner *et al.*, 1971)

A—Oxygenator

B—Perfusion via the umbilical artery

C—Artificial amniotic fluid

D—Fetal perfusate collected via the other umbilical artery

E—Perfusate collected from the blood bathing the placenta

F—Container with fetal perfusate

G—Heater

H—Infusion pump for radioactive material

I—Container with the placenta submerged in maternal blood

ovaries. It has been discovered, however, that the human placenta, at mid-pregnancy, is an incomplete endocrine organ because it lacks some of the essential enzyme systems that are required for steroid synthesis. Many of these are present in the fetus.

Workers in the field had perfused the isolated placenta and isolated fetus on many occasions *in vitro* (for review see Lerner, Saxena & Diczfalusy, 1971), but no satisfactory apparatus for perfusion *in vitro* of the intact feto-placental unit had been evolved until Lerner and Diczfalusy produced the apparatus shown in Fig. 6.2. It consists of a tank containing 5·5% (w/v) glucose in isotonic saline kept at a temperature of between 36° and 37°C. The fetus, still attached to the placenta, is placed in this tank. The placenta is placed in a smaller separate tank submerged in the artificial amniotic fluid, and is bathed in the same blood as that used to perfuse the system.

The oxygenator is filled with heparinized, oxygenated, fresh Rh-negative group-O blood diluted with 15% (v/v) isotonic saline. A catheter is inserted into one umbilical artery in the direction of the placenta and another one in the other artery towards the fetus. The complete feto-placental unit is then perfused via the first catheter. The blood that bathes the placenta is continuously replaced by fresh blood from the oxygenator, and a perfusate—the 'maternal' perfusate—is collected. A second perfusate is collected from the fetus. In studies in which radioactive material is administered, this can be introduced into the oxygenated blood at a constant rate by some form of infusion pump. In this system the perfused material reaches the fetus via the placenta and the umbilical vein; as both arteries are cannulated no labelled material is recirculated from the fetus to the placenta.

Diczfalusy (1969) has modified this system for certain experiments by the introduction of a T-catheter into the umbilical vein. This makes it possible to introduce labelled material directly into the fetal compartment of the feto-placental unit during a perfusion experiment. No blood pump is involved in this system—the only pump in the system is the fetal heart. Using this technique Lerner *et al.* (1971) have maintained the feto-placental system in good condition for 2–3 h.

A third type of apparatus has been described by Chamberlain (1968) and is more elaborate than the two types described above. It was intended for use in the development of work designed to maintain alive a premature baby suffering from respiratory distress syndrome, and is not so suitable for work on smaller fetuses. Chamberlain maintained a 980-g male fetus on his system for 5 h 8 min. In the smaller fetuses—fetuses of 300–800 g were perfused—blood flow was poor. The most difficult problem was to establish a return flow to the circuit via the umbilical arteries. Negative pressures, papaveretum and oxygenated warm saline were all tried but the best results followed proper placing of the catheters so that their tips were in the larger arteries. The longest survival in the series came with the largest fetus.

The apparatus consists of a gas exchanger, the most satisfactory being a cellophane membrane wound round a spacing frame. With this equipment high oxygen tensions could be obtained at good flow rates. An auxiliary circulating pump was incorporated in the system. It was of variable output and capable of delivering up to 100 ml/min. The circuit was much the same as that used in adult heart–lung work, the blood coming from the fetus to a reservoir. From the reservoir the blood passed through the pump to the oxygenator and then back into the fetus via the umbilical vein, the flow being regulated by a clip and monitored by a drop counter.

This apparatus differs from the other two mainly by the introduction of an external pump. This is necessary since the blood is being recirculated from an external container. This system is really more suitable for larger fetuses since the umbilical arterial flow rates in the smaller fetuses used in perfusions (Diczfalusy, 1962; Greig & Macnaughton, 1967; Coutts & Macnaughton, 1969) are so sluggish that the blood volume involved is relatively small, and that the use of a pump would rapidly lead to fetal engorgement and cessation of the perfusion.

The apparatus described by Chamberlain would, therefore, seem to be best suited for use in the maintenance of the premature baby, rather than for experiments on fetuses of gestation periods of 20 weeks or less. Chamberlain found that the fetuses under 500 g in his system did not survive any longer than would have been expected if no perfusion had been performed.

Temperature

In many of the early experimental studies on the pre-viable human fetus, the perfusion temperature was at 20°C (Westin et al., 1958). These workers thought that this was the optimum temperature for perfusion. They based this on the fact that when a pre-viable fetus was kept at room temperature, the heart rate, systolic blood pressure, body temperature and oxygen tension of the blood rapidly decreased, while excessive amounts of carbon dioxide accumulated in the blood. The majority of these fetuses show no signs of life about 3–4 h after birth. When the fetus was kept at 37°C, the period of survival was considerably limited and this was thought to be due to higher metabolism and a limited store of oxygen, glucose and other important nutritive elements.

As a result of this evidence much of the early experimentation by Diczfalusy and his colleagues in Stockholm on the metabolism of steroid hormones in the isolated fetus was performed at 20°C (for reviews of this work see Klopper & Diczfalusy, 1969). It has been questioned, however, whether the perfusion of the pre-viable fetus at 20°C is really equivalent to experimental conditions *in vivo* because of the low temperature. This might well affect the enzyme systems that are involved in steroid synthesis and metabolism in the fetal organism. More recently, therefore, work of this type has usually been performed at 37°C, as in the author's method mentioned above.

Preparation of perfusion fluid

The perfusion fluid used is adult blood. This is not ideal since at the fifth month the normal adult form of haemoglobin constitutes only about 10% of the total haemoglobin of the fetus. Before this time the blood contains practically only fetal haemoglobin. The fetal haemoglobin (haemoglobin F) binds more oxygen than adult haemoglobin, and this is part of the reason why the oxygen dissociation curve of fetal red blood cells, rich in haemoglobin F, lies to the left of that of normal red blood cells which contain primarily haemoglobin A. However, fetal blood in sufficient quantities is not available for this work. Adult blood may also differ in osmolality, protein and electrolytes.

Perfusion is, therefore, performed with group-O Rh-negative adult blood. This should preferably have been recently drawn from the donor and treated with heparin to prevent coagulation.

One of the great problems encountered in preparing perfusion fluid is the lack of knowledge of the physiological values in the fetus. Several investigators have reported on pH, P_{O_2} and P_{CO_2} values of maternal blood and umbilical-cord blood at varying stages of gestation. Table 6.1 shows

Table 6.1. *Average pH, P_{O_2} and P_{CO_2} values of umbilical cord blood at different stages of gestation as reported by Yamuda (1970)*

Period of gestation (weeks)	Umbilical artery			Umbilical vein		
	pH	P_{O_2}	P_{CO_2}	pH	P_{O_2}	P_{CO_2}
15–20	7·35	29·2	36·3	7·28	18·6	50·6
21–28	7·37	29·3	37·7	7·30	18·7	46·0
29–38	7·37	27·3	39·1	7·31	17·9	48·8
39–42	7·36	26·5	38·6	7·31	17·9	47·8

some of the average pH, P_{O_2} and P_{CO_2} values of maternal and umbilical blood at different stages of gestation as reported in the literature. These are reported to be normal values, but it is well known that in order to obtain blood from fetuses various non-physiological factors have been introduced which may well affect the results. Maternal anaesthesia and fetal manipulation have both occurred before blood could be obtained for these values, and as soon as the umbilical cord is touched spasm begins so that the values must be regarded with caution. The figures in the non-manipulated fetus may be different from those shown in the above Table. Most of the perfusion fluids used have not, in fact, very closely resembled what is considered to be the physiological situation.

Oxygenation

This may be done in three ways: (a) By a single-bubble oxygenator. This gives good oxygenation but causes cell fragmentation and careful debubbling is necessary. (b) Screen oxygenation. This allows a thin film of blood to be exposed to circulating oxygen. The method is used by the author. (c) Membrane oxygenator. This imitates the lung in separating blood from the environment, and oxygen and carbon dioxide counter diffuse through the thin membrane. This method is not very suitable for the very small volumes required in a fetus of less than 20 weeks' gestation, and is more suited to larger fetuses (900 g or so) or to premature babies (Chamberlain, 1968).

With the screen oxygenator, two long wide-bore needles are inserted through the rubber stopper of a blood transfusion bottle containing 200 ml of donor blood. The bottle is then inverted and a mixture of 95% oxygen and 5% carbon dioxide is passed in one needle, circulated in the bottle and passes out through the second needle. The bottle is moved in such a way as to form a fine film of blood on the sides. In this way good oxygenation is obtained in about 5 min. This method is similar to that provided with the oxygenator shown in Fig. 6.2. The oxygenator is a more elaborate device but may give more accurate results. Lerner et al. (1971) found that oxygenation of the blood perfused with carbogen (93·5% oxygen and 6·5% carbon dioxide) resulted in a low pH, highly elevated P_{O_2} and almost normal P_{CO_2} values, while oxygenation with air and carbon dioxide (air + 23·1% − 3·7% CO_2) gave a normal pH and almost normal P_{O_2} and P_{CO_2} values.

It is necessary, if the carbogen mixture is used, to raise the pH to physiological levels by addition of 4% (w/v) sodium bicarbonate solution (Coutts & Macnaughton, 1969), but the P_{CO_2} and P_{O_2} tend to remain high in this perfusion fluid.

Blood gas values

Table 6.2 shows the average pH, P_{O_2} and P_{CO_2} values of perfused blood and of the perfusates obtained during the perfusion of mid-term fetuses by the method involving the screen oxygenator.

Lerner *et al.* (1971) found that when fetuses were perfused with blood oxygenated with carbogen, the already low pH of the perfusate diminished further, the P_{O_2} values became elevated and the P_{CO_2} values

Table 6.2. *Average pH, P_{O_2} and P_{CO_2} values of the perfused blood and of the perfusates obtained during mid-gestation experiments*

Samples	Carbogen oxygenation			Author
	pH	P_{O_2}	P_{CO_2}	
Perfused blood	7·26	319·7	46·4	Lerner *et al.*, 1971
Perfused blood	7·39	210·0	180·0	Macnaughton (unpublished)
Perfusate– umbilical artery	6·99	53·8	104·6	Lerner *et al.*, 1971
Perfusate– umbilical artery	7·30	95·8	166·6	Macnaughton (unpublished)

remained too high. When air and carbon dioxide were used for oxygenation in fetal perfusions, the pH of the perfusate increased, the P_{O_2} became higher and the P_{CO_2} values were significantly diminished. These workers considered that the air and carbon dioxide mixture resulted in better experimental conditions than those obtained when carbogen was used. This mixture should, therefore, be used for oxygenation of the blood in this type of experimental work.

Flow rates

One of the serious problems in the perfusion of the pre-viable human fetus is achieving an adequate flow rate. Great variations in flow rates are recorded in the literature. In a fetus weighing 275 g Westin *et al.* (1958) recorded a flow rate of 14·2 ml/min. This is rather greater than that found by Lerner *et al.* (1971). They found the mean blood flow in the fetal perfusions to be 4·3 ml/min, when an air and carbon dioxide mixture was used for oxygenation, and 2·6 ml/min when carbogen was used. Flow rates in our own laboratory would agree with those from Diczfalusy's laboratory, but occasionally higher rates of 6–7 ml/min can be recorded.

Varying concentrations of gases in the perfusion fluid may produce different flow rates in different species. Lawn and McCance (1964) found in the pig fetus, that when the concentration of oxygen in the gas mixture was reduced the flow rate fell. These workers found that this effect on the flow rate was nearly always obtained. It is clear that reproducible results are difficult to obtain. The heart rate in these fetuses is little guide to the flow rate, and the work of Nyberg and Westin (1962) on the human fetus has been confirmed by Lawn and McCance in the pig fetus. The heart rate may remain the same while the flow rate alters considerably.

In general terms, the bigger the fetus the greater is the chance of achieving a high flow rate. Chamberlain (1968), in an experiment where the fetus weighed 980 g, achieved flow rates varying from 15 to 40 ml/min, but this magnitude of flow rate does not occur with fetuses of under 500 g, which are commonly used in fetal perfusions.

It is evident that the flow rates achieved in these isolated human fetuses are probably not physiological, but the data are really inadequate for a final conclusion to be made. Rudolph (1971) made some measurements in human fetuses in mid-gestation while they were still attached to the placenta during hysterectomy. He found the measurements were not very different from those of the lamb, and considered that the reported flow rates of 2–3 ml/min in a 150-g fetus were probably about 6% of the normal flow. This worker also found that in the fetal lamb the proportion of cardiac output distributed to the placenta was higher at mid-term than at term, and he considered that this was probably true of the human fetus as well.

Since the flow rates achieved in human perfusion experiments are unphysiological, the various biochemical data that have been obtained in these experiments have to be viewed with caution. Some progress has been made towards improving the conditions of the perfusions, but further studies are required before more physiological conditions can be obtained. It is of particular importance to establish experimental conditions under which the acidosis of the fetus can be rapidly eliminated.

Application in the human subject

The main application in man has been in the study of various interactions in the metabolism of steroid hormones in the fetus and feto-placental unit. It is not appropriate in this chapter to discuss these in detail, but just to indicate the broad aspects of the studies. For full details readers should consult 'Foetus and Placenta' edited by Klopper and Diczfalusy (1969).

In this general approach tracer doses of radioactive steroids have been injected into the various compartments of the feto-placental unit in normal and adrenalectomized fetuses. After a period of time, usually 1-2 h, the organs are dissected and the steroid metabolites are isolated and characterized. A pattern of metabolism of steroid hormones and their inter-relationships has been built up by this method.

Cassmer (1959) was the first to show that the fetus was important in steroidogenesis in the feto-placental unit. He ligated the umbilical cord in patients undergoing legal abortion and left the placenta and dead fetus *in situ* for 3 d. Interruption of the feto-placental circulation resulted in a very large and immediate drop in the urinary excretion of the three 'classical' oestrogens—oestrone, 17β-oestradiol and oestriol—whereas pregnanediol excretion showed only a slight decrease, suggesting that for placental oestrogen synthesis fetal precursors are more important than maternal precursors.

Perfusion of pre-viable fetuses has been used to study oestrogen metabolism (for reviews, see Beling, 1971, and Diczfalusy and Mancuso, 1969), and the isolation, formation and metabolism of neutral steroids (for reviews, see Younglai & Solomon, 1969, and Solomon & Fuchs, 1971) and androgen metabolism (for review, see Gandy, 1971).

This information has led to a greater understanding of hormone production and metabolism during human pregnancy and in the case of oestrogen metabolism to the development of an assay method for oestriol, which is used widely as a monitor of fetal growth and development in clinical practice (Macnaughton, 1967; Macnaughton, 1969; and Klopper, 1968).

Experiments are now in progress to determine the control of steroid synthesis by trophic hormones and by steroids elaborated with the feto-placental unit.

It is hoped that from this the further significance for fetal viability of steroids in maternal urine and blood will be determined, and, the use of steroids of uniquely fetal origin, and metabolites other than steroids, may be developed as indicators of fetal viability.

References

BELING, C. G. (1971) In *Endocrinology of Pregnancy*, ed. Fuchs, F. & Klopper, A. Ch. 3, pp. 32–65. New York and London: Harper and Row.

CASSMER, O. (1959) Hormone production of the isolated human placenta. *Acta Endocrinologica* (Kbh.), Suppl. 45: 66.

CHAMBERLAIN, G. (1968) An artificial placenta. *American Journal of Obstetrics and Gynecology*, **100**, 615–626.

COUTTS, J. R. T. & MACNAUGHTON, M. C. (1969) The metabolism of [4-14C]cholesterol in the pre-viable human foetus. *Journal of Endocrinology*, **44**, 481–488.

DICZFALUSY, E. (1962) Endocrinology of the foetus. *Acta Obstetrica et Gynecologica Scandinavica* (Suppl. 1), 41: 45–85.

DICZFALUSY, E. (1969) Steroid metabolism in the foeto-placental unit. *Excerpta medica* (Amst.) *International Congress Series*, **183**, 65–109.

DICZFALUSY, E. & MANCUSO, S. (1969) In *Foetus and Placenta*, ed. Klopper, A. & Diczfalusy, E. Ch. 5, pp. 191–248. Oxford and Edinburgh: Blackwell Scientific Publications.

GANDY, H. M. (1971) In *Endocrinology of Pregnancy*, ed. Fuchs, F. & Klopper, A. Ch. 5, pp. 101–154. New York and London: Harper and Row.

GREIG, M. & MACNAUGHTON, M. C. (1967) Radioactive metabolites in the liver and adrenals of the human foetus after administration of [4-14C]progesterone. *Journal of Endocrinology*, **39**, 153–162.

KLOPPER, A. (1968) The assessment of feto-placental function by estriol assay. *Obstetrical and Gynecological Survey*, **23**, 813–838.

KLOPPER, A. & DICZFALUSY, E. (1969) *Foetus and Placenta*. Oxford and Edinburgh: Blackwell Scientific Publications.

LAWN, L. & McCANCE, R. A. (1964) Artificial placentae. *Acta Paediatrica*, **53**, 317–325.

LERNER, U. & DICZFALUSY, E. (1968) A new method for the 'in vitro' perfusion of the human foeto-placental unit at mid-pregnancy. *Excerpta medica* (Amst.) *International Congress Series*, **170**, 19.

LERNER, U., SAXENA, B. N. & DICZFALUSY, E. (1971) Extra corporeal perfusion of the human foetus, placenta, and foeto-placental unit. *Karolinska Symposia on Research Methods in Reproductive Endocrinology*, 4th Symposium. Perfusion Techniques, 310–325.

MACNAUGHTON, M. C. (1967) Hormone excretion as a measure of fetal growth and development. *American Journal of Obstetrics and Gynecology*, **97**, 998–1019.

MACNAUGHTON, M. C. (1969) In *Modern Trends in Obstetrics 4*, ed. Keller, R. J. pp. 110–134. London: Butterworth.

NYBERG, R. & WESTIN, B. (1962) An experimental study of the pre-viable human foetus. *Journal of Obstetrics and Gynaecology of the British Commonwealth*, **69**, 831–835.

PEEL, J. (1972) The use of fetuses and fetal material for research (1972): Report of the Advisory Group. London: H.M. Stationery Office.

RUDOLPH, A. M. (1971) In *Karolinska Symposia on Research Methods in Reproductive Endocrinology*, 4th Symposium. Perfusion Techniques, p. 328.

SOLOMON, S. & FUCHS, F. (1971) In *Endocrinology of Pregnancy*, ed. Fuchs, F. & Klopper, A. Ch. 4, pp. 66–100. New York and London: Harper and Row.

THOMSON, A. M., BILLEWICZ, W. Z. & HYTTEN, F. E. (1968) The assessment of foetal growth. *Journal of Obstetrics and Gynaecology of the British Commonwealth*, **75**, 903–916.

WESTIN, B., NYBERG, R. & ENHORNING, G. (1958) A technique for perfusion of the previable human foetus. *Acta Paediatrica*, **47**, 339–349.

YAMUDA, N. (1970) Respiratory environment and acid-base balance in the developing fetus. *Biology of the Neonate*, **16**, 222–242.

YOUNGLAI, E. V. & SOLOMON, S. (1969) In *Foetus and Placenta*, ed. Klopper, A. & Diczfalusy, E. Ch. 5, pp. 249–298. Oxford and Edinburgh: Blackwell Scientific Publications.

7 Extracorporeal Circulation for the Study of the Pre-term Fetus

COLIN H. M. WALKER and
B. J. N. Z. DANESH

The fact that economic prosperity and improved standards of medical care go together is well illustrated by the revolution in the management of the newborn infant in the Western world over the last half century. The success of methods of resuscitation and of sustaining life in these small patients is reflected in the steady fall in perinatal mortality, and while this is rewarding and reassuring it is not enough. Having taken it upon ourselves to increase survival of the 'at risk' fetus and newborn infant we are now obliged to do everything possible to ensure that the life that is to follow is a normal happy one, both for the child and his family. But we still lack much fundamental knowledge about fetal development during late pregnancy and about progress during labour and delivery, knowledge that is required for the optimum management of the complicated case and for the treatment of the prematurely born or abnormal baby.

The present concentration of energy and resources on a field as narrow as perinatology must appear to some, especially those belonging to the lesser privileged communities, to be luxury medicine, if not to be frankly wasteful. However, the need for knowledge and the potential benefits of research can be readily recognized by a visit to a cerebral palsy clinic or congenital abnormality clinic or to a mental deficiency hospital.

There is no argument. Infants who would have died a few years ago now survive. We therefore have an obligation to intensify, whenever possible, the study of perinatal physiology and improve methods of life support in the newborn. It was with the accomplishment of these ends in mind that the system described here was developed.

209

Historical review

Searching the medical archives can be a very humbling experience. To those of us who think of an artificial extracorporeal circulation as something relatively new it may come as a surprise to know that this was first conceived in 1813 by Le Gallois.*

During the first half of the nineteenth century the value of artificially infused oxygenated blood for preserving tissue viability was established. The latter part of the 1800s saw the advent of systems for continuous organ perfusion as illustrated by the sophisticated extracorporeal circuit attributed to von Frey & Gruber (1885), which appears on the cover pages of the excellent treatise on 'Heart–Lung Bypass' by Galletti & Brecher (1962). The deleterious effects of blood as a priming medium were suspected by Prevost and Dumas as early as 1821, and a pump oxygenator circuit not requiring blood for priming was reported by Brodie in 1903. The importance of such factors as aeration, composition and temperature of perfusate, continuity of flow, infusion and pulse pressures were all recognized by 1920, yet it was not until 1937 that techniques had advanced to the point where Gibbon proposed the use of extracorporeal circulation during cardiac surgery in man. During the seven-year period from 1948 workers all over the world designed and tested diverse systems, and about 1955 these efforts culminated in the clinical application of extracorporeal perfusion in open-heart surgery. The many problems, mechanical and physiological, theoretical and practical, that such systems present have been ably described by Galletti & Brecher (1962), but it was not till more recent years that Galletti and others turned their attention to the problems now before us— that of designing a safe system for use in subjects of very low body weight.

The use of partial, as opposed to total, cardiopulmonary bypass was recorded by Melrose as early as 1953 but it was not for some years that long-term extracorporeal bypass for assisting circulatory collapse was used clinically.

In parallel with this came the development of systems sufficiently well controlled for use in small animals and ultimately for the perfusion of the pre-term human fetus. Following the publications of Kolff & Effler (1956) and Westin, Nyberg & Enhorning (1958) there have appeared reports from some 20 European and North American centres in which

* Early historical data abstracted from 'Heart–Lung Bypass' (Galletti & Brecher, 1962).

various techniques have been developed. Some have designed systems for the maintenance of the fetus immersed in a simulated liquor amnii solution (Westin et al., 1958; Lawn & McCance, 1964; Callaghan & Angeles, 1961; Alexander, Britton & Nixon, 1966) while others have directed their efforts to the maintenance of life after the fetus has established independent existence (Krasna et al., 1962; Schramel et al., 1961; Walker, Rose & Simons, 1962; Galletti et al., 1963; Rashkind et al., 1965; Chamberlain, 1968; Dorson et al., 1969a; Kolobow et al., 1970; White et al., 1971). During the same period there was a gradual accumulation of data which indicated that exchange methods involving membranes, which avoid a direct blood-gas interface, were less injurious to blood. It is largely for this reason that disc and bubble oxygenators have become less popular for long-term perfusion. A great deal of effort has been expended in finding materials with exceptionally good gas-transfer properties. The Teflon membranes used for a number of years following the work of Kolff & Effler (1956) were superseded by membranes of silicone rubber. Kolobow et al. (1969) used silicone-rubber membrane in the form of a spiral tube, while Dorson et al. (1969a, b) found multiple parallel silicone tubes provided very efficient gas transfer, as did Vervloet, Edwards & Edwards (1970) and Boyd, Moran & Clark (1972) with spaced sheet membranes.

Much effort has also been directed towards miniaturization for the purpose of limiting or perhaps eliminating blood for priming. While this is of obvious importance in a system that can be primed with clear fluid, there is little virtue in achieving a total system-priming volume of much less than one unit if blood has to be used. It is the presence of donor blood, and not the volume used, that is the main factor of importance. Nevertheless, miniaturization of the circuit can lead to more accurate control of blood volume and flow, and many of the systems already referred to contain modifications directed to this end.

Methods have been devised in which the subject's own blood pressure activates flow through the extracorporeal system (Lawn & McCance, 1964; Rashkind et al., 1965; Dorson et al., 1969b), though most circuits incorporate some form of pump. The earlier workers frequently used a digital sigma-motor pump, but this was gradually replaced by the roller type. The point of most physiological significance is whether or not peristaltic flow was necessary (Wilkens, Regelson & Hoffmeister, 1962). The peristaltic pump is more complex and while there is some evidence that it provides a more physiologically efficient arterial flow (Trinkle et al., 1970) it does not seem to be essential to survival after long-term perfusion. Its use is even less necessary, and may even be contra-indicated in bypass circuits returning blood to the venous system.

Finally, in this tale of technical evolution, there is the way in which the system to be described below would seem to be unique. Systems with separate gas-exchanger and dialyser in series (Lawn & McCance, 1964), with dialyser coils (Chamberlain, 1968) and membrane assemblies suitable for either gas or dialysate transfer (Landé *et al.*, 1967), have been described, but ours would seem to be the first attempt to study the functions of an artificial oxygenating lung and kidney dialyser with two types of membrane in the one multiple-membrane assembly. Admittedly such a system is complex and it has not yet been possible to incorporate much miniaturization. But it has the great advantage of being able to correct both the respiratory and metabolic components of cardio-pulmonary distress—an advantage that we feel may prove to be worth the price of complexity.

A striking feature in the world literature on this subject is the limited clinical application of systems that have shown such promise in the experimental laboratory. Only now are reports appearing in the literature of successful long-term partial bypass of up to 5 d in adults (Hill *et al.*, 1972). Survival during partial bypass of 7 d in lambs (Kolobow *et al.*, 1970) and 10 d in an infant (White *et al.*, 1971) has also now been reported. It must be remembered, however, that the lambs were well and had normal pulmonary function to assist in long-term survival, but that it is a very different matter treating a desperately ill newborn infant with pulmonary insufficiency. The problems are many and varied as we shall see, but in the field of fetal and neonatal maintenance it may be that too much attention has been paid to the support of lung function and too little to that of the kidney. The desire to miniaturize has perhaps taken precedence over methods that compensate for their complexity by providing more comprehensive biochemical control, and in which there is the added advantage of the blood flow control to the fetus being independent of the control of the flow through the oxygenator and/or dialyser. An example of the latter type of lung–kidney (Klung) system is what follows.

The combined lung–kidney (Klung)

The concept of a combined multimembrane lung–kidney unit was intro-duced by Galletti and Brecher in 1962, having been evolved from the

Pierce (1960) modification of the Clowes artificial membrane lung (Clowes, Hopkins & Neville, 1956). To our knowledge, however, the unit has been used only as a lung (Galletti *et al.*, 1963) or as a kidney (Someren *et al.*, 1963), but never as both simultaneously, probably because of the difficulty of obtaining success with such a complex membrane assembly. Nevertheless, this seemed an ideal system with which to treat the severe combined metabolic and respiratory acidosis of the newborn infant in cardio-pulmonary distress and after much experimentation and practice a successful procedure has been achieved.

Description

Artificial kidney and lung assembly
The gas and electrolyte exchange area consists essentially of an assembly of several layers of lung membrane (Teflon 0·5 mil = 0·0005 in) and kidney

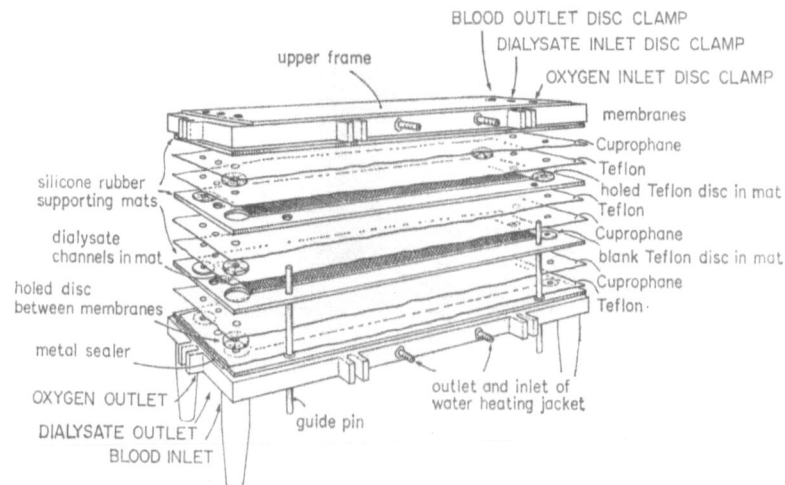

Fig. 7.1. Diagram illustrating placement of mats and membranes in 'Klung' assembly

membrane (Cuprophane PT 150) laid on 'multiple cone' soft silicone rubber or polyvinyl chloride mats (Fig. 7.1). Oxygen is directed between the Teflon and the mat, and dialysing fluid between the Cuprophane and the mat. The purpose of the cones is to permit plastic moulding of the membranes, thus achieving a larger exchange surface area than the basic mat area of 5 × 20 in ($\frac{1}{16}$ m^2 approx.). The assembly is such that as blood passes between two membranes it is oxygenated over one surface and dialysed over the other.

213

The membrane–mat assembly is held firm by an upper and lower aluminium cast frame held tightly by six screw clamps tightened to a strength of 12 lbf/ft² by a torque wrench. The frames are hollow and warmed by water circulation thermostatically controlled at 38°C.

At each end of the mats there are three $\frac{7}{16}$-in ports. These are paired for gas, dialysing fluid (dialysate) and blood circulation, and with a series of moulded P.T.F.E. Teflon distribution discs (Someren *et al.*, 1963) the flow through the three compartments is established, yet all three remain isolated (Fig. 7.2). These distribution discs are secured by a torque

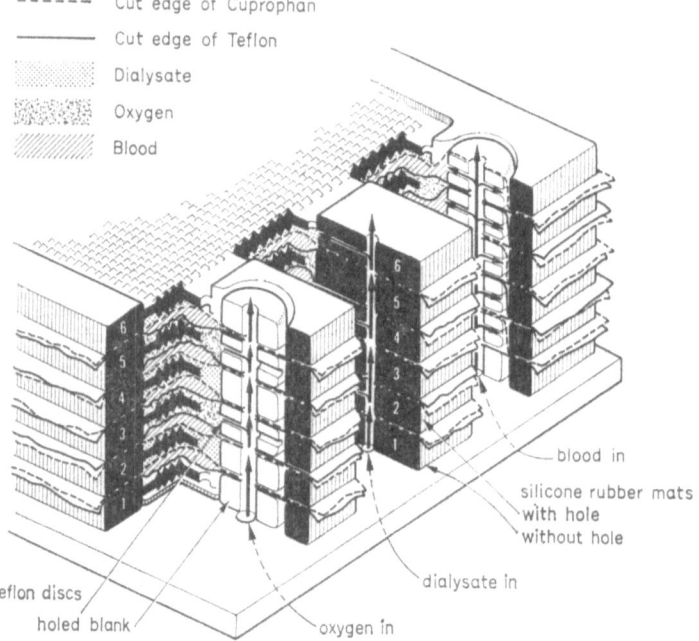

------ Cut edge of Cuprophan

———— Cut edge of Teflon

    Dialysate

    Oxygen

    Blood

blood in

silicone rubber mats
with hole
without hole

teflon discs

holed blank

dialysate in

oxygen in

Fig. 7.2. Diagram illustrating distribution of gas, dialysate and blood by P.T.F.E. Teflon discs

wrench set at 2 lbf/ft². The disc distribution is such that blood and gas are directed diagonally across the assembly, and dialysate from centre to centre port, in counter-current fashion. For the purposes of both the basic studies and the clinical trials of the system, a total of one square metre exchanging area was used. This consisted of five pairs of Teflon and three pairs of Cuprophane membranes, i.e. 16 surfaces each of $\frac{1}{16}$ m².

The blood-pressure gradient across the assembly is monitored by two aneroid manometers and the inlet pressure of dialysate by a third.

Plate 7.1. Infant incubator and perfusion assembly:

1. Cantilever weighing beam
2. Shock absorber
3. Incubator and suspension tray
4. Servo-controlled incubator power pack
5. Solenoid safety valve
6. Micro-screw valve
7. Filter, bubble trap
8. Klung
9. Klung warmer water tank
10. Infant blood-flow pump
11. Klung circuit Sarns pump
12. Dialysate pump
13. Dialysate tank
14. Oxygen supply and suction outlet

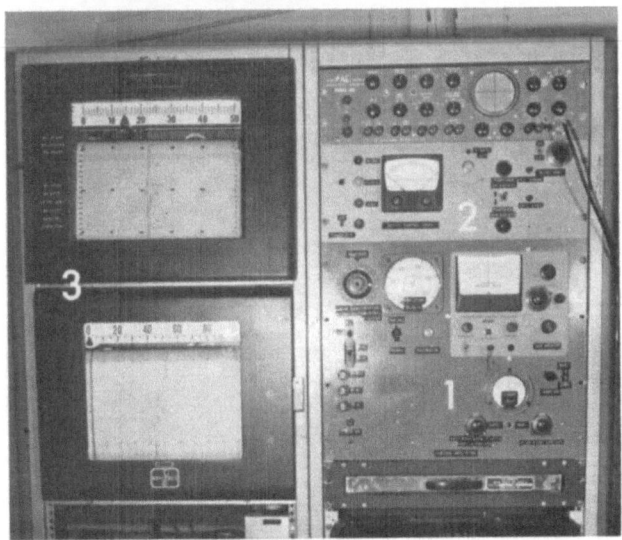

Plate 7.2. Control console:
1. Transducer amplifier and infant-pump control
2. Safety panel with contact alarm meter
3. Recorders for body weight, and temperature of points on blood circuits

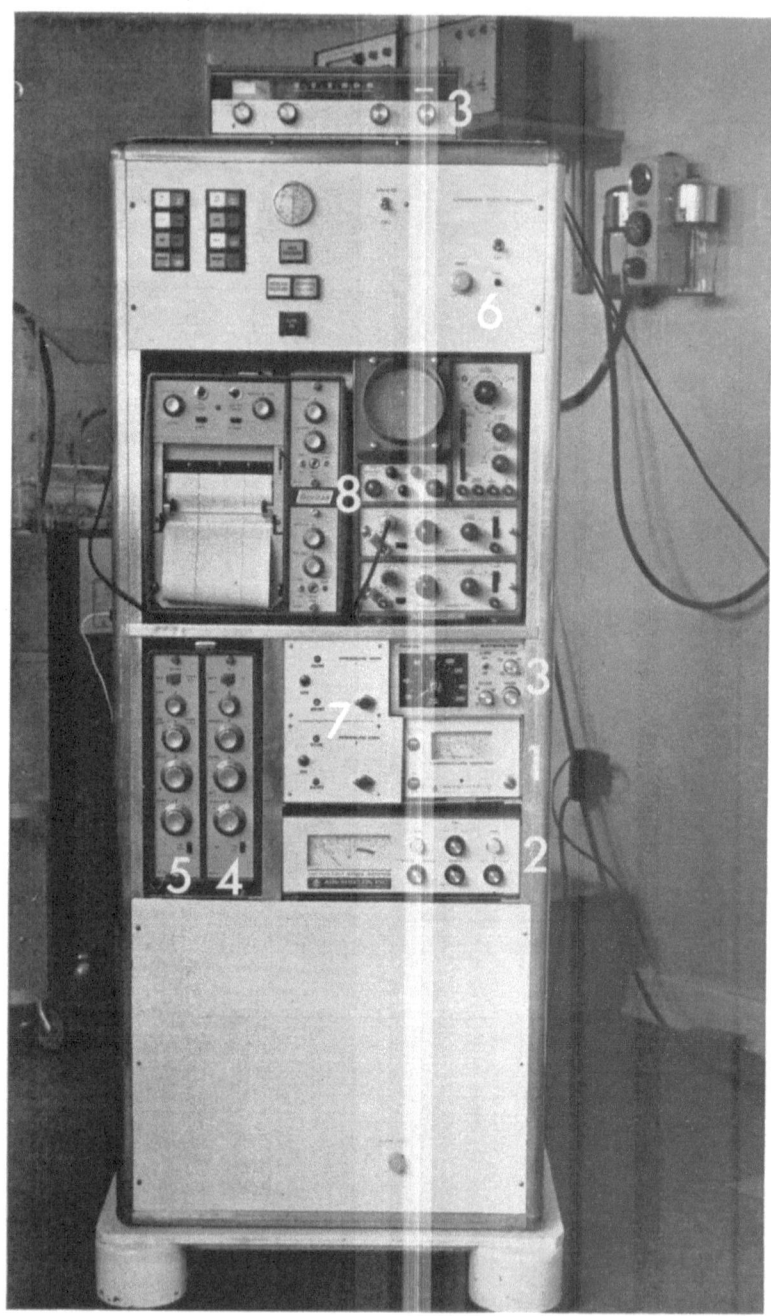

Plate 7.3. Recently completed console for:

1. Temperature
2. Respiratory rate and alarm
3. Heart rate (including FM transmitter/receiver) and alarm
4. ECG (standard leads)
5. EEG
6. Surface-impedance blood pressure
7. Intravascular pressures
8. Oscilloscope and recorder

The Klung circuit

The Klung circuit (Fig. 7.3 and Plate 7.1) is completed by tubing connecting: (*a*) the Klung blood phase to a blood reservoir (thermostatically controlled at 37°C) and back to the Klung via a Sarns roller pump; (*b*) the dialysate phase proximally to the dialysate tank (thermostatically controlled at 37°C) and distally the dialysate is discarded; (*c*) the gas phase to a cylinder of 100% oxygen or 96% oxygen and 4% CO_2 and distally the gas is allowed to escape into room air.

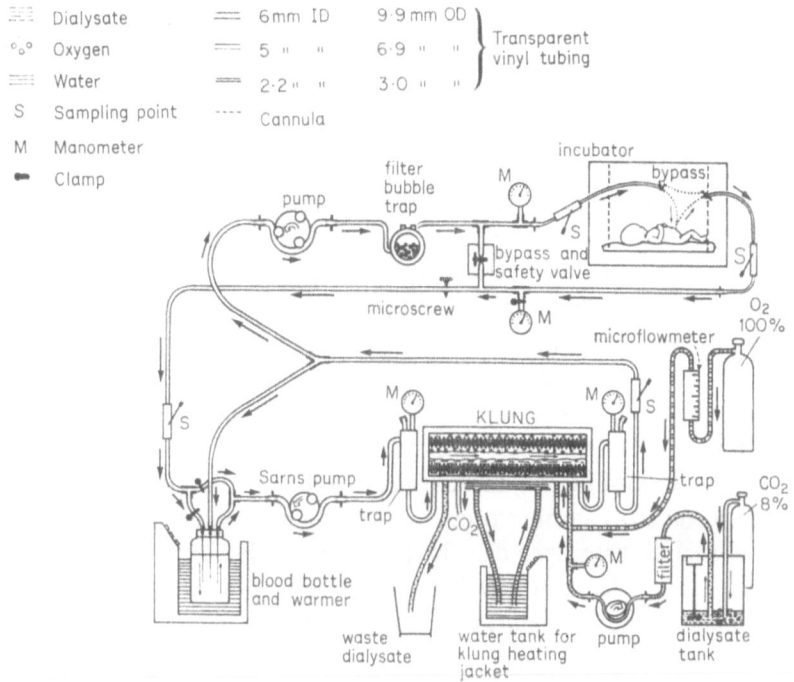

Fig. 7.3. Diagram of the 'Klung' and infant circuits

This system is leak tested after slowly priming the blood phase with saline at 180 ml/min so as to gently distend the space and mould the membrane over the mats. The exit tube is clamped and pumping continued till a pressure of 140 mmHg is obtained. The inlet tube is then clamped and if the pressure is sustained the membranes are known to be intact. Saline is then circulated in the dialysate phase, and the blood-phase saline is again circulated, and with gas flowing at 500 ml/min the pressures are checked. Values for an intact system are known and if these are not obtained a further check is made by introducing 0·5 g Evans blue into the blood phase. If this appears in the dialysate a leak is confirmed; if not, the dye can be readily washed out of the system. In any event the pressure in the blood phase is always maintained at 20–30 mmHg higher than that of the dialysate phase so that, in the unlikely event of a leak developing during perfusion, the direction of loss would be safe, i.e. blood to dialysate, and would be immediately recognized in the discarded dialysate.

The Klung unit has a priming volume of 100–120 ml/m², increasing to 140–150 ml with further gradual moulding deformation of the membranes over one hour, and the blood flow through the Klung is set between 180 and 200 ml/min m². This ensures sufficient flow to avoid pooling of blood between the membranes, and it can be increased to 400 ml/min without untoward increase in pressure within the Klung. The inflow blood pressure at the lower flow rate, however, may occasionally rise from 80 to 200 mmHg over the period of a five-hour perfusion. This is probably due to gradual occlusion of the flow channel by aggregations of blood cells.

The dialysate flow averages 70 ml/min per 0·37 m², with an inflow pressure of 60–100 mmHg. Again this gradually rises during perfusion but this is acceptable provided there is a positive gradient from blood to dialysate of 20–30 mmHg. The oxygen flows at 0·5–1·0 l/min.

The assembly time for one-square-metre membrane-exchange area is about 30 min. To this must then be added leak testing and sterilization time (as described below), making a total of about two hours. After some practice, success can be expected in 80% of assemblies. The problems arising include improper distribution disc positioning, damage due to piercing or wrinkling of membranes, application of asymmetrical pressure when securing mats and discs, and too rapid priming causing perforation of a membrane.

When the assembly is shown to be satisfactory, it is sterilized by circulating 0·1% HCl through the blood phase for 30 min, then washing out by circulating sterile saline through both the blood and dialysate phases. This is a quick and efficient method but tends to leave the cuprophane membrane slightly brittle. A simpler method requires the use of an ethylene oxide chamber, but this has the disadvantage of a 24-h delay while all the gas escapes from the silicone-rubber mats. Formaldehyde may be used but the wash-out procedure is cumbersome and time consuming.

The combined Klung and Infant circuits are shown in Fig. 7.3. Their point of contact is the blood reservoir which is maintained at 37°C in a bottle warmer, and from which an occlusive Sarns pump draws blood and passes it via a bubble trap (and pressure manometer) into the Klung. From the Klung exit it passes through another bubble trap (and manometer), past a sampling site and back to the reservoir. From a point on this last stage, treated blood is drawn off by a branch line to the infant, only when the infant pump is activated.

The advantages of the Klung circuit being independent of the infant circuit are that: (a) it may be primed and equilibrated without influencing infant blood flow; (b) it can operate at an optimum rate for gas and electrolyte exchange independently of the infant flow rate and (c) samples can be taken as often as required without altering infant blood volume.

The infant circuit

From the branch line of the outlet of the Klung, the blood is drawn off by a three-arm De Bakey-type roller pump (under infant-weight servo-control) and passed through a steel mesh filter bubble trap, through a bypass safety valve and into the infant by way of an umbilical or femoral cannula. Blood is withdrawn from the infant via similar blood vessels, either by the hydrostatic head of pressure between infant and reservoir, or by arterial pressure when withdrawal of blood is from an artery, or by both. If flow is inadequate gentle sucking may be exerted by opening the clamp on the link line just above the reservoir and closing the line to the reservoir. In this way the Sarns pump encourages direct drainage into the Klung. Both the 'in' and 'out' lines pass through the bypass safety valve so that, in the event of an unacceptable change in weight the infant is automatically isolated by the stopping of infant flow, while the remainder of the extracorporeal circulation can continue in readiness for restarting perfusion. After the bypass valve the 'out' line passes through a manual microscrew flow-control clamp and on down to the reservoir. Further manometers and sampling sites may be attached as indicated and disposable tubing systems are now assembled to our specifications and packed and sterilized by Portex Ltd. specifically for this work.

Mechanical control and safety circuit

The earlier methods of controlling blood volume during extracorporeal circulation frequently depended upon crude level measurements in reservoirs or in oxygenators, and are insufficiently accurate for use in subjects of less than several kilograms in weight. Miniaturization reduces the degree of error, as does more precise electronic methods of level detection (Walker, Blackwell & Massengale, 1961), but it is still debatable whether measurements of extracorporeal blood volume are sufficiently accurate when subjects below 1500 g are being perfused at relatively high flow rates. For example, a sudden total blockage of a return-line cannula may deplete an infant of 1·0 kg, perfused at a rate of 40 ml/min, of almost one quarter of his blood volume in 30 s. For these reasons the method of control in this system is based upon the subject's body weight, changes in which control the pump speed by means of an electronic proportional servo-control circuit (Walker *et al.*, 1962, 1963a, 1963b).

The infant is nursed on a perspex tray suspended from a cantilever beam by a shock absorber (Plate 7.1). Displacement of this beam is detected by a load-cell transducer standardized so that, no matter what the initial load between 0 and 10 kg may be, an increase in weight of 1 g will effect a movement of about 1×10^{-5} in. The signal generated by this movement is amplified and used to increase or decrease the infant blood-pump-motor speed in such a way that the body weight of the subject is maintained at ± 5–10 g from the initial weight, despite fluctuations in the rate of blood flow (Walker *et al.*, 1963b).

Initiation of perfusion
Priming of the whole system requires about 400 ml of fresh heparinized blood, and this is first circulated in the Klung System till equilibration of gases and electrolytes is achieved. The infant circuit is then filled by opening the safety valve and priming the line up to the small link attached to the infant cannula. When ready to perfuse, the safety valve is closed, and the infant cannula opened to the main lines. The infant thus pressurizes the lines, and a note is taken of the 1- to 2-g weight loss that usually occurs. The safety valve is then opened and flow started by slowly opening the microscrew valve. As blood flows into the reservoir a weight loss is detected by the suspension system. This activates the infant pump at a loss of under 10 g, and weight is restored automatically by the servo-control system which balances weight (blood) loss with infant-pump-flow rate. The microscrew valve is gradually opened to its full extent, the pump automatically keeping pace to within 10 g body weight, and once full flow is established the pump will match the outflow to within ± 5 g of the initial body weight.

If errors in flow control are not detected and corrected immediately, the infant must at once be isolated from the pump circuit. This has been achieved by a safety circuit which is 'fail-safe' in respect to power or circuit failure, and which isolates the subject by activating the solenoid safety valve whenever the body weight exceeds preset limits for a predetermined time (Plate 7.2). The limits are chosen in accordance with the weight of the subject, e.g. 5 g for a 1-kg infant and a 10- to 15-s delay is all that is permitted. In this way the infant is protected from any mechanical or electrical breakdown, and experience has shown this circuit to be most reliable. Once isolated from the pump circuit at the safety-valve level, the infant can have any excessive volume error corrected manually and the source of error can be detected at leisure. Manual volume correction is rarely necessary, however, as the proportional servo-mechanism responds in such a way that pump speed will cease when the infant's weight increases by 10–15 g, or will greatly increase with decreases in weight of this order.

Biochemical control

Many of the recommendations for biochemical control of extracorporeal systems have in the past assumed that the patient, usually an adult, has the capacity to correct some of the imbalances that occur during perfusion. In respect to the artificial lung, for example, the almost total loss of partial pressure of carbon dioxide (P_{CO_2}) when 100% oxygen is used in the oxygenator is soon corrected by the adult. In respect to the artificial kidney, for example, no attempt is made to control the loss of low-molecular-weight amino acids, and the metabolism of the adult is also relied upon to convert the acetate, used to control the dialysate pH, to bicarbonate. Such licence cannot, however, be taken with the very small subject whose blood volume is usually considerably less than the extracorporeal blood volume, and whose metabolic activity may well be immature and, as in newborn infants with cardio–pulmonary distress, defective.

That it is necessary to pay much greater attention to this aspect of total bypass for surgery in infants has already been recognized (Turina et al., 1972). Even more so is it important in the long-term perfusion of the fetus and newborn, and thus basic studies were required to assess membrane equilibrium ratios and determine the interaction of blood gases, proteins and electrolytes in the combined lung–kidney. There follows a summary of the results of such studies, some of which have already been published (Danesh, Walker & Mathers, 1968, 1970).

Blood gases

The great advantage of this system is that blood circulates through the Klung at least four times as often as through the infant. This provides considerable biochemical advantage despite the relatively large priming volume. There is, therefore, no difficulty in maintaining optimum oxygenation and elimination of carbon dioxide. Initially it may take up to 60 min to achieve normal gases in the heparinized donor blood because, no matter how fresh, it has on priming a partial pressure of oxygen (P_{O_2}) of around 50 mmHg and a P_{CO_2} well above 100 mmHg. The P_{O_2} is corrected in 5–10 min, but the P_{CO_2} takes longer (Fig. 7.4), because the blood is being dialysed against a solution of 'normal' P_{CO_2}, for reasons that will be discussed later. This equilibration could be hastened by increasing the number of kidney membranes or increasing the flow rate. The former adds to the assembly time and priming volume of the Klung, however, and the latter slightly increases the risk of haemolysis. As there is usually time available during the cannulating and preparation of the subject, neither of these modifications have been adopted. It should be added that it is unwise to circulate blood and gas without also circulating dialysate, as it is important to establish both gas exchange and electrolyte equilibrium prior to perfusion.

Provided the dialysate contains P_{CO_2} in the quantities described below, it is not necessary to use carbon dioxide in the oxygenator gas phase. If the system is used as an oxygenator only, however, this must be added in a concentration of about 4% to maintain a normal P_{CO_2} during perfusion. The gas does not need to be humidified, as it is with disc or bubble oxygenators, as direct contact with the cell does not occur.

The importance of the control of pH in respect to electrolyte balance will be detailed below. Suffice it to say that normal gas content can readily be achieved and maintained by the methods described, despite wide fluctuations in perfusion flow rate.

Electrolytes

It has been recognized for some considerable time that electrolyte disturbances may follow repeated renal dialysis. In the mid 1960s many reports of hypercalcaemia and soft tissue calcification appeared in the literature, and these were followed by reports of hypermagnesaemia. Study of these and other reports (Danesh *et al.*, 1970) revealed a surprisingly wide range in the calcium and magnesium content of the dialysate, and it was realized that the equilibrium ratios of the basic ions would have to be re-established for the Klung as constructed for perfusion of the newborn. Such studies were designed to consider the effects of some of the many factors inherent in a membrane transfer system, e.g. binding of ion by the membrane, haemolysis, denaturation of protein, protein binding of ions, changing hydrostatic pressure, changing pH, and

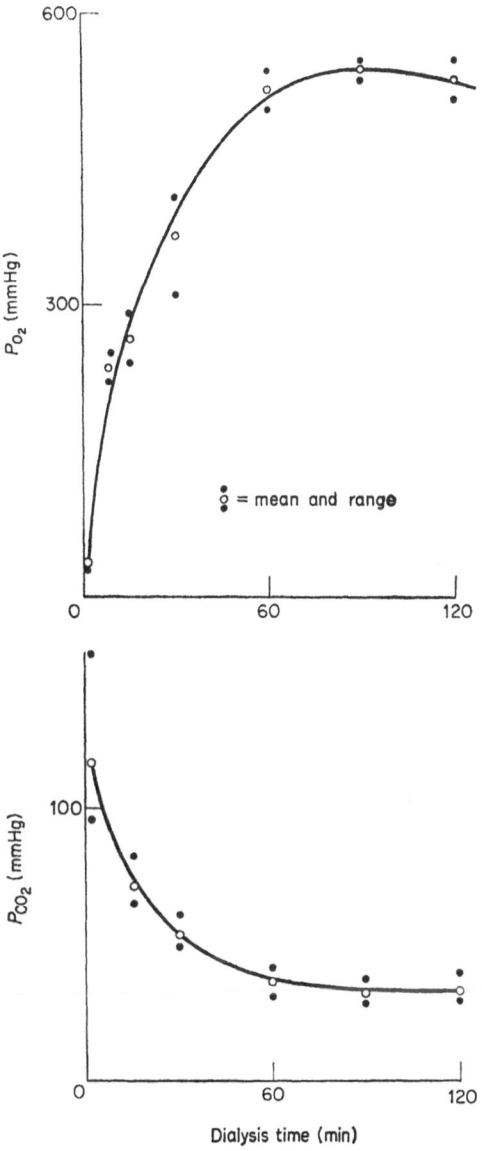

Fig. 7.4. Equilibration of oxygen and carbon dioxide in the perfusing blood with 100 % oxygen and with dialysate containing a P_{CO_2} of 30–40 mmHg

the loss of less significant diffusible substances such as sulphate, citrate, lactate, low-molecular-weight amino acids and urea, many of which are not routinely incorporated in dialysate. The comments that follow summarize the work of Danesh and his co-workers (Danesh *et al.*, 1968, 1970; Danesh, 1972).

The first aim of these experiments was to assess the degree of imbalance of ions in stable equilibrium which is produced on each side of the semi-permeable membrane by the two types of interaction that occur between proteins and electrolytes: (*a*) A weak interaction in which the electrolytes remain active and diffusible—known as the Donnan effect. (*b*) A strong binding of ion by protein so that the ions become ionically inactive and non-diffusible.

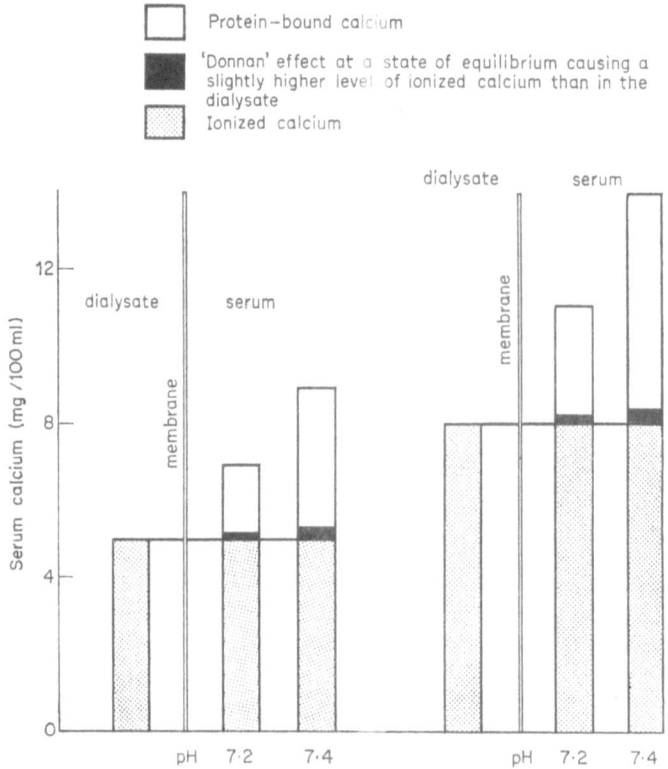

Fig. 7.5. Effects of dialysate calcium level and pH on equilibrium dialysis levels of serum calcium. (Reproduced by permission of *The Lancet*)

The effect of these two interactions is shown in Fig. 7.5. The ionized fraction equilibrates as expected, but the presence of protein in serum attracts additional ionized calcium (Donnan effect) and, as there is a constant ratio between diffusible and bound ion at any given pH, a further increase occurs due to the subsequent extra ion binding by protein. This figure also illustrates the considerable effect on postdialysis serum levels caused by changes in dialysate levels of calcium and by changes in pH.

Not only are similar changes seen with magnesium but these, and presumably other ions, compete for protein-binding sites. For example, the relationship between dialysate and post-dialysis plasma-protein-bound calcium at different levels of dialysate magnesium is shown in Fig. 7.6.

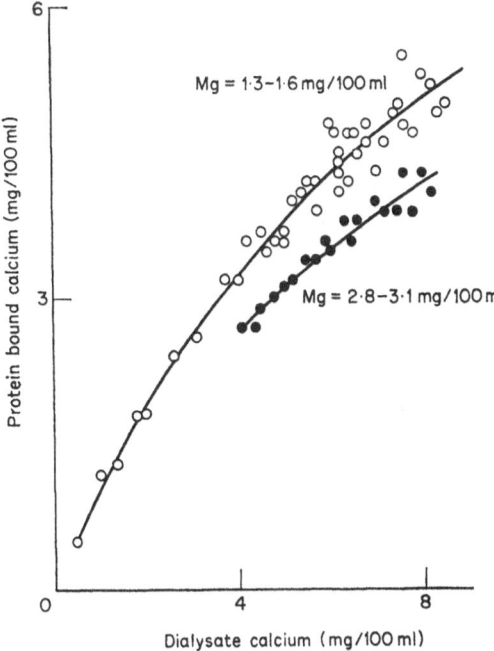

Fig. 7.6. Relationship between dialysate and protein-bound calcium at different levels of dialysate magnesium

Extensive studies were performed in order to establish equilibrium ratios at normal blood-gas tensions, and so evolve a dialysate that provided homeostatic levels of electrolytes in the serum. A list of ratios, i.e. the dialysate value/plasma value at equilibrium, is given in Table 7.1 and the composition was: Na^+, 143; K^+, 4; Mg^{++}, 1–1·5; Ca^{++}, 2·5–3·0; HCO_3^-, 27–30; Cl^-, 105 and $PO_4^=$, 2·6 mEq/l at a pH of 7·35–7·40. If the patient should require calcium or magnesium levels beyond the usual range, the dialysate concentrations required to produce these may be found from the nomograms calculated from these *in-vitro* studies (Fig. 7.7).

The signal difference between this dialysate and that used in routine renal dialysis in older patients is the presence of bicarbonate (HCO_3^-). This is usually provided by the addition of acetate which is converted

Table 7.1. *Ion values equilibrated at normal* pH

	Total dialysate concentration (mEq/1 H₂O)	Total plasma concentration (mEq/1 H₂O)	Equilibrium ratio (ER)
CATIONS (dialysate/plasma)			
Na	140	151·4	0·93
K	4·4	5·0	0·88
Mg	1·3	2·0	0·64
Ca	2·9	5·4	0·54
ANIONS (plasma/dialysate)			
Cl	112·2	109·8	0·98
HCO₃	27·2	26·7	0·98
PO₄	2·4	2·8	1·18

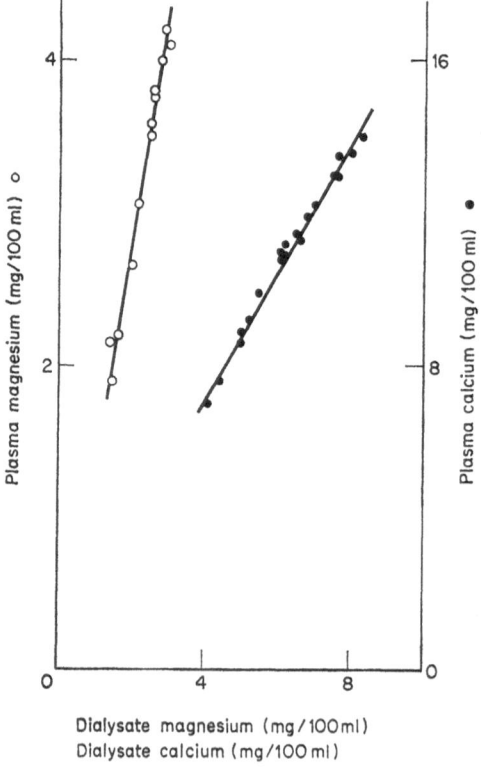

Fig. 7.7. Nomograms for the prediction of post-dialysis levels of calcium and magnesium in plasma. (Reproduced by permission of *The New England Journal of Medicine*)

to HCO_3^- in the Krebs-cycle metabolism of the patient. In the debilitated and immature fetus or newborn, however, we considered this an unacceptable metabolic requirement and sought to provide HCO_3^- directly in the dialysate.

In order to prevent precipitation of calcium carbonate when adding HCO_3^- and calcium together to the dialysate 3 ml/l of normal HCl was added to the final solution. As a result of the subsequent liberation of carbon dioxide (CO_2) and fall in pH from 8·0–7·4, calcium chloride could be added without the risk of precipitation. In most dialysate reservoirs, even though covered, there is a continuous loss of carbon dioxide and a tendency for a rise in pH. This, and the consequent precipitation, was obviated by the continuous bubbling of 8% carbon dioxide in oxygen into the dialysate solution maintained at 37°C. This is the reason why the dialysate must have the P_{CO_2} referred to previously.

Table 7.2. *Equilibrium ratios of amino acids*

Amino acid	Ratio plasma/dialysate	Amount added to 15 l of dialysate (mg)*
Alanine	0·76	581
Glutamic acid	1·90	60
Glycine	0·79	491
Isoleucine	0·73	107
Leucine	0·82	174
Methionine	0·65	102
Phenylalanine	0·74	264
Proline	0·68	470
Tyrosine	0·74	255
Valine	0·73	329

The following were also added though the ratios were not studied:†

Arginine	141
Aspartic acid	17
Butyric acid	42
Cystine	220
Histidine	179
Lysine	440
Ornithine	181
Serine	258
Threonine	388
Tryptophane	98

* Amount required for maintenance of normal values over 5 h dialysis.

† These values probably underestimate the required concentration.

Amino acids and proteins

It would appear that in adult renal-dialysis units little attention is paid to the loss of the low-molecular-weight diffusible amino acids which cross the semi-permeable membranes of artificial kidneys. Again, we considered this unacceptable practice for the pre-term fetus and new-born, and so a series of dialysis experiments was performed to establish

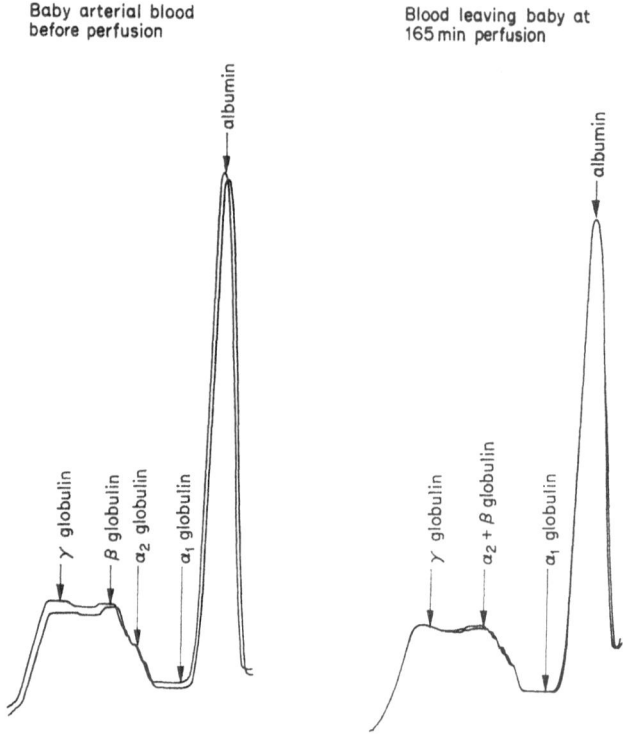

Baby arterial blood
before perfusion

Blood leaving baby at
165 min perfusion

Fig. 7.8. Scans of plasma-protein electrophoresis strips showing minimal effects of short-term perfusion on protein electrophoresis

equilibrium ratios for those amino acids obtainable for addition to dialysate solutions. The equilibrium ratios of those tested, and the appropriate amounts added to the 15-l dialysate tank are shown in Table 7.2.

The denaturation of protein caused by extracorporeal circulation, as shown in the change of appearance on simple paper electrophoresis, has been known for some time. It has also been agreed that this is less marked when blood and gases are separated by membranes (Fig. 7.8), though it may still occur. It is our contention that too little attention has been paid to the numerous other plasma constituents to which proteins are linked or otherwise related. While this matters little in short-term bypass for cardiac surgery, it may have considerable importance in the long term or repeated extracorporeal perfusions required for life support, and even more when such systems are used to study metabolic processes in the human subject of any age.

Miscellaneous factors

To date no attempt has been made to study the equilibrium properties of citrates, or sulphates. The dangers of cardiac arrest and convulsions following rapid infusions of blood anti-coagulated with citrate are well known. These dangers are greater in the presence of hepatic or renal dysfunction, hypothermia or when hepatic blood flow is diminished during surgical procedures. The effects of citrate were therefore studied further during these *in-vitro* experiments, and it was found that citrates not only decreased ionic calcium and magnesium by forming citrate salts but they also combine with the protein-bound ions from which the ionic pool would otherwise have been replenished. For these reasons, only fresh heparinized donor blood was considered safe to use for infant perfusions, though just outdated acid–citrate–dextrose blood was used for experimental work and, when necessary, it was reconstituted with calcium and heparinized. Allowance had to be made for this, and especially for the increased tendency for this older blood to haemolyse.

Partial perfusion

Perfusion of puppies

Techniques

The early stages of this work were directed principally towards the development of the electronic servo-control flow system already described. The accuracy of control was aided further by the monitoring of blood volume in the oxygenator with a capacitance level detecting device (Walker *et al.*, 1961). By these means the puppy weight could usually be maintained within ± 5 g of the starting value at varying flow rates.

The circuit was less complex initially, consisting of a similar suspension-weight transducer system, but with a small $4\frac{1}{2}$-in, 15-disc Kay–Cross oxygenator with an uptake capacity of 37·5 ml O_2/min (Walker *et al.*, 1962). The puppies were anaesthetized by intraperitoneal Nembutal and were kept under sedation throughout the perfusion. A small but important factor was the amount of sedation they received from the donor priming blood. The donors were initially sedated with pheno-barbitone but latterly, to avoid the risk of overdose, they were sacrificed by electrocution and then bled into heparinized containers. Biochemical control was at this time limited to the blood gases, and for these animal perfusions humidified 100% oxygen or 4% carbon dioxide in 96% oxygen was used in the oxygenator. As more attention was paid to the biochemical parameters it became obvious that there was a considerable loss of bicarbonate during perfusion, and this was added to the blood directly as sodium bicarbonate. Various blood-flow routes in the puppies were established with combinations of femoral artery and vein catheters, and superior vena cava cannulation via the jugular vein. Blood pressure was monitored by another arterial catheter, and the electrocardiogram and electroencephalogram were recorded.

During the development of this system reports of the value of haemo-dilution during total bypass for cardiac surgery began to appear. As histological studies of the liver and lungs of the puppies that died sug-gested that even partial perfusion could result in post-perfusion syn-dromes similar to those seen clinically after total bypass, the perfusate was diluted with isotonic saline to an haematocrit of 30–35% (Walker *et al.*, 1963b). This resulted in great improvement in the mortality rate and in the post-perfusion clinical condition, and the beneficial effects of

Table 7.3. *Progress in perfusions of puppies of various weights*

Perfusate	Below 1·0 kg		1·0–1·5 kg		Over 1·5 kg	
	Total	Survived	Total	Survived	Total	Survived
Whole blood (June 1960–June 1961)	2	1(50%)	7	1(14%)	18	12(67%)
Whole blood (mechanical improvement) (June 1961–Jan. 1962)	5	1(20%)	3	2(67%)	2	2(100%)
Whole blood with bicarbonate and CO_2 (Jan. 1962–Aug. 1962)	6	1(17%)	4	3(75%)	2	2(100%)
Haemodilution with bicarbonate and CO_2 (Aug. 1962–Dec. 1963)	10	5*(50%)	11	8(73%)	5	5(100%)

* One puppy sacrificed at 7 d because of neurological complication.

haemodilution were also evident histologically (Rowlands & Walker, 1964). Attempts at priming extracorporeal circuits with clear fluid in such volumes as to cause a fall in haematocrit to values ranging from 10–20% have not proven very successful in dogs (Depp & Hughes, 1971), though some success has been claimed with this method when associated with deep hypothermia for congenital heart surgery (Laver & Buckley, 1972).

The improvement in mortality these changes brought about over the years is seen in Table 7.3, the full experience with a total of 95 puppies.

Problems

The hazard already mentioned of the over-sedation of the puppy with the high levels of anaesthetic in the donor blood is peculiar to this type of study. It occurs because of the relatively large priming volume of donor blood as compared with the small blood volume and relatively inefficient liver function of the nearly newborn puppy. Avoiding this by the use of electrical methods of sacrificing the donor has the added advantage of reducing the clotting tendency of the donor blood. Heparinization was, however, necessary for long-term perfusion experiments and repeated additions were required to correct for metabolic loss within the puppy.

Any system with a sensitivity of 1 g such as this has the disadvantage of requiring very careful setting in respect to the weight imposed by cannulae, monitoring cables and artificial respirator tubing, but, provided these are adequately tethered and tared, they pose no great problem. More difficult to control are the extraneous signals generated by movement of the subject—puppy or infant. After about 2 h of perfusion, some puppies became sufficiently restless and mobile that false weight-change signals were detected by the transducer. Shivering is common in dogs in which the effects of barbiturate are wearing off and was often seen. This is counteracted by frequency filters which can recognize certain false signals, and by incorporating a 5-s delay before the system will respond to the weight-change signal. There are times, however, when further sedation of the subject may be required.

The main disadvantage of using body weight as a guide to volume and flow control is that sequestration is not detected (Lefemine, Fosberg & Harken, 1968) nor pooling of blood with subsequent diminished circulating-blood volume. Dogs are particularly prone to splanchnic pooling (Halley, Reemtsma & Creech, 1959; Gianelli *et al.*, 1960; Tanaka *et al.*,

1962), owing possibly to activity of the sympathetic nervous system (Brooksby & Donald, 1971), and the puppies in these experiments were no exception. The redistribution of blood to various organ systems during total bypass indicates increased flow to the mid-brain, brain stem and splanchnic area and reduced flow to the heart, cerebral hemispheres and muscles (Lees *et al.*, 1971). The reduced 'effective' blood volume and hypotension that followed sequestration at times required additional donation of blood to restore normal blood pressure. A more consistent finding was that of metabolic acidosis, and this manifestation of reduced effective circulation and tissue hypo-perfusion frequently required buffering with bicarbonate or TRIS (THAM). It is our suggestion that the acidosis may be greatly reduced by using venovenous bypass, thus preventing the localized ischaemic acidosis caused by either arterial drainage or the reduced cardiac output of venous drainage.

On the other hand this method of control ensured that hypervolaemia was not the cause of the hepatic and pulmonary congestion seen in the puppies that died. The hepatic findings may be explained by the sequestration phenomenon, and the lung findings by the changes in cardio-pulmonary dynamics that occur during bypass (Yong *et al.*, 1965). Perfusion of an immobile or hypoventilated lung inevitably causes congestion (Dodrill, 1958; Cartwright *et al.*, 1962), and transudation of fibrinous fluid into the alveoli may even lead to the development of classical hyaline membranes—the very lesion the effects of which the system may be used to alleviate. It has been suggested that the pulmonary lesions are in part due to loss of surfactant material during bypass (Gardner *et al.*, 1962), though our limited studies of pressure–volume curves of puppy lungs after perfusion failed to support this hypothesis. The lung lesions resemble those found in animals suffering from hypo-volaemic shock (Schramel *et al.*, 1963) with intracapillary red-cell and/or platelet aggregation. Congestion of this kind can be greatly reduced, if not eliminated, by haemodilution and by avoiding the vaso-constrictive effects of acidosis.

The pulmonary haemorrhage seen occasionally could either have been due to the more severe form of congestion induced by changes in pulmonary dynamics, or to frank haemorrhagic disease, which is caused by loss of platelets and other haemostatic factors known to occur during bypass and which is discussed in more detail later. Though the entity of disseminated intravascular coagulation was not well recognized at the time of the puppy studies, it is not unlikely that this at times played a part in causing death. Another 'popular' diagnosis, and again one more common in dogs than human subjects, was endotoxic shock, though the precise mechanism of this is still not well understood.

It soon became clear that perfusion via peripheral-vessel cannulation yielding flows of 15–35 ml/kg min provided sufficient oxygen to maintain cardiovascular function despite total respiratory arrest. Changes in the EEG unfortunately do not correlate closely with haemodynamic or metabolic disturbance during perfusion (Hopf, Galletti & Goldstein, 1961), and the voltage changes that occurred with changing levels of sedation made interpretation of the tracings difficult. The EEG and the progressive acidosis did, however, indicate that the peripheral oxygenation was inadequate, and that either increased flow from central vessel cannulation or some additional form of ventilation would be required if the puppy was to be totally dependent on artificial respiratory support for long periods of time.

Perfusion of infants

Techniques
The first newborn infant to receive a period of partial perfusion was one with total respiratory arrest caused by severe and prolonged hypoxia sustained during an arrested delivery. At that time the disc oxygenator was still in use and the system was similar to that used in the puppy experiments. The infant was maintained for 6 h but never achieved spontaneous respiration. Two infants were dialysed with a single-coil Kolff artificial kidney unit in place of the oxygenator, and linked via a common reservoir to the perfusion control system. Details of the technique and results of one of these have been published (Walker *et al.*, 1963b). The experience of these three initial cases confirmed the suitability of the control system for use in the small patient and gave impetus to the development of the combined Klung system already described.

To date five infants have been treated with the Klung system, one receiving two periods of perfusion. A summary of the clinical data of these infants is presented in Table 7.4.

The assembly of the Klung and infant circuits has already been described. The various routes of flow used include umbilical artery and vein, femoral artery and vein and combinations of these. In addition to the various blood studies required during preparation for perfusion, the following parameters were carefully monitored during the procedure (Plate 7.3):

(1) *Temperature.* A thermistor probe taped to the thorax over the

Table 7.4. *Clinical data of infants perfused*

	Age (d)	Weight (g)	Diagnosis	Duration of perfusion (min)	Route	Maximum flow rate (ml/min)	Outcome	Comments
1. S.G. A.	4	820	Severe prematurity with apnoea and atelectasis	12	AV (both umbilical)	15	Survived Improved	Discontinued for technical reasons
B.	17	940		35	AV (umbilical to femoral)	15	Survived Unchanged Died at 19 d	Successful run
2. S.M.	49	2000	Large occipital cervical meningo-coele. Cerebellar dysgenesia	30	AV (femoral)	20	Died during perfusion, having been almost mori-bund on starting	Successful run Autopsy—old right atrial thrombosis caused by Pudenz CSF valve. Long standing pneumonia
3. J.B.	4	3000	Anencephalic	180	AV then VA Femoral artery Umbilical vein	50	Survived for 7 d	Successful run
4. D.C.	21	1380	Severe prematurity Severe respiratory distress. Heart failure. Probable patent ductus	90	AV (femoral)	20	Survived Little change	Successful run Probable increase in pulmonary oedema and in liver size
5. P.T.	10	2060	Inoperable tracheo-oesophageal fistula, with pulmonary complications	52	AV (femoral-umbilical)	25	Cardiac arrest during pro-cedure Resuscitated but died 4 h later	Autopsy revealed pulmonary haemorrhage suspected pre-perfusion

sternum activates a servo-control circuit for the heater in the special incubator power pack (Air-Shields U.K.). Theoretically, the thermostatic control should be set at neutral temperature for pre-term infants, i.e. skin temperature of 36·0°C–37·0°C (Silverman & Sinclair, 1966), and thus the additional oxygen requirements of infants who become hypothermic may be avoided. In practice, the setting must be considerably higher (40°C or more) for reasons discussed below.

(2) *Pulse, blood pressure and electrocardiogram.* A continuous ECG was displayed on an oscilloscope for visual monitoring and was recorded at intervals on a Schwartzer four-channel physiological recorder. The pulse was found to vary remarkably little but ECG-wave-form changes were observed if the infant became unduly hypoxic.

Ideally, the blood pressure should be monitored continuously by means of an indwelling arterial catheter. This is neither easy to achieve nor is it desirable to use a second artery in the newborn. Many surface techniques with electronic sensors have been devised, and an impedence plethismographic method designed for this work (Walker & West, 1972) was used in these studies.

The EEG proved to be of limited value as an index of progress in puppies but, due to problems of rapid and reliable electrode application, it has yet to be evaluated in infants.

(3) *Blood chemistry.* Samples were taken initially at 5-min intervals and later every 10 or 15 min. They were obtained from the sampling sites in the circuit lines entering and leaving the infant and at the two ends of the Klung. The various parameters detailed above were assayed and, if necessary, appropriate corrections were made to the gases or dialysate. An example of the extent of biochemical monitoring of these infants is seen in Table 7.5 and Fig. 7.9, and summary of selected data appears in Table 7.6.

(4) *Blood coagulation.* The infant was initially heparinized intravenously with 100 u/kg body weight and the priming blood was anticoagulated with heparin on donation. As heparin is lost across the dialysis membrane and to a lesser extent by liver metabolism in the infant, it was necessary to add 1000 u hourly to the Klung circuit during perfusion. At the end of the procedure the infant was given a slow injection of protamine sulphate, at the rate of 1 mg/100 u heparin, if bleeding from the wound continued and if the clotting time did not return to a value of less than 20 min.

Table 7.5. *Biochemical data*

Birth weight: 1420 g
Age: 4 weeks

Details	Perfusion time (min)	pH	P_{O_2} (mmHg)	P_{CO_2} (mmHg)	T_{CO_2} (mMol/l)	Base excess (mEq/l)	HCO_3 (mEq/l)	Cl (mEq/l)	PO_4 (mEq/l)
Donor's blood (heparinized)	0	7·34	44	48	26	−1	25	108	1·7
DIALYSATE		7·41	122	48	31	+5	30	108	2·3
Priming blood (equilibrated)	60	7·42	360	38	25	0	24	99	2·0
Preperfusion baby art. blood	0	7·30	58	64	32	+3	30	96	2·3
Blood leaving baby	30	7·31	44	64	33	+4	31	100	2·4
Blood leaving baby	60	7·35	96	52	29	+2	28	100	2·6
entering baby		7·43	305	45	30	+5	29	100	2·3
Blood leaving baby	90	7·30	56	52	26	−2	25	100	2·5
entering baby		7·39	245	45	28	+2	27	99	2·6
Dialysate	90	7·58	255	40	30	+6	29	108	2·2
Baby art. blood 15 hours after perfusion	−	7·34	64	48	26	−1	25	96	
28 hours after perfusion	−	7·25	70	70	32	+2	30		

(5) *Clinical evaluation.* It is not possible to examine the infant during perfusion as any handling will cause an erroneous change in the transducer weight detection and thus an inappropriate change in infant blood-pump speed. Should examination become necessary or should the infant require aspiration of mucus from the airway, or if, for example, adjustment of catheters or monitoring electrodes is required, the weight status is noted and the flow to the infant stopped by activating the solenoid safety valve. Thereafter the system is re-set with allowance for any previous weight deviation, and flow is restarted as before.

Problems

The difficulties encountered in a system as complex as this are plentiful and fall into three groups—clinical, technical and biochemical.

The *clinical problems* have been mentioned first because at this time they present the most difficulty.

The first of these is timing. No great difficulty is encountered in the preparation of the system in time for the elective delivery of a pre-viable fetus deliberately selected for perfusion. This can be done during the day when the clinical and laboratory staff are all available. It is a very different matter, however, when an unexpected pre-term fetus is de-

on a perfused infant

Perfused, using combined kidney and lung
Perfusion route: Arterio-venous (femoral-femoral)
Perfusion flow: 20–25 ml/min

Na (mEq/l)	K (mEq/l)	Ca (mg/100 ml)	Mg (mg/100 ml)	Lactate (mg%)	Urea (mg%)	Glucose (mg%)	Osmolality (mOsm kg)	Protein (g/100ml)	PCV	Cortisol (μg%)
136	3·8	7·1	1·7	32	21	525	318	6·5	39	
138	4·6	5·7	1·6							
135	4·7	9·8	1·9	11·0			288	6·5	43	
131	4·1	9·1	2·1	34	10	195	393	5·9	42	27
135	3·2	8·8	2·1						40	
137	3·6	9·8	2·1	39		130			44	60
138	3·9	9·8	2·1						44	
136	3·6	9·4	2·1	57	8	124	288	6·7	43	107
138	4·1	9·7	2·1	24	5	155	288	6·9	44	100
139	4·6	5·6	1·5			188	289			
124	3·1									

livered, one whose progress may either be totally uneventful or may be such as to require full intensive care, and one who, failing to respond to this, may become a candidate for therapeutic perfusion at short notice. The deterioration may be so rapid that valuable time is lost before perfusion can be started, and these events so often happen at night or over weekends that only a full 24-h cover by the perfusion team will provide an adequate service. The team has to be specially trained and cannot, with the possible exception of the clinical chemistry technician, be routinely employed elsewhere. These factors impose considerable restrictions on this as a therapeutic procedure, and indeed it has been difficult, even under research conditions, for the team to be available when an infant presents for perfusion.

To this is added the further difficulty of the choice of infant for perfusion. Until the system has been proven to be of value one is obliged to treat infants for whom there is virtually no chance of survival, and special permission must be obtained from the parents of such infants. This has, quite naturally, not always been given and the accumulation of clinical experience has therefore of necessity been extremely slow.

The second major clinical problem is flow rate. There is always reluctance to cannulate major vessels in the neck, though it is only in this way that flows providing enough oxygen for total body metabolism may be

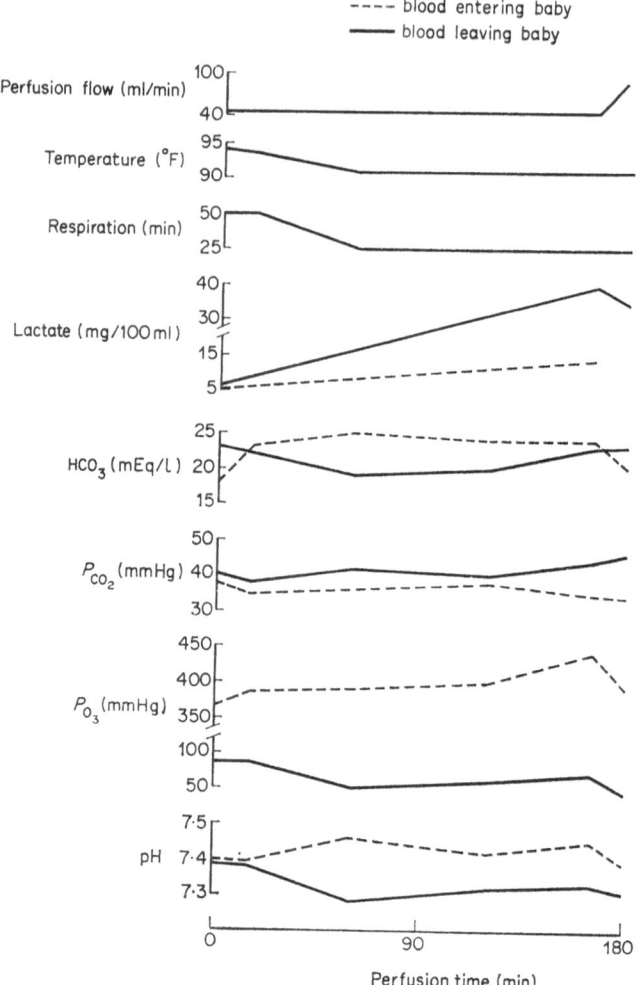

Fig. 7.9. The parameters studied in a 3020-g anencephalic infant with severe hypoventilation (unassisted) illustrate the problem of hypoxia and lactic acidosis poorly controlled by low extracorporeal blood flow. The first five samples were taken from the femoral artery and the last from the umbilical vein, following reversal of the direction of flow.

obtained. Over 35 ml/kg min can be obtained from the umbilical artery of a newborn, but this represents only about 25% of the cardiac output and femoral-vessel flow provides considerably less. Various combinations of umbilical and femoral vessels were tried in order to determine which would provide the greatest blood flow, and the biochemical findings were studied in retrospect to see which route produced least lactic

Table 7.6. Biochemical data on infant perfusion

Infants perfused	Perfusion time (min)	Perfusion flow (ml/min)	Source of blood	Blood to or from baby	pH	P_{O_2} (mmHg)	P_{CO_2} (mmHg)	HCO_3 (mEq/l)	Blood lactate (mg %)	Resp. rate /min	Heart rate /min	Atm. O_2	Babies temp. (°F)
J.B. 3020 g (Anencephalic 4 days old)	0	–	F. artery	Pre-perfusion	7·39	86	40	23	5·6	50	120	20 l/min	94
AV (F-U)	165	40–50	F. artery	Bl. leaving	7·33	68	44	23	39	25	120	air	91
			Extracorp	Bl. entering	7·45	390	38	24	13	–	–	0·5 l/min	91
VA (U-F)	180	95	U. vein	Bl. leaving	7·31	44	46	23	35	25	120	20 l/min	
S.G. 960 g (Premature 3 days old)	0	–	U. artery	Pre-perfusion	7·23	80	47	19	28	44	150	R=100% O_2	96
AV (U-U)	12	20–30	U. artery	Bl. leaving	7·28	84	44	20	39	44	150	R=100% O_2	93·4
			Extracorp	Bl. entering	7·62	425	20	21	18	–	–	0·5 l/min	–
(16 days old) 2nd perfusion	0	–	U. artery	Pre-perfusion	7·32	86	40	20	13	48	180	R=100% O_2	97
AV (U-F)	30	20–30	U. artery	Bl. leaving	7·42	94	38	22	34	54	180	R=100% O_2	97·8
			Extracorp	Bl. entering	7·46	344	34	24	19	–	–	0·5 l/min	–
S.M. 3900 g (o.c. meningocele 6 weeks old)	0	–	U. artery	Pre-perfusion	7·33	60	60	31	10	100	180	20 l/min	100·5
	30	20–30	U. artery	Bl. leaving	7·33	64	58	30	32	100	180	20 l/min	99·3
AV (U-U)			Extracorp	Bl. entering	7·4	420	42	25	12	–	–	0·5 l/min	–
D.C. 1420 g (Premature, Heart failure Pulm. fibrosis) 4 weeks old	0	–	F. artery	Pre-perfusion	7·30	58	64	30	34	52	166	R=100% O_2	96
AV (F-F)	90	20–25	F. artery	Bl. leaving	7·30	56	52	25	57	54	180	R=100% O_2	95·4
			Extracorp	B. entering	7·39	245	45	27	24	–	–	0·5 l/min	–
P.T. 2060 g (irreparable T.O.F., 4 days old)	0	–	F. artery	Pre-perfusion	7·42	58	50	26	11·0	45	120	20 l/min	95
AV (F-U)	55	25–30	F. artery	Bl. leaving	7·39	86	52	31	26	28	180	20 l/min	95
			Extracorp	Bl. entering	7·47	385	37	26	9·4	–	–	0·5 l/min	–

F = Femoral U = Umbilical Extracorp = Extracorporeal R = On respirator AV = arteriovenous (perfusion route)
Bl = Blood

acidosis. Flow was insufficient to sustain a normal biochemical state by any route if the infant's own respiratory effort was poor (Fig. 7.9). This may in part be related to the sequestration phenomenon already mentioned, though fortunately this occurs to a lesser extent in man than in animals (Litwak et al., 1963).

As an alternative to neck-vessel cannulation (White et al., 1971), this problem could perhaps be answered by the cannulation of both umbilical arteries but this has yet to be tried. The success of venovenous perfusion in raising arterial oxygen saturation was described by Krasna et al. as early as 1962, and it is interesting that, in an attempt to avoid peripheral ischaemic acidosis, workers are continuing to favour this route (Boyd et al., 1972). The simultaneous artificial ventilation with a respirator has been used as a means of augmenting oxygenation. Although the ventilator tubes are cumbersome and add to the problem of setting up stable conditions for weight detection, this method has proved successful and it has the added advantage of maintaining adequate lung movement, the importance of which has already been recorded (Cartwright et al., 1962).

There is a considerable volume of literature describing systems that rely, not on weight, but on an extracorporeal volume control for determining pump speed, and some systems do not employ any pumps. Fortunately the sequestration syndrome seen so commonly in perfused dogs, is much less common in human subjects, though its detection necessitates monitoring blood pressure as well as blood volume which, in our system, is mirrored by body weight. We feel, however, that extracorporeal volume monitoring is often insufficiently accurate for use in very small infants, e.g. around 1000 g, whose total blood volume is only about 90 ml and in whom an error of say 10 ml (over 10% of the blood volume) would be physiologically unacceptable.

Those who use pumpless systems use only an oxygenator and do not incorporate kidney dialysis which we now feel offers considerable advantages. Moreover, they rely on arterial pressure to activate flow through the artificial lung (Rashkind, 1965; Kolobow, 1968), and it has not yet been shown that the myocardium of the very ill premature infant can sustain this load. Besides, we have, as has already been stressed, come to question the suitability of arterial drainage, because of the localized peripheral ischaemia and subsequent serious lactic acidosis for which the infant's metabolism cannot compensate. Thus it would appear that, except when a respirator is used simultaneously, the cannulation of two umbilical arteries or major neck vessels is obligatory.

Limited haematological studies were performed on four of the per-fused infants (Table 7.7). Post-perfusion haemorrhage has occurred, though only once was this prolonged and apparently unresponsive to protamine sulphate. The changes in blood-coagulation factors appear to be the same for partial perfusion as they are for total bypass. There is an initial loss in platelets (Brinsfield *et al.*, 1962; Wisselink *et al.*, 1970) and white blood cells, but the former stabilize at a low level and the latter are

Table 7.7. *Haematological data on infant perfusion.*

Infants perfused	Blood sample	Hb %	PCV	RBC million/mm^3	WBC/ mm^3
S.G.					
First perfusion	after perfusion	90	38	4·72	7400
Second perfusion	before perfusion	85	41	3·92	4200
	after perfusion	100	38	3·59	4700
S.M.	before perfusion	60	27	2·87	10,100
	after perfusion	88	39	4·33	7800
D.C.	before perfusion	102	45	4·82	6600
	after perfusion	101	43	4·53	8000
P.T.	before perfusion	91	37	3·62	19,800
	after perfusion	101	41	4·76	4700

soon replenished by myeloid tissue stimulation (Brinsfield *et al.*, 1962). Recent studies have failed to establish a relationship between serotonin and the endothelium-stabilizing effects of platelets (Wisselink *et al.*, 1970). The prothrombin, proaccelerin and thromboplastin generation deficiencies are usually restored on neutralization of heparin (Philips, Malm & Deterling, 1963), though the severe post-perfusion haemorrhage sometimes seen may be attributable to fibrinogenopenia and increased fibrinolytic activity (de Vries *et al.*, 1961; Philips *et al.*, 1963). Some suc-cess in limiting the loss of platelets by coating the extracorporeal circuit with albumen prior to perfusion has recently been reported by Depp & Hughes (1971). The hypotensive dangers of protamine have been well documented (Gourin *et al.*, 1971), and infants are not given this unless bleeding from the cutdown wound is excessive. The blood-platelet loss that occurs during perfusion with this system is detectable but not excessive, but we have not yet had the opportunity to study the effects on factors related to disseminated intravascular coagulation.

The maintenance of temperature may be regarded as both a clinical and a technical problem. Despite all efforts to prevent hypothermia, exposure of the very small, unwell infant during cannulation may lead to mild or moderate loss of heat, and the dangers of 'cold injury' in infants are well documented. Once within the incubator and on the suspension tray, the servo-control system can take over and the environment then will assist in prevention of heat loss. However, the blood lines to and from the Klung are long, and while the Klung and reservoir are thermostatically controlled, a considerable amount of heat may be lost by blood in transit. It is desirable to prevent changes greater than 1–2° as these are said to encourage denaturation of protein. To this end, heating coils are wound round the tubing at places of greatest loss. The lines are also covered with aluminium foil wherever possible and thermocouples are attached at various places to check the temperature fluctuations.

Finally, the clinician must accept the small risk of donor-blood-borne infection. The risks of hepatitis are well known and to these must now be added the recently confirmed and small additional hazard of cytomegalovirus infection (Lang *et al.*, 1968; Kantor & Johnson, 1970). It may be that some of the other non-specific post-perfusion syndromes (Reyman, 1966; Riemenschneider & Moss, 1966) are also viral in origin. However, when these are weighed against the advantages of the Klung, the use of priming blood would seem fully justified.

In regard to purely *technical problems*, reference has already been made to the error in weight detection imposed by infant movement. The careful attachment of blood lines, monitoring cables and respirator tubes to the suspension tray is essential, but in practice these cause surprisingly little difficulty. However, much electronic circuitry is required, and to avoid interference and distortion of the physiological monitoring signals a new multipurpose monitoring console has been built (Plate 7.3). A search for faulty electrode application or inadequate grounding usually reveals the problem, but this may delay the start of perfusion, and to obviate such episodes constant maintenance of such equipment is obligatory. Further consideration will also have to be given to the recent innovation of using Dacron wool (Hill *et al.*, 1970; Ashmore *et al.*, 1972) or polypropylene (Patterson & Twitchell, 1971) filter material, as these have been shown to reduce the risk of post-perfusion fat embolism (Miller *et al.*, 1962).

The *biochemical problems* have been left till last as they, in fact, present the least difficulty. Once the system has been primed and equilibrated, the dual lung and kidney function of the Klung maintains the blood in excellent biochemical condition (Fig. 7.10). Despite all the above precautions, however, some denaturation of protein probably still occurs, and it will not be without much more study that the cause and remedy of this will be found.

In addition to a few estimations of plasma haemoglobin, the tendency for the plasma potassium to fall (rather than rise as it would in the presence of significant red-cell destruction) provided indirect evidence that there was negligible haemolysis during perfusion. Any haemolysis

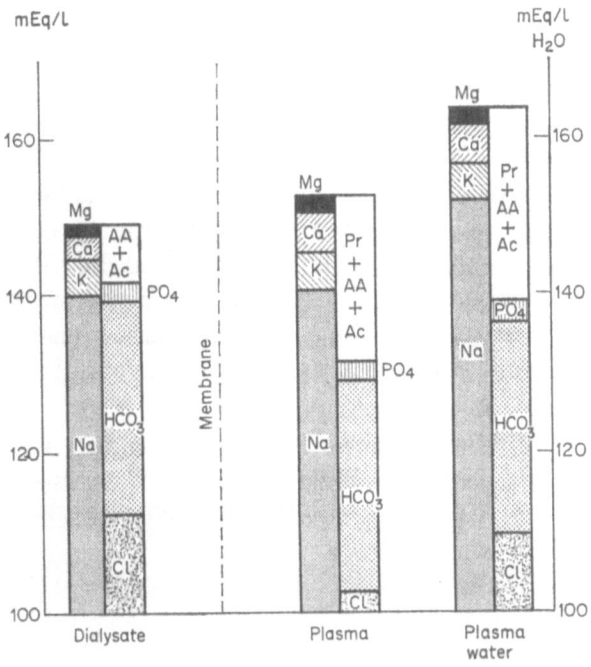

Fig. 7.10. Equilibration data (at normal pH) showing effect of dialysis on various constituents of plasma and plasma water. AA =amino acids. Ac =acetate. Pr = protein changes.

that does occur would seem to be well within the tolerance limits of the infant kidney, and in the event of unexpectedly severe haemolysis the infant is protected from hyperkalaemia by the dialysing function of the Klung.

Apart from the influence of the plasma environment on the survival of the red cell, the haemolysis rate is related to the mechanical stress to which the cell is exposed. After considerable debate in the early years of perfusion work, most authorities agreed that with the 'De Bakey'-type roller pump, which incorporates a stator, complete occlusion was less traumatic to red blood cells than partial occlusion. Our experience supported the contention that if a roller acts against a stator the pump should probably be totally occlusive, but recent evidence suggests that this need not be so, especially when the roller acts purely on the chamber and does not compress it against a hard surface. Both a non-stator, non-occlusive pump of the Holter variety, and a stator 'Sarns' pump with precise non-occlusion of $0.001-0.0015$ inch in a 0.6-in internal diameter polyurethane pump chamber have proved satisfactory (Kolobow *et al.*, 1970).

Why the red cell survives a limited amount of 'leak-back' in any pump is not clear, because our studies indicate that the cell is most vulnerable when exposed to negative pressure. Pressures up to 30 lbf/in^2 were deliberately exerted upon red cells in a perfusion line, yet no appreciable rupture occurred. This was presumably because, by the time the cells reached the end of a long high-resistance small-bore catheter, the pressure was dissipated and the cell emerged gently from the end. When the same cells were subjected to brief vigorous suction imposed by a sampling syringe, haemolysis of considerable degree occurred. It is therefore necessary in all perfusion systems to ensure that there are no sites in which sudden release of pressure or areas of turbulent vortices occur.

At the present time, however, we would suggest that, if perfusion is required for long periods, it would seem safer to reprime the system with fresh blood every 6 h, perhaps with protein electrophoresis or plasma haemoglobin as a guide.

Discussion

The degree of complexity of the system we have described, when compared to some infant systems and to those required for adult extracorporeal perfusion, must appear unnecessary to some people, but we believe that it is yet not enough. The immature ill newborn infant has little capacity to maintain homeostasis at the outset, let alone to compensate for any metabolic imbalance imposed by the perfusate. It will have already become evident that many more constituents of blood require study, not the least of these being the pulmonary vasoactive agents (Swan & Meagher, 1971), and that the task is formidable indeed. How far one must go in biochemical control is not yet known, but we suggest further work along these lines will improve the quality of the perfusate and possibly lessen the complication rate in both infants and adults, and add to the reliability of experimental studies of the fetus. Some of the topics worthy of study that immediately come to mind include the value of pressurized oxygen, the control of citrates and sulphates, the estimation of changes during perfusion of fibrinogen-degradation products, triglycerides, phospholipids and free fatty acids, and further estimation of protein denaturation in relation to the more adequate control of these components of plasma.

The rise in plasma cortisol and related fall in potassium noted in these experiments suggests that the infant is responding to a stressful influence, and the effect of this on other steroids has not yet been assessed. It must by now be abundantly evident that we place a great deal of importance on the adequate preparation of any blood perfusate for infants and that, if this is not done, the results of studies that claim to reveal the 'normal' metabolic activity of an infant or fetus cannot be accepted without reservation.

In the last analysis we are obliged to return to the objectives stated in the introductory remarks—where does all this effort and expensive research take us? Is it really necessary?

As a means of maintaining life in the pre-term newborn infant this technique, and others like it, may have limited use because of difficulties such as timing and patient selection already discussed. These, however, are not insurmountable, and if an infant is going to survive it is the obligation of the clinician to give him or her every possible chance of becoming a normal child and adult. What we need, therefore, are better means of determining the potential for survival—and then we shall know which infant to treat, and avoid the accusation of being meddlesome.

The system has also been used to perfuse dog cadavers with hypothermic blood prior to donation of organs for transplant, and it has the degree of mechanical and biochemical control suited to the maintenance of viability of single isolated organs prior to transplant.

As a method of studying the changes that occur during the extracorporeal circulation of blood under closely controlled conditions, this system offers more than most. Combined kidney and lung control of blood has not been studied adequately, and many of the lessons learned so far are applicable and can be of value to those treating adults requiring circulatory support. Examples recorded quite recently of difficulties still facing those using oxygenators alone include problems of magnesium control and its relevance to post-perfusion cardiac dysrhythmia (Sheinman *et al.*, 1971) and of sodium and water imbalance (Cohn, Angell & Shumway, 1971).

The system reported here was designed for the pre-term infant who had established respiration and is thus essentially different from those 'artificial placenta' systems that allow for total immersion of the fetus while being perfused. Nevertheless, it is perfectly possible to study the metabolism of infants during perfusion as the blood taken for repeated sampling is automatically and immediately returned to the infant by the servo-control, eliminating all blood-loss risk to the infant. The ethics of the use of any injected material, e.g. isotopes, must be judged individually on the merits of the test and its possible value to the infant—always remembering that infants so treated will only provide data that is essentially non-physiological, because they started with the failing metabolic state which necessitated the use of artificial support.

Acknowledgements

We wish to express our thanks to the medical, nursing and technical staffs of the University of Colorado Medical Center, Denver, Colorado, and the University and Royal Infirmary, Dundee, Scotland, too numerous to mention individually, who have assisted in this project. We are particularly grateful to Dr. C. Henry Kempe, Dr. Julie Wershing and Mr. Sanford L. Simons of Denver; to Professor P. D. Griffiths (Clinical Chemistry), Professor E. G. Cullwick (Electrical Engineering), Mr. L. A. MacKenzie (Regional Physics), Dr. C. Cameron (Blood Transfusion Service) and their members of staff in Dundee for invaluable help; to Mrs. M. Smernicki, Mr. A. R. Whytock for more recent technical assistance; and to Mrs. E. Forsyth and Mrs. J. Duncan for secretarial help. We also thank Messrs. J. P. Bemberg A.G., West Germany, for donating the Cuprophane membranes, and Air-Shields U.K. for contributing certain items of equipment to this project.

This was work supported by Grants from the University of Colorado, Colorado Heart Association, National Institutes of Health (Grants H 5481 and HE 05481-04), Scottish Hospitals Endowment Research Trust (Grant 206) and the Board of Management, Dundee General Hospitals.

References

ALEXANDER, D. P., BRITTON, H. G. & NIXON, D. A. (1966) Maintenance of the isolated foetus. *British Medical Bulletin*, **22**, 9–12.

ASHMORE, P. G., SWANK, R. L., AMBROSE, P. & PRICHARD, K. H. (1972) Effect of Dacron wool filtration on the microembolic phenomenon in extracorporeal circulation. *The Journal of Thoracic and Cardiovascular Surgery*, **63**, 240–248.

BRINSFIELD, D. E., HOPF, M. A., GEERING, R. B. & GALLETTI, P. M. (1962) Hematological changes in long-term perfusion. *Journal of Applied Physiology*, **17**, 531–534.

BOYD, J. C., MORAN, J. F. & CLARK, R. E. (1972) An analysis of the operating characteristics of the 0·25 M^2 Travenol infant membrane oxygenator. *Surgery*, **71**, 262–269.

BROOKSBY, G. A. & DONALD, D. E. (1971) Dynamic changes in splanchnic blood flow and blood volume in dogs during activation of sympathetic nerves. *Circulation Research*, **29**, 227–238.

CALLAGHAN, J. C. & ANGELES, J. D. (1961) Long-term extra corporeal circulation in the development of an artificial placenta for respiratory distress of the new born. *Surgical Forum*, **12**, 215–217.

CARTWRIGHT, R. S., LIM, T. P. K., LUFT, U. C. & PALICH, W. E. (1962) Pathophysiological changes in the lungs during extracorporeal circulation. *Circulation Research*, **10**, 131–141.

CHAMBERLAIN, G. (1968) An artificial placenta. *American Journal of Obstetrics & Gynaecology*, **100**, 615–626.

CLOWES, G. H. A., Jr., HOPKINS, A. L. & NEVILLE, W. E. (1956) An artificial lung dependent upon diffusion of oxygen and carbon dioxide through plastic membranes. *Journal of Thoracic Surgery*, **32**, 630–637.

COHN, L. H., ANGELL, W. W. & SHUMWAY, N. E. (1971) Body fluid shifts after cardiopulmonary bypass. *The Journal of Thoracic & Cardiovascular Surgery*, **62**, 423–430.

DANESH, B. J. N. Z., WALKER, C. H. M. & MATHERS, N. P. (1968) Predicting serum-calcium after haemodialysis. *The Lancet*, **2**, 433–435.

DANESH, B. J. N. Z., WALKER, C. H. M. & MATHERS, N. P. (1970) The relation of post-dialysis plasma calcium and magnesium to the dialysate levels and to changes in blood pH. *New England Journal of Medicine*, **282**, 771–775.

DANESH, B. J. N. Z. (1972) *Thesis for Doctor of Philosophy*. University of Dundee.

DEPP, D. A. & HUGHES, R. K. (1971) Venovenous perfusion with a membrane oxygenator. *The Journal of Thoracic & Cardiovascular Surgery*, **62**, 658–662.

DODRILL, F. D. (1958) The effects of total body perfusion upon the lungs. In *Extracorporeal Circulation*, ed. Allen, J. G. Springfield: Charles C. Thomas.

DORSON, W., BAKER, E., COHEN, M. L., MAYER, B., MOLTHAN, M., TRUMP, D. & ELGAS, R. (1969a) A perfusion system for infants. *Transactions. American Society for Artificial Internal Organs*, **15**, 155–160.

DORSON, W. J., BAKER, E., HULL, H., MOLTHAN, M., MEYER, B., FARGOTSTEIN, R. & COHEN, M. (1969b) A long-term partial bypass oxygenation system. *The Annals of Thoracic Surgery*, **8**, 297–311.

GALLETTI, P. M. & BRECHER, G. A. (1962) *Heart-Lung Bypass, Principles and Techniques of Extracorporeal Circulation*. New York: Grune and Stratton.

GALLETTI, P. M., MARTINEZ, F. J., BRINSFIELD, D. E. & PEIRCE, E. C. II. (1963) A pediatric perfusion system. *Transactions. American Society for Artificial Internal Organs*, **9**, 244–250.

GALLETTI, P. M., SADLER, J. H., BARBOUR, B. & ORRELL, L. (1968) The miniklung and the minikiil. Laboratory and clinical evaluation of two low prime, high transfer rate haemodializers. In *Proceedings of the European Dialysis and Transplant Association*. Vth Conference, Dublin, June 1968. Excerpta Medica International Congress Series No. 179.

GARDNER, R. E., FINLEY, T. N. & TOOLEY, W. H. (1962) The effect of cardiopulmonary bypass on surface activity of lung extracts. *Bulletin de la Société Internationale de Chirurgie*, **21**, 542–551.

GIANNELLI, S., MAHAJAN, D. R., NAVARRE, J. R. & PRATT, G. H. (1960) Studies of blood volume changes during cardiopulmonary bypass in dogs. *Annals of Surgery*, **152**, 190–196.

GOURIN, A., STREISAND, R. L., GREINEDER, J. K. & STUCKEY, J. H. (1971) Protamine sulfate administration and the cardiovascular system. *The Journal of Thoracic & Cardiovascular Surgery*, **62**, 193–204.

HALLEY, M. M., REEMTSMA, K. & CREECH, O. (1959) Hemodynamics and metabolism of individual organs during extracorporeal circulation. *Surgery*, **46**, 1128–1134.

HILL, J. D., OSBORN, J. J., SWANK, R. L., AGUILAR, M. J., LANEROLLE, P. & GERBODE, F. (1970) Experience using a new dacron wool filter during extracorporeal circulation. *Archives of Surgery*, **101**, 649–652.

HILL, J. D., O'BRIEN, T. G., MURRAY, J. J., DONTIGNY, L., BRAMSON, M. L., OSBORN, J. J. & GERBODE, F. (1972) Prolonged extracorporeal oxygenation for acute post-traumatic respiratory failure (shock-lung syndrome). *The New England Journal of Medicine*, **286**, 629–634.

HOPF, M. A., GALLETTI, P. M. & GOLDSTEIN, L. (1961) Electroencephalographic observations during long-term partial heart-lung bypass. *Transactions. American Society for Artificial Internal Organs*, **7**, 231–236.

KANTOR, G. L. & JOHNSON, B. L. (1970) Cytomegalovirus infection associated with cardiopulmonary bypass. *Archives of Internal Medicine*, **125**, 488–492.

KOLFF, W. J. & EFFLER, D. B. (1956) Disposable membrane oxygenator (heart-lung machine) and its use in experimental and clinical surgery while heart is arrested with potassium citrate according to Melrose technique. *Transactions. American Society for Artificial Internal Organs*, **2**, 13–17.

KOLOBOW, T., ZAPOL, W. & PIERCE, J. (1969) High survival and minimal damage in lambs exposed to long-term (1 week) veno-venous pumping with a polyurethane chamber roller pump with and without a membrane blood oxygenator. *Transactions. American Society for Artificial Internal Organs*, **15**, 172–177.

KOLOBOW, T., ZAPOL, W. M., SIGMAN, R. L. & PIERCE, J. (1970) Partial cardiopulmonary bypass lasting up to seven days in alert lambs with membrane lung blood oxygenation. *The Journal of Thoracic & Cardiovascular Surgery*, **60**, 781–788.

KRASNA, I. H., STEINFELD, L., KREEL, I. & BARONFSKY, I. D. (1962) Studies in prolonged veno-venous perfusion with oxygenation in hypoxia of respiratory origin. *The Journal of Thoracic & Cardiovascular Surgery*, **43**, 135–138.

LANDÉ, A. J., DOS, S. J., CARLSON, R. G., PERSCHAN, R. A., LANGE, R. P.,

SONSTEGARD, L. J. & LILLIHEI, C. W. (1967) A New Membrane Oxygenator-Dialyzer. *Surgical Clinics of North America*, **47**, 1461–1470.

LANG, D. J., SCOLNICK, E. M. & WILLERSON, J. T. (1968) Association of cytomegalovirus infection with the post-perfusion syndrome. *The New England Journal of Medicine*, **278**, 1147–1150.

LAVER, M. B. & BUCKLEY, M. J. (1972) Extreme hemodilution in the surgical patient. In *Hemodilution, Theoretical Basis and Clinical Application*, ed. Messmer, K. & Schmid-Schonbein, H. Basel: Karger.

LAWN, L. & McCANCE, R. A. (1964) Artificial Placentae. *Acta Paediatrica*, **53**, 317–325.

LEES, M. H., HERR, R. H., HILL, J. D., MORGAN, C. L., OCHSNER, A. J., THOMAS, C. & VAN FLEET, D. L. (1971) Distribution of systemic blood flow of the rhesus monkey during cardiopulmonary bypass. *The Journal of Thoracic & Cardiovascular Surgery*, **61**, 570–586.

LEFEMINE, A. A., FOSBERG, A. M. & HARKEN, D. E. (1968) Prolonged partial extracorporeal perfusion. *American Heart Journal*, **75**, 531–536.

LEFEMINE, A. A. & HARKEN, D. E. (1971) Extracorporeal support of the circulation by means of venoarterial bypass with an oxygenator. *The Journal of Thoracic & Cardiovascular Surgery*, **62**, 769–780.

LITWAK, R. S., SLONIM, R., WISOFF, B. G. & GADBOYS, H. L. (1963) Homologous-blood syndrome during extracorporeal circulation in man. II. Phenomena of sequestration and desequestration. *The New England Journal of Medicine*, **268**, 1377–1382.

MILLER, J. A., FONKALSTRUD, E. W., LATTA, H. L. & MALONEY, J. V. (1962) Fat embolism associated with extracorporeal circulation and blood transfusion. *Surgery*, **51**, 448–451.

PATTERSON, R. H. & TWICHELL, J. B. (1971) Disposable filter for microemboli. *The Journal of the American Medical Association*, **215**, 76–80.

PEIRCE, E. C. II. (1960) A modification of the Clowes membrane lung. *The Journal of Thoracic & Cardiovascular Surgery*, **39**, 438–448.

PHILLIPS, L. L., MALM, J. R. & DETERLING, R. A. (1963) Coagulation defects following extracorporeal circulation. *Annals of Surgery*, **157**, 317–326.

RASHKIND, W. J., FREEMAN, A., KLEIN, D. & TOFT, R. W. (1965) Evaluation of a disposable plastic, low volume, pumpless oxygenator as a lung substitute. *The Journal of Pediatrics*, **66**, 49–102.

REYMAN, T. A. (1966) Postperfusion syndrome. *American Heart Journal*, **72**, 116–123.

RIEMENSCHNEIDER, T. A. & MOSS, A. J. (1966) Postperfusion syndrome. *The Journal of Pediatrics*, **69**, 546–552.

ROWLANDS, D. T. & WALKER, C. H. M. (1964) Effects of haemo-dilution on pathological changes in small dogs following partial perfusion. *British Journal of Experimental Pathology*, **45**, 450–457.

SCHEINMAN, M. M., SULLIVAN, R. W., HUTCHINSON, J. C. & HYATT, K. H. (1971) Clinical significance of changes in serum magnesium in patients undergoing cardiopulmonary bypass. *The Journal of Thoracic & Cardiovascular Surgery*, **61**, 135–140.

SCHRAMEL, R., SCHMIDT, F., DAVIS, F., PALMISANO, D. & CREECH, O.

(1963) Pulmonary lesions produced by prolonged partial perfusion. *Surgery,* **54,** 224–231.

SILVERMAN, W. A. & SINCLAIR, J. C. (1966) Temperature regulation in the newborn infant (concluded). *The New England Journal of Medicine,* **274,** 146–148.

SOMEREN, T., GEERING, R. B., KERN, G., MARTINEZ, F. J. & GALLETTI, P. M. (1963) A simple, no-prime, pumpless artificial kidney: A preliminary report. *Transactions. American Society for Artificial Internal Organs,* **9,** 73–89.

SWAN, H. & MEAGHER, D. M. (1971) Total body bypass in miniature pigs. *The Journal of Thoracic & Cardiovascular Surgery,* **61,** 956–967.

TANAKA, T., HAYASHI, H., CHIBA, C., OSAWA, M., SHIMIZU, T., IWA-MATO, J., HASHIMOTO, A., MATSUMURA, G. & SAKAKIBARA, S. (1962) Extracorporeal circulation in dogs. *The Journal of the American Medical Association,* **179,** 708–716.

TRINKLE, J. K., HELTON, N. E., BRYANT, L. R. & GRIFFEN, W. O. (1970) Pulsatile cardiopulmonary bypass: Clinical evaluation. *Surgery,* **68,** 1074–1077.

TURINA, M., HOUSEMAN, L. B., INTAGLIETTA, M., SCHAUBLE, J. & BRAUNWALD, N. S. (1972) An automatic cardiopulmonary bypass unit for use in infants. *The Journal of Thoracic & Cardiovascular Surgery,* **63,** 263–268.

VERVLOET, A. F. C., EDWARDS, M. J. & EDWARDS, M. L. (1970) Minimal apparent blood damage in Lande–Edwards membrane oxygenator at physiologic gas tensions. *The Journal of Thoracic & Cardiovascular Surgery,* **60,** 774–780.

DE VRIES, S. I., VAN CREVELD, S., GROEN, P., MULLER, E. & WETTER-MARK, M. (1961) Studies on the coagulation of the blood in patients treated with extracorporeal circulation. *Thrombosis et Diathesis Haemorrhagica,* **5,** 426–446.

WALKER, C. H. M., BLACKWELL, L. & MASSENGALE, O. (1961) A new electronic blood level indicator. *Journal of Applied Physiology,* **16,** 925–927.

WALKER, C. H. M., ROSE, R. L. & SIMONS, S. L. (1962) Closed partial perfusion in puppies. *American Journal of Diseases in Children,* **104,** 296–301.

WALKER, C. H. M., WERSHING, J. M., SIMONS, S. L., HOLMES, J. H., SITPRIJA, V. & O'BRIEN, D. (1963a) Hemodialysis in infantile nephrotic syndrome. *American Journal of Diseases of Children,* **106,** 479–483.

WALKER, C. H. M., WERSHING, J. M., SIMONS, S. L., HOLMES, J. H. & SITPRIJA, V. (1963b) A technique for partial perfusion and hemodialysis in the newborn. *Transactions. American Society for Artificial Internal Organs,* **9,** 86–90.

WALKER, C. H. M. & WEST, P. J. (1972) Indirect estimation of systolic and diastolic blood pressure in the newborn. *Pediatrics,* **50,** 387–394.

WESTIN, B., NYBERG, R. & ENHORNING, G. (1958) A technique for perfusion of the previable human fetus. *Acta Paediatrica,* **47,** 339–349.

WHITE, J. J., ANDREWS, H. G., RISEMBERG, H., MAZUR, D. & HALLER, J. A. (1971) Prolonged respiratory support in newborn infants with a membrane oxygenator. *Surgery,* **70,** 288–296.

WILKENS, H., REGELSON, W. & HOFFMEISTER, F. S. (1962) The physiologic importance of pulsatile blood flow. *The New England Journal of Medicine*, **267**, 443–446.

WISSELINK, P., FEOLA, M., ALFREY, C. P., SUZUKI, M., ROSS, J. N. & KENNEDY, J. H. (1970) Prolonged partial cardiopulmonary bypass and the integrity of small blood vessels. *The Journal of Thoracic & Cardiovascular Surgery*, **60**, 789–795.

YONG, N. K., EISEMAN, B., SPENCER, F. C. & ROSSI, N. (1965) Increased pulmonary vascular resistance following prolonged pump oxygenation. *The Journal of Thoracic & Cardiovascular Surgery*, **49**, 580–587.

8 Physiological Principles Involved in the Care of the Pre-term Human Infant

EDMUND HEY

The word mature is derived from the Latin word *maturus*, meaning ripe, and a mature baby may be considered to be one who is ripe for the change from intrauterine life. All such maturity is of course relative, since the tissues and organs of the body progress towards final physiological maturity at very different rates, and maturation continues for many years after birth. An assessment of the functional development of each of the major organ systems of the body is necessary in assessing a baby's ripeness for delivery. The competence of the cardio–respiratory system and the digestive tract, the ability of the kidney and liver to maintain a reasonably constant *milieu interieur* once the placenta ceases to fulfil this role, and the ability of the body to resist infections and maintain body temperature in a hostile and sometimes cold environment—these will be the main topics to be discussed in this chapter, because these are the problems that most frequently confront the Paediatrician faced with the care of a small and immature infant. Much has now been learnt about the maturation of the various endocrine glands, the musculo–skeletal system and the nervous system, but less attention will be paid to these because there is, as yet, less evidence that their immaturity ever jeopardizes life after birth.

It is, of course, next to impossible to set up a totally satisfactory definition of maturity when the various body systems are developing at different and probably quite unrelated rates. Birthweight is perhaps the simplest criterion by which to judge a baby's maturity, and over 50 years ago the famous Finnish Paediatrician Ylppö suggested that babies who weighed less than 2·5 kg should be classified as immature. The American Academy of Pediatrics formally adopted the same definition in 1935, and two years later the International Medical Committee of the League of Nations recommended its general acceptance. A similar definition was

251

suggested by the First World Health Assembly in 1948 and endorsed in 1950 by the World Health Organization Expert Group on Prematurity who wrote: 'for the purpose of this classification an immature infant is a liveborn infant with a birth weight of $5\frac{1}{2}$ lb (2500 grams) or less, or specified as "immature". In some countries however, this criterion will not be applicable. If weight is not specified, a liveborn infant with a period of gestation less than 37 weeks or specified as "premature" may be considered as the equivalent of an immature infant.'

The alternative of defining maturity in terms of gestational age failed to gain general acceptance at this time. This was partly because it was realized that babies might develop at different rates *in utero*, and partly because it was realized that reliable menstrual data might often be unobtainable. The limitation of a definition based entirely on birthweight has, however, become more and more apparent in the last 20 years. Some developing countries were shown to have unusually high 'prematurity' rates, even though the proportion of mothers delivered more than three weeks before term was not excessive and perinatal mortality was low. More recently it has become plain that the problems that face a full-term baby of low birth weight are very different from those that face a baby of the same weight that is born before term; undergrown full-term babies have been variously termed pseudomature, dysmature, small-for-dates and light-for-dates, while they have occasionally been referred to (somewhat derogatively) as 'stunts' or 'runts', and their low birth weight has been ascribed to fetal malnutrition or chronic fetal distress.

A start was made towards rationalizing the increasingly confused terminology in 1961 when the World Health Organization Expert Committee on Maternal and Child Health recommended that babies who weighed 2·5 kg or less at birth should, in future, be termed 'low birthweight' instead of 'premature'. Soon after this the policy of classifying babies by both gestational age and birthweight became widespread (American Academy of Pediatrics Committee on Fetus and Newborn, 1967), and it has now been suggested that all babies born more than 21 days before term should be called 'pre-term', while babies born more than 14 days after term should be called 'post-term' (European Association of Perinatal Medicine, 1970). 'Term' for these purposes is considered to be 280 days (40 weeks) after the first day of the mother's last period (as long as the cycle is reasonably regular and of normal duration), and use of the term 'premature' is deliberately avoided. A further subclassification according to whether the baby is or is not of average weight for its gestational age is also of value, and various sex- and parity-specific charts indicating the range of birthweight to be expected at any given gestational age have recently been published. Perhaps the most comprehensive of these is one based on an analysis of legitimate singleton births in Aberdeen, Scotland, between 1948 and 1964 (Thomson, Billewicz & Hytten, 1968).

Most human infants born five weeks before term survive without handicap, although at this gestation perinatal mortality is ten times what it is for a baby born at term. Even when a baby is born three-quarters of the way through pregnancy there is at least an even chance that the child will survive, given good nursing care. We know less about the consequences of premature delivery in other animal species. The piglet and the Rhesus monkey can survive premature birth if given appropriate care, but the lamb and the rabbit rarely survive if delivered much before term (mortality approaches 100% if pregnancy is interrupted at 0·9 of term). These wide differences make any direct comparison between species difficult if not hazardous, and underline the continuing need for direct study of the pre-term human infant.

In all that follows, therefore, it is the human infant that will be under discussion and the handicap imposed by incomplete development that will be under review. Functional maturity cannot be measured by means of any single yardstick (save perhaps by outcome), and the functional development of each of the major body systems will therefore be reviewed in turn. Special attention will be paid to the physiological handicaps facing babies of less than 31 weeks' gestation for these babies usually weigh only half as much as babies born at term, and have, at the moment, less than a 50% chance of surviving the perinatal period. There is, however, no clear cut dividing line between what can be considered viable and what should be considered pre-viable. 10% of babies born at less than 28 weeks gestation, and 5% of babies weighing less than 1 kg at birth, survived the perinatal period in Britain at the time of the National Perinatal Mortality Survey in 1958, and there is no doubt that the prognosis has improved since then. However, no very dramatic fall in mortality is likely to be achieved until the problems associated with immaturity of the lung have been mastered.

Respiration

Momentous and fundamental changes in cardiorespiratory function occur with great rapidity when the lungs take over the respiratory function of the placenta at birth. The mechanisms involved in this transition have been authoritatively reviewed by Dawes (1968), and a description of these changes is not relevant to the theme of the present chapter, except in so far as delivery before term influences the actual process of transition.

Maturation of lung structure

Our understanding of the structural organization of the developing lung has advanced very materially in the last 15 years, and existing knowledge has been concisely summarized in a book edited by Emery (1969).

Cellular architecture

It has become traditional to identify three major stages in the development of lung structure. The primary lung bud forms on the ventral aspect of the foregut and grows into the pleuro-peritoneal space until, by the time the fetus is 30 mm long, the pattern of adult lobes and fissures has already been fully established. During the first 16 weeks, primitive branching tube-like bronchi lined by pseudostratified columnar epithelium ramify rapidly through the mesenchymal tissues. These epithelial lobes are surrounded by a thin basement membrane, and aggregates of mesenchyme that are probably destined to form peribronchial muscle surround some of the larger tubes. Narrow capillary blood vessels lined by thick endothelium are evenly distributed throughout the mesenchyme but are not seen in apposition to the primitive bronchi. This has been termed the *glandular period* of development because of the tissue's appearance when viewed under the light microscope, but it might perhaps more appropriately be called the period of bronchial development.

Branching of the primitive bronchi continues at a diminished rate after formation of the main bronchial tree is complete at about 16 weeks' gestation, and primitive bronchioles form slowly throughout the remainder of fetal life. Cartilagenous support for the major bronchi develops rapidly, and by the end of the twenty-fourth week the pieces of cartilage have nearly all taken on the shape and distribution they will retain throughout adult life. The newly formed primitive respiratory bronchioles are readily identifiable because they are lined with largely undifferentiated cuboidal epithelium; the bronchioles are still separated from each other by abundant mesenchyme, but capillary vessels are now seen in increasing numbers adjacent to the bronchiolar basement membrane. This period of early bronchiolar development has usually been called the *canalicular period*, but the term is perhaps a little unfortunate since it rather implies that canalization is occurring at the time; this is not so because at no stage are the developing bronchial and bronchiolar ducts solid.

The final stage of structural maturation begins at a relatively early gestational age in man. Differentiation of the primitive epithelium lining the terminal air spaces of the lung results in the formation of a large number of flattened (type I) cells with attenuated cytoplasm, and in the formation of a number of large rounded (type II) cells with conspicuous inclusion bodies, plentiful mitochondria, a well-developed rough endoplasmic reticulum, and numerous microvilli on the luminal surface. Early evidence of this differentiation can often be detected by 20 weeks' gestation, and the change is usually well advanced by 26 weeks; very little cuboidal epithelium can usually be found in babies of more than 34 weeks gestation. The septa become progressively thinner during this period, and an increasing proportion of all the capillaries are now found coursing close to the surface of the terminal air spaces; indeed, in the fully mature and non-expanded lung at term, capillaries distended with red cells often seem to bulge into the lumen. The thin-walled endothelium of the capillary becomes closely applied to the basement membrane of flattened type I epithelial cells, and, as a result, capillary blood is separated from air by no more than a thin barrier two cells thick. As maturation proceeds the epithelial cells come to look less metabolically active and to contain progressively less glycogen, while the type II cells appear to become proportionately less numerous.

Visible connective-tissue septa containing large lymph channels separate the lung of the newborn baby into small segments 1–2 mm in diameter, and the basic pattern of this septation is already present by about 18 weeks' gestation. Collagen constitutes the principal connective-tissue element throughout the lung in early fetal life, but elastic tissue comes to predominate round the terminal air spaces during the last third of pregnancy. These elastic fibres appear to form a lattice through which the developing alveoli grow, and a constraining network that balances the negative pull from the pleural surface. This final stage of fetal development is normally termed the *alveolar* period; nevertheless the alveolar region undergoes so much further growth and segmentation after birth that some workers prefer to describe the terminal air spaces present in a baby born at term as 'clusters of alveolar saccules', rather than true 'alveoli' (Boyden & Thompsett, 1965). Whatever terminology one chooses to adopt, the notable thing is that the lung has developed sufficiently to be a competent organ for gaseous exchange by 26 to 30 weeks' gestation in the human infant.

Lung fluid

The potential airways of the lung contain fluid until the time of birth; we do not know how much fluid the human lung contains, but the lung of the fetal lamb has up to 30 ml per kg body weight, a volume approximately equal to the Functional Residual Capacity for air after birth (that is the volume of air present at the end of quiet expiration), and there is indirect evidence that the human lung contains a similar amount of fluid at birth. It used to be thought that this fluid was derived from the surrounding amniotic fluid, but it has now been shown that the fluid is actively formed and secreted by the fetal lung (Adamson *et al.*, 1969a). Early estimates suggested that this fluid was formed at a remarkably high rate, but the technique of measurement was open to question (Boston *et al.*, 1968) and it now seems that, in the lamb at least, the fluid is formed at a rate of about 2 ml per kg body weight per hour (Normand *et al.*, 1971).

Surfactant

Most body fluids generate a high surface tension and Radford pointed out in 1954 that if such a pressure were to be applied, unmodified, across the internal surface of the alveoli it would cause total lung collapse. Not long after this Pattle showed that there was indeed a substance within the lung with remarkable surface-active properties. Within a period of 10 years it was shown that various phospholipids (of which lecithin is the most important) are responsible for this 'surfactant' activity. It is now certain beyond all reasonable doubt that this material is manufactured in the type II epithelial cells of the terminal air spaces, and then stored as visible osmiophilic inclusion material within the cell's lysosomes prior to its extrusion into the lumen of the airway by a process of evagination. Osmiophilic granules can usually be detected in babies of more than 20 weeks' gestation, and they become more numerous as gestation advances; these granules are, however, largely intracellular in babies of less than 26 weeks' gestation, no active surfactant can be detected by the Wilhelmy balance, and the lung fails to remain stable when inflated with air (Gandy, Jacobson & Gairdner, 1970).

A detailed and systematic study of surfactant synthesis by Gluck and his co-workers has shown that lecithin is synthesized by two separate and independent enzyme systems. Synthesis of dipalmitoyl lecithin by means of the phosphocholine transferase reaction is the major pathway; this normally only matures a few weeks before term, but precocious maturation appears to occur some days after premature delivery. Synthesis of palmitoyl myristol lecithin by means of a methyl transferase reaction is, however, active in man (but not in most other mammals) many weeks before term. Methylation is readily compromised by acidosis, hypoxia and hypothermia, and this may well explain why respiratory distress due to surfactant deficiency is only seen in babies born before term, and is also commoner in babies subjected to perinatal stress (Gluck et al., 1972). Gluck's studies highlight the dangers inherent in relying on an animal 'model' when studying a human disease.

In summary, the lung needs to develop adequate terminal air spaces with a capillary blood supply to its surface and adequate reserves of functional surfactant before it can be considered a competent organ for gaseous exchange.

Maturation of lung function
The lung functions with quite remarkable efficiency from the moment of birth in those babies who have enough surfactant to acquire stable air-filled terminal airways at birth, and most tests of pulmonary function indicate that the lung of the full-term baby behaves very like that of an adult within a few hours of birth if due allowance is made for the great difference in body size (Nelson, 1966).

The thorax is normally compressed as the baby passes through the birth canal (Johnson, 1962), and recoil usually results in some air entering the chest during delivery. Functional residual capacity (FRC) then rises rapidly as the infant takes its first few breaths, and stabilizes at a value equivalent to between 40 and 50% of total lung capacity within an hour or two of birth. Compliance rises rather more slowly during the first day of life as lung fluid is slowly absorbed by the lymphatics (Chu et al., 1964). The amount of air in the functional airways and alveoli of the lung at end expiration (FRC) is frequently less than the total volume of gas within the thorax and respiratory tract (TGV) as measured by the method of Dubois in the first few days of life (Nelson et al., 1963a; Krauss & Auld, 1971), and this must indicate that there are some areas of 'trapped gas' within the thorax at this time.

Despite this, nitrogen enters the respiratory tract rapidly and evenly after a period spent breathing pure oxygen, and this indicates that the various regions of the lung are equally and efficiently ventilated even on the day of birth (Strang & McGrath, 1962). There is, however, some indication that lung function does improve perceptibly in this respect during the first three days of life (Hanson & Shinozaki, 1970). Since the alveolar-arterial gradient for carbon dioxide and nitrogen is small (Nelson *et al.*, 1962; Nourse & Nelson, 1969; Krauss, Soodalter & Auld, 1971), it must be assumed that these virtually unventilated areas of 'trapped gas' are also virtually unperfused; were this not so the overall match between ventilation and perfusion immediately after birth could not be as good as it is. Arterial oxygen tension is, however, low in the first few days of life, and there is a large alveolar-arterial oxygen gradient. The fact that this gradient persists when the baby is given 100% oxygen to breathe indicates that the gradient is almost certainly due to a 'true' veno–arterial shunt (Nelson *et al.*, 1963b; Thibeault, Poblete & Auld, 1967); the absence of any other evidence of mismatch between ventilation and perfusion provides further confirmation for this view. As much as a quarter of the cardiac output may pass through these shunts in the first two days of life, and 10% of the output of a small but healthy pre-term baby may continue to be shunted throughout the whole of the first month of life (Orzalesi *et al.*, 1967). Some of this blood may be shunted through the foramen ovale and ductus arteriosus, and some through small areas of atelectasis within the lung. Because of this, mean arterial P_{O_2} in most healthy babies of short gestation is about 70 ± 14 mm Hg in the first two days of life and about 80 ± 12 mm Hg (mean $\pm$ SD) during the remainder of the first month of life.

The nature and location of the trapped gas so often present in babies of short gestation immediately after birth remains a mystery, particularly as there is usually no radiological evidence of pneumothorax, pneumo-mediastrium or cyst formation within the lung. Since there is good evidence that gas trapping can persist for many days, the trapping is probably occurring in alveoli that are only being ventilated intermittently when the infant takes a particularly deep breath. If this is so it seems remarkable just how accurately perfusion usually manages to match this irregularity of ventilation.

Babies of less than 30 weeks' gestation
Lung function is unfortunately rather less perfect in many babies who weigh less than 1·5 kg at birth. The degree of malfunction is very variable, but the most seriously affected babies develop the clinical and radiological features of the 'Wilson–Mikity' syndrome (Burnard *et al.*, 1965). These babies have signs of chronic respiratory distress with chest recession, moderate cyanosis and a mild respiratory acidosis; a

few die at this stage, while others make a slow recovery over a period of weeks or months. (A few also die suddenly and unexpectedly later in the first year of life after appearing to make an adequate clinical recovery.) X-rays reveal a fine reticular granular pattern early in the disease process, with later thickening of the perihilar pulmonary markings; classically there is at least some transient evidence of 'bubbly' focal emphysema. While all the clinical and radiological features can on occasion be ascribed to delayed recovery from hyaline membrane disease (p. 262) or to secondary oxygen toxicity (p. 274), the full syndrome frequently develops quite unexpectedly in a small but otherwise healthy infant 1–3 weeks after birth, in the absence of any previous history of respiratory distress and even though there has been no significant prior use of supplemental oxygen (Hodgman *et al.*, 1969). Lung-function studies reveal that the chest recession is due to a fall in dynamic compliance, but that static compliance remains normal; this apparently paradoxical situation is in fact to be expected when ventilation of the lung is grossly uneven (Aherne *et al.*, 1967). Thoracic gas volume is often particularly low in the second and third week of life even in asymptomatic babies, and the low lung volume at end expiration is probably at least partly due to high chest-wall compliance. There is also evidence of severe ventilation-perfusion mismatch with a large alveolar-arterial gradient for carbon dioxide, oxygen and nitrogen in the more severely affected infants, but there is no evidence of a shunt within the heart (Krauss *et al.*, 1970). There may in addition be evidence of pulmonary hypertension.

It is tempting to relate the increasing respiratory difficulty in the second and third week to the very low thoracic gas volume, especially as the alveolar-arterial carbon dioxide and oxygen gradients fall when lung volume is increased by the use of a constant transthoracic pressure; however, this is unlikely to be the whole explanation since ventilation-perfusion difficulties frequently persist long after thoracic gas volume has risen once more (Thibeault, Poblete & Auld, 1968). The fact that respiratory function is normal at birth, that respiratory difficulty only develops after a number of days, and that it becomes slowly and progressively more troublesome over a period of weeks, makes it conceivable that the disturbance is caused by respiratory movement damaging the delicate structure of the immature lung. Collagen constitutes the principal connective element in the lung during early fetal life, and elastic tissue only begins to make its appearance during the last 12 weeks of gestation. The bronchioles are very long in the pre-term baby

259

(Boyden & Tompsett, 1965) and the airways so small and compliant that quite a small rise in transpleural pressure might cause their collapse during expiration, with consequent trapping of gas in distended alveoli (Burnard *et al.*, 1965). Nevertheless the presence of 'trapped gas' appears to be such a common finding in asymptomatic babies that it can hardly be the full explanation. Repeated distension and collapse of these airways might, however, damage the bronchioles, modify cellular maturation, and distort growth, causing lobular emphysema, atelectasis and uneven ventilation. Symptomatic treatment is all that is available at the present time.

Respiratory drive

Repeated episodes of apnoea sometimes complicate management in babies of short gestation in the first month of life. The presence of foreign material in the pharynx provokes an extremely powerful inhibitory reflex in the newborn baby, and a small quantity of regurgitated food will sometimes be discovered in the back of the throat following the onset of an unexpectedly severe episode of apnoea. A similar phenomenon has been studied in the lamb, where the sensory receptors in the larynx can apparently differentiate between saline solutions and other fluids such as water, milk and glucose (Johnson, Dawes & Robinson, 1972). Recurrent apnoea can be the first sign of impending respiratory collapse in a baby with severe respiratory distress, while the emergence of episodes that cannot be influenced by gentle stimulation may be the first sign of intraventricular haemorrhage. An attack of apnoea is also on occasion the only readily detectable outward manifestation of a generalized convulsion. However, in very many cases no underlying cause or precipitating factor can be found, and careful observation reveals that the infant has developed a form of periodic breathing in which some of the apnoeic pauses become inexplicably prolonged to the point where cyanosis, limpness and bradycardia become apparent. While these episodes are not in themselves thought to be of serious significance, there is a real risk of progressive brain damage from hypoxia if they go unchecked (Bacola *et al.*, 1966).

Periodic breathing occurs in 20–40% of healthy babies of short gestation in the first 10 days of life, but clinical concern arises because a number of otherwise healthy babies develop periods of apnoea that last long enough to cause significant hypoxaemia. The respiratory centre appears to respond normally to oxygen and to carbon dioxide in these babies, and blood-gas values are not usually abnormal (except, of course, during the periods of apnoea). However, newborn babies in the first week of life only show a transient response to hypoxia before ventilatory depression sets in (Brady & Ceruti, 1966; Ceruti, 1966) and this abnormal response may be linked in some way with the risk of apnoea (Rigatto & Brady, 1972). Episodes of apnoea can be provoked by minor external stimuli such as handling, feeding or a sudden change in environmental temperature (Perlstein, Edwards & Sutherland, 1970), and there is also some evidence that attacks may be more common when the environmental temperature is unduly high (Daily, Klaus & Meyer, 1969) or the ambient humidity is unduly low (Belgaumkar & Scott, 1972). Burnard & Grauaug (1965) have shown that the babies who develop frequent apnoea have usually lost more weight than average and have a raised haematocrit; the problem of water balance in these small babies probably deserves more attention than it has received in the past (see p. 323).

Some episodes seem to be vagal in origin and are associated with an initial tachycardia or with an early bradycardia that can be abolished by atropine (Siassi et al., 1972). A fairly modest increase in the oxygen content of the air will often reduce the incidence of periodic breathing (Graham et al., 1950) and the total number of apnoeic attacks, but it has been suggested that the residual episodes last longer and that the cumulative total time that the baby remains apnoeic is only marginally reduced (Miller, Behrle & Smull, 1959). Troublesome recurrent apnoea can also be abolished on occasion by correcting an existing anaemia, and it may be that improved tissue oxygenation is the common factor here. Apnoeic attacks are particularly frequent in babies of less than about 34 weeks gestation during the first 10 days of life (Fig. 8.1), and sighing and periodicity may be linked with the low thoracic-gas volume that many of these babies possess at this time. Thibeault et al. (1968) showed that periodicity could be reduced by applying a constant transthoracic pressure to increase functional residual capacity, and this approach merits further study with the equipment now available for providing a continuous transthoracic airway pressure (Gregory et al., 1971). Whether the decreased periodicity is due to abolition of inhibitory stimuli coming from pulmonary mechanoreceptors or to improved oxygenation is not known.

Many clinicians feel that supplementary oxygen can be of help in this baffling condition, but such an approach is fraught with danger in the absence of significant hypoxaemia (see p. 272). Continuous monitoring together with gentle but early tactile stimulation will reduce the number of attacks that have to be terminated by bag and mask resuscitation

(Johnson & Babson, 1967), and this would appear to be the most logical method of management. A wide range of monitoring devices are now available. Where changing chest impedence is used to monitor respiration (Walker & Hanwell, 1968) the same pair of electrodes can be used

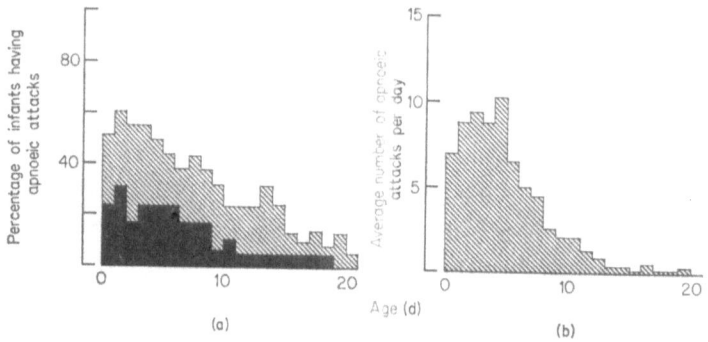

(a)   (b)

Fig. 8.1 (a). The incidence of apnoea in babies of less than 34 weeks' gestation in the first 3 weeks of life. Babies were classified as having apnoea if more than two episodes of apnoea were recorded, lasting more than 10 s in a single 24-h period. The black area indicates the number of babies who had more than 10 such episodes each day (Princess Mary Maternity Hospital data, Newcastle) (b). The average number of apnoeic attacks of more than 10-s duration witnessed each day in a series of 30 babies of 28–40 weeks' gestation, subjected to continuous monitoring because of recurrent episodes of apnoea. (From the data of Blake *et al.*, 1970)

to record the electrocardiogram. A cheap but effective apnoea-alarm mattress has recently become commercially available (Lewin, 1969), while radar is used to detect respiration in another recently designed monitoring device (Caro & Bloice, 1971; Bloice, 1972).

Hyaline membrane disease

Respiratory difficulty provides by far the commonest reason for anxiety over a pre-term baby's progress during the first few days of life. Distress usually becomes apparent soon after birth and is manifest by cyanosis, by abnormal retraction of the chest wall with every breath, and by either an increased respiratory rate or an audible grunt during expiration. Causes for such distress are multiple and include pneumonia, aspiration of foreign material into the lung, pulmonary haemorrhage, pneumothorax and various rare congenital malformations (Avery, 1968). Sometimes, indeed, what appears at first sight to be a respiratory problem turns out to be a cardiac or a metabolic problem, and damage to the central nervous system can also simulate pulmonary disease. However, by far the commonest cause for respiratory distress in the first week of life, and the commonest cause of neonatal death, is the syndrome now normally known as hyaline membrane disease.

No condition has excited the interest of those involved in the care of the newborn in the last 20 years more than the challenge posed by the existence of this syndrome; the interested reader is referred to an authoritative article on epidemiology by Fedrick and Butler (1970a) and to articles reviewing aetiology and management by Nelson (1970), Reynolds (1970) and Scopes (1971). Farber and Sweet described the presence of hyaline material in the lungs of 18 babies in 1931, and suggested that the membranes were caused by aspirated vernix from the baby's skin. This continued to be the accepted explanation for the condition until Miller and Hamilton pointed out in 1949 that hyaline material was never seen in the lungs of stillborn babies; from a reappraisal of the facts then known they concluded that 'the hyaline-like membrane or "vernix membrane" is not aspirated vernix but represents a tissue reaction to injury of the bronchioles, alveolar ducts and alveoli'. Today, nearly 25 years later, it is now generally agreed that the eosinophilic membranes are largely composed of fibrin and only form over a period of hours as a result of damage to the underlying epithelium; yet although this is now agreed we still do not know what causes the original injury.

The most important single contribution to our understanding of this problem was the demonstration that infants dying of the disease in the first two days of life lack lung surfactant (Avery & Mead, 1959). Lack of surfactant can account for the observed change in the mechanical properties of the lung, the atelectasis, and the maldistribution of gas within the lung, while blood flow through unaerated lung appears to be largely responsible for the severe central cyanosis that develops (Murdock et al., 1970). Lack of surfactant may also contribute to the delayed absorption of lung fluid through the lymph channels of the lung at birth, but it seems that damage to the underlying epithelium is more likely to be responsible for the appearance of plasma proteins in the alveoli (Normand, Reynolds & Strang, 1970). Since the risk of hyaline membrane disease is extremely small in infants of more than 37 weeks' gestation, and inversely related to gestational age, it is clear that lung immaturity must contribute to the aetiology of the disease in some fairly fundamental way. Babies of 30 weeks' gestation have about a 50% chance of developing features of the disease, and those who do so run about a 30% risk of dying of the disease (Usher, Allen & McClean 1971).

We have seen, however, that most infants of more than 26 weeks' gestation have stable lungs that contain functional surfactant at birth. Furthermore many immature infants who go on to die of hyaline membrane disease appear to be in good condition at birth, and then develop progressive respiratory difficulty over a number of hours. Surfactant has a relatively short half-life, and it has therefore been suggested that the disease develops in these infants because the surfactant present at birth is lost or inactivated, in babies of short gestation, faster than it can be replaced (Boughton, Gandy & Gairdner, 1970). However, it has recently been shown that the primary act of lung aeration at birth is abnormal in many infants who eventually die of hyaline membrane disease (Hey & Hull, 1971), even though overt evidence of respiratory difficulty only develops some hours later; inadequate surfactant reserve cannot be the explanation for this, and changes due to intrapartum asphyxia are possibly more important than is generally acknowledged.

Perhaps the most convincing single piece of evidence for this view has come from the study of twin births. Where, as is often the case, only one of the pair dies of hyaline membrane disease, it is four times as likely to be the second twin as the first twin (Table 8.1). These records show

Table 8.1. *Incidence of death from hyaline membrane disease (HMD) in twin pregnancy.*

| | Number of pregnancies in which | | | Increased risk |
Country	both babies died with HMD	only first born died with HMD	only second born died with HMD	to second twin if only one twin dies
Czechoslovakia	32	7	25	3·5:1
England	5	3	12	4:1
United States of America	12	2	12	6:1

fairly conclusively that perinatal stress potentiates the risk due to preterm delivery, and that the hazard is demonstrable in babies of 30–34 weeks' gestation, as well as in more mature babies. It is, of course, conceivable that the apparent increase in the incidence of hyaline membrane disease in second twins is spurious, that these babies died for other reasons, and that the presence of hyaline membranes would have gone undetected if the infant had not died from some unrelated cause. While this possibility is difficult to refute, it should be pointed out that the incidence of severe but *non-lethal* respiratory distress is equally increased in the second-born twin (Neligan, Robson & Hey, 1969).

Fetal asphyxia is known to cause intense pulmonary vasoconstriction and, as a result, almost all the blood returning to the right side of the heart is shunted through the foramen ovale and the ductus arteriosus. Although no significant changes appear to be detectable during or immediately after a period of prenatal asphyxia (Humphreys & Strang, 1967) the fluid content of the lung rises during the subsequent period of recovery, presumably as a result of a temporary change in permeability (de Sa, 1968, 1969). Marked perivascular and interstitial oedema is a consistent and early feature in hyaline membrane disease and Gandy *et al.* (1970) found dilated lymphatics, interstitial oedema and alveolar epithelial-cell necrosis, as well as loss of surfactant, in babies dying within an hour of birth, well before hyaline membrane formation was detectable. de Sa has argued that pulmonary-oedema formation is the initial event, and that loss of surfactant is a secondary consequence; this, however, fails to account for the fact that surfactant appears to function normally in babies with hydrops fetalis or pulmonary haemorrhage (Reynolds, Roberton & Wigglesworth, 1968). Gandy on the other hand has argued that the oedema and cell necrosis are secondary consequences of surfactant deficiency developing after lung aeration; this, however, fails to account for the evidence of abnormal lung function immediately after birth. The suggestion that continuing pulmonary ischaema after birth is the primary event responsible for the development of surfactant deficiency (Chu *et al.*, 1967) is equally unsatisfactory, because the right-to-left shunt across the lung is often no greater in the early stages of hyaline membrane disease than it is in babies with other forms of respiratory difficulty (Reynolds *et al.*, 1968). Hypoxia and acidosis are clearly capable of making loss of surfactant a self-perpetuating problem, and the discovery that surfactant reappears after about 4 days, despite the continued existence of a massive right-to-left shunt in even the most severely affected infant, was a further completely unexplained feature of the disease process until Gluck *et al.* (1972) showed that there were two synthetic pathways for surfactant production.

It will be seen from this extremely brief resumé that there is, as yet, no unanimity as to the root cause of this scourge of the special-care nursery. It is not that we lack facts; indeed the disease process comes to affect almost every system of the body, and Nelson (1970) in his review of the available evidence suggested that the overall fully developed syndrome was one of 'neonatal shock'. This may well be true, but such a form of words does nothing to clarify our understanding of the basic aetiology, and, while treatment of the established syndrome remains as unsatisfactory as it is at the present time, this must continue to be our aim.

Surfactant deficiency apparently develops if the available reserves are utilized or inactivated faster than they can be replenished by the type II epithelial cells of the immature lung, and the resulting atelectasis is responsible for most of the clinical features of the disease process; a plasma exudate also forms within the terminal air spaces at much the same time as a result of a change in the permeability of the epithelium, and this produces the hyaline membranes seen at autopsy in babies more than a few hours old. The available evidence is at least consistent with the hypothesis that inhibition of palmitoylmyristol lecithin production in babies of short gestation (Gluck *et al.*, 1972) and pulmonary oedema (de Sa, 1969) are secondary and interrelated consequences of perinatal stress, prior to or shortly after aeration of the lungs at birth. Such an hypothesis has at least the merit of focusing attention on the contribution that can be made by good preventive medicine.

Prevention

Prevention is better than cure, and it is of fundamental importance to avoid delivery before 37 weeks' gestation whenever possible. There is no complete substitute for an accurate knowledge of the date of the last menstrual period, supplemented by a clinical assessment of uterine size towards the end of the first trimester. Gestational age and fetal size are difficult to assess by abdominal palpation later in pregnancy, but serial measurements of biparietal diameter with ultrasound can serve to separate the pregnancy complicated by fetal growth retardation from the pregnancy in which the dates are in error, if measurements are started early enough. If the need to confirm gestational age does not become apparent until late in pregnancy, amniocentesis and examination of the liquor can help to determine fetal age (Lind & Billewicz, 1971), while Gluck *et al.* (1971) have recently suggested that estimation of the lecithin:sphingomyelin ratio in the liquor can also determine whether the enzyme systems necessary for surfactant production are fully mature. Bhagwanani, Fahmy & Turnbull (1972) have found this a promising approach, but others have found the change in the lecithin:sphingomyelin ratio more gradual than was claimed by Gluck (Gusdon & Waite, 1972), and the data published by Nelson (1972) indicate that it may be better to measure amniotic-fluid lecithin concentration directly.

A comparative study of the differing analytic techniques currently in use is urgently needed. It is possible to assess the risk of a baby developing respiratory distress after birth fairly precisely from an accurate knowledge of gestational age; measurement of liquor lecithin will need to increase the accuracy of this assessment if it is to achieve its objective and become a widely used clinical tool. The technique is, however, an elegant and logical approach to a clinical problem of great importance.

Another approach of great theoretical interest has stemmed from the observation that lung maturation was enhanced by injecting dexamethasone into the fetal lamb (Liggins, 1969). Precocious lung maturation has now been documented in the rabbit as well, and correlated with premature appearance of lung surfactant after fetal injection with thyroxine (Wu et al., 1971) or 9-fluoroprednisolone (Kotas & Avery, 1971). Steroid treatment appeared to have a specific effect on lung maturation without having any effect on skeletal maturation or skin maturation (Kotas et al., 1971). These steroids provoked premature delivery as well as precocious lung maturation; methyl prednisolone, however, provoked premature delivery without enhancing lung maturation (Tanesch, Avery & Sugg, 1972). Animal experiments have not, as yet, produced any good evidence that these steroids are of value if given to the mother rather than the fetus (Motoyama et al., 1971), and it is not yet known whether steroid treatment significantly accelerates surfactant production in species (like man) that are capable of synthesizing palmitoyl myristol lecithin well before term. Nevertheless, Liggins & Howie (1972) have just completed a controlled trial that seems to show that treatment with betamethasone can reduce both perinatal mortality and the incidence of clinical respiratory distress in babies of less than 32 weeks' gestation, as long as *the mother* is started on treatment at least 24 h before delivery. Liggins gave betamethasone as soon as there was evidence of spontaneous premature labour: treatment begun after delivery does not appear to be of value (Baden et al., 1972). More work will need to be done to confirm the effectiveness of this form of treatment, and to assess whether there are any untoward complications, before recommending the widespread adoption of what is essentially a prophylactic measure. It would be exceedingly rash to embark on a preventive measure involving the treatment of a large number of children in order to save just a few from death, before a reasonable assessment can be made of the inherent short-term and long-term risks.

There is growing evidence that the mode of delivery has an influence on mortality from hyaline membrane disease. The risk that respiratory distress is increased by ante-partum haemorrhage has long been recognized and this has often been invoked to explain the increased incidence of respiratory distress following Caesarian section. Evidence nevertheless continues to accumulate that respiratory distress is commoner following delivery by Caesarian section (Usher *et al.*, 1971), and that section before the onset of labour further increases the risk of death from hyaline membrane disease (Fedrick & Butler, 1972). An intriguing speculation is that changes in fetal steroid production associated with the onset of labour might enhance surfactant production in some way.

There can be little doubt that careful assessment throughout labour, and meticulous care in the first hour of life, can help to reduce mortality from hyaline membrane disease (although it is extremely difficult to get direct statistical proof of this). Adequate oxygenation is extremely important, and mortality is much higher in those babies who have central cyanosis on arrival in the special-care nursery, despite what was thought to be adequate resuscitation at birth (Table 8.2). This cyanosis is

Table 8.2. *The relation between clinical condition on arrival in the special-care nursery and the incidence of clinical respiratory distress or death with hyaline membrane disease HMD, at the Princess Mary Maternity Hospital, Newcastle, 1960–1967 (Neligan, Omer & Marr, 1972).*

Birthweight and clinical condition on arrival in the nursery within 30 min of birth		Clinical respiratory distress (%)	Death from HMD (%)	Death from other causes (%)
137 babies of 1000–1500 g birthweight	Pink on admission ...	48	19	8·0
	Cyanosed on admission ...	50	22	32
328 babies of 1501–2000 g birthweight	Pink on admission ...	19	3·1	2·7
	Cyanosed on admission .	73	39	12

probably due partly to the fact that lung function is already compromised at the time of birth in some of these babies (Hey & Hull, 1971); there is, however, no doubt that adequate oxygenation can abolish central cyanosis in almost every baby in the first few hours of life, and produce a measurable decrease in the right-to-left shunt across the heart (Sinclair, Engel & Silverman, 1968). Early correction of the acidosis present at birth is also of value (Gupta, Dahlenburg & Davis, 1967;

Russell & Cotton, 1968; Marini *et al.*, 1970), and there are important theoretical reasons for minimizing cold stress at this time (see p. 298). It is also advisable to allow babies to receive at least a small transfusion of placental blood at delivery (p. 280). A case can be made out for intubating and artificially ventilating almost every small pre-term baby at birth, when this can be achieved quickly, gently and without trauma, in order to aid expansion of the lung, correct any existing respiratory acidosis with the minimum of delay, and secure and maintain adequate oxygenation during transfer to the special-care nursery. The importance of adequate care immediately after delivery has been highlighted by a recent analysis of the causes of perinatal mortality in the province of Quebec. Mortality from hyaline membrane disease in maternity hospitals with their own neonatal intensive-care facilities was half what it was elsewhere in the province; mortality in those hospitals that referred their sick infants elsewhere for intensive care after birth was only marginally better than in those hospitals that made no use of intensive-care facilities (Usher, 1971).

Cure

Additional oxygen, adequate warmth, intravenous glucose and great gentleness are the mainstay of supportive care at the present time in the baby with signs of established respiratory distress. Care should be taken to avoid rough handling or any sudden fall in the ambient oxygen concentration, as there is some evidence that this can occasionally cause a rise in the right-to-left shunt across the lung that cannot always be reversed. A slow infusion of sodium bicarbonate can help to combat acidosis, and there is no doubt that significant anaemia should always be corrected. Whether the hypotension so often present is due to hypovolaemia is doubtful, and there is, as yet, no clear indication that vasoconstrictor agents are called for.

More aggressive therapy with drugs has been singularly unrewarding to date, and several forms of treatment have proved positively harmful. In all too many instances no adequate attempt was made to judge the value of the new form of treatment by controlled trial. Sinclair reviewed the evidence adduced in favour of 30 different forms of treatment in 1966, and concluded that none had been shown to be unequivocally beneficial; subsequent reports on the use of intravenous tolazoline, bradykinin, heparin, and urokinase-activated human plasmin, on the use of a continuous infusion of acetylcholine through a catheter positioned within the pulmonary artery, and on the use of intramuscular ethyl adrianol (a synthetic analogue of noradrenaline) have been equally discouraging. Exchange transfusion with fresh adult blood in order to improve arterial oxygen tension (Delivora-Papadopoulos, Roncevic & Oski, 1971) and peritoneal dialysis (Boda et al., 1971) have been advocated, but the benefits derived are marginal and the techniques themselves not without risk.

Intravenous infusions of glucose and sodium bicarbonate have been used in the treatment of respiratory distress for more than a decade (Usher, 1959), and several controlled trials have confirmed that mortality can be reduced by this form of treatment (Savignoni et al., 1969). A slow continuous intravenous infusion of glucose prevents hypoglycaemia, as well as decreasing the hyperkalaemia and the other signs of tissue catabolism that develop in babies with respiratory distress (Auld, Bhangananda & Mehta, 1966). The trial undertaken by Kitchen & Campbell (1971) suggests that infusion of sodium bicarbonate is the more important of the two ingredients of this regime, but there is at the present time no proof that continued slow infusion of bicarbonate after correction of the initial acidosis is of any value. A great deal of trouble is taken in many hospitals to measure blood pH at regular intervals so as to estimate the amount of sodium bicarbonate required with greater precision, but it seems possible that these efforts are sometimes misguided, as there is growing evidence that continued treatment with sodium bicarbonate causes a serious rise in P_{CO_2} if respiratory function is already restricted, and that pH, in consequence, changes very little (Evans et al., 1970).

Indeed it seems at least theoretically possible that the could point be reached where further treatment with sodium bicarbonate would actually cause a fall in pH (Ostrea & Odell, 1972). Over-treatment with sodium bicarbonate can cause a dangerous change in the osmolal concentration (Kravath et al., 1970) and, while there is no doubt that the rapid infusion of a single dose of bicarbonate soon after birth produces a measurable improvement in effective pulmonary blood flow (Russell & Cotton, 1968), and a demonstrable improvement in prognosis (Marini et al., 1970; Hobel et al., 1971), there is also good evidence that mortality rises if too many attempts are made to correct acidosis abruptly (Usher, 1967). The advantages of using tris(hydroxymethylaminomethane) instead of sodium bicarbonate, to correct persistent acidosis (Strauss, 1968) need further assessment, even though there is a risk of transient respiratory depression after its rapid administration (Roberton, 1970). Whether continued buffer administration to babies with established respiratory distress is ever of more than transient benefit after the first 12–24 hours of life remains to be shown.

Since alveolar collapse and atelectasis are prominent and early features in hyaline membrane disease, causing a large decrease in functional residual capacity (Krauss & Auld, 1970), it is logical to attempt to maintain a permanent pressure gradient across the thorax to counter the tendency for alveoli to collapse and close at the end of expiration. The audible grunt that many of these distressed babies develop is caused by expiration against a partially closed glottis, and the consequential increase in transthoracic pressure appears to cause a rise in arterial oxygen tension (Harrison, Heese & Klein, 1968). Recent reports have shown that maintenance of an increased transthoracic pressure with an endotracheal tube (Gregory *et al.*, 1971), a pressurised head box or a negative-pressure body tank (Chernick & Vidyasagar, 1972) also improves arterial oxygen tension. Early use of this form of treatment could therefore possibly bring about a measurable decrease in overall mortality.

If all these therapeutic approaches fail, artificial ventilation provides the only hope. Success is rare in a baby weighing less than 1·5 kg, because small size augments the magnitude of the technical problems, but increased experience is bringing its due reward (Minkowski, Monset-Couchard & Amiel-Tison, 1970). Nasotracheal intubation is usually preferred to orotracheal intubation if artificial ventilation is required for any length of time, because it is easier to secure the tube in place, but Gregory (1972) has put forward reasons for preferring orotracheal intubation. Care should be taken to position the end of the tube midway between the vocal cords and the carina (see Table 8.5, p. 322). Tracheostomy is not indicated.

Nevertheless it is clear that any attempt to maintain respiratory function by mechanically ventilating a severely damaged lung can only hinder the vital process of repair. To subject such a delicate organ as the lung to high pressure, a high oxygen tension, poor humidification, constant trauma and a greatly increased risk of bacterial invasion can only delay recovery. Ideally the damaged organ should be rested rather than flogged to death on a mechanical respirator! For the hard core of children who continue to develop this common and frequently lethal condition despite careful resuscitation immediately after birth, the development of an efficient pump oxygenator probably offers the best long-term hope for the future. Such an 'artificial placenta' would need to be capable of maintaining adequate respiratory exchange for at least 2 or 3 d without undue hazard to the circulation.

The problem of oxygen toxicity

The potential danger of oxygen toxicity has been appreciated for many years: pure oxygen at more than 2 atm pressure causes convulsions, while pulmonary symptoms develop after a number of hours at rather lower partial pressures. Toxicity in adult man appears to be related not only to the partial pressures of the gas but also to the length of time for which it is administered (Welch, Morgan & Clamann, 1963). Many adult laboratory animals develop severe and even fatal lesions when exposed to between 60 and 100% oxygen at atmospheric pressure for more than a few days (Bean, 1945; Haugaard, 1968), but newborn rats seemed to be relatively resistant to this damage, and it was at one time thought that the human infant shared the same relative immunity.

Retinal damage

Faith in the safety of administering supplementary oxygen was, however, shattered by the discovery that its administration could cause retrolental fibroplasia in babies of short gestation, and that the more immature the infant the greater was the risk of retinal destruction (Fig. 8.2(a)). Retinal-artery vasoconstriction in response to a high arterial oxygen tension seems to be the factor that initiates the pathological process that culminates in retrolental fibroplasia, so there are good grounds for giving supplemental oxygen to babies with central cyanosis, and no grounds for thinking that a high ambient oxygen concentration is ever a significant danger to the retina as long as arterial P_{O_2} is kept below 100 mmHg.

Retinal damage similar to that seen in man can be produced in the eyes of newborn kittens, and here there is undoubted evidence that oxygen concentration and duration of exposure are both factors of importance (Ashton, Ward & Serpell, 1954), and that intermittent exposure reduces the hazard (Ashton, Garner & Knight, 1971). While the dose–time relationship has never been studied in detail, retrolental fibroplasia never seems to develop in the kitten as long as the ambient oxygen concentration is kept below 35%, and many paediatricians came to believe that the same was true of the human infant. Indeed so confident did many people become that this was the 'safe limit' that manufacturers started to design incubators in which oxygen administration was mechanically regulated at a level that precluded the concentration rising above 35%. Belief in the existence of this 'safe level' was fostered by reports of two controlled trials in America (Patz, Hoeck & De la Cruz, 1952; Lanman, Guy & Dancis, 1954). These showed that a quarter of all the babies of less than 1·8 kg nursed in 60–70% oxygen for a minimum

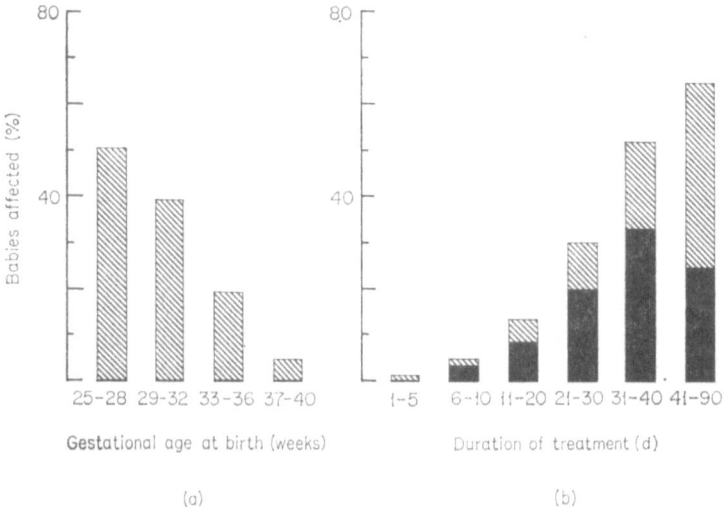

Gestational age at birth (weeks)

Duration of treatment (d)

(a)   (b)

Fig. 8.2, (a). The relation between gestational age and the incidence of retrolental fibroplasia in infants of less than 2 kg admitted to the Babies Hospital, New York, in 1950 and 1951 who were discharged home alive. (From the data of Silverman *et al.*, 1952)

(b). The relation between the duration of treatment with oxygen and the incidence of subsequent retrolental fibroplasia in the MRC United Kingdom survey of 1095 babies of less than 1·8 kg, born between October 1951 and May 1953 and alive at 2 months. The proportion found to have progressive retinal damage sufficient to cause severe visual handicap or complete blindness is indicated by the black area in this histogram

of 2 weeks developed cicatricial retrolental fibroplasia, whereas there was not a single case of progressive damage in the two comparable groups of babies who were never given more than 40% oxygen. The results of a third less strictly controlled trial published by Gordon, Lubchenco & Hix (1954) showed a similar reduction in retinal damage when measures were taken to reduce ambient oxygen concentration. Unfortunately it was not generally noticed that in all three of these trials the babies who were allowed more than 40% oxygen to breathe also received supplemental oxygen for very much longer, and the important findings of the British Medical Research Council survey published in 1955 were almost entirely overlooked (Fig. 8.2(b)).

Once it became known that the use of supplemental oxygen could cause blindness, it became standard practice to measure ambient oxygen concentration at regular intervals and keep records of the amount given. Since it quickly became clear that the cyanosis of some severely ill babies could only be relieved by giving more than 40% oxygen, it also became standard practice in most large centres to monitor arterial oxygen tension, particularly in babies receiving more than 40% oxygen. Repeated blood samples can be obtained from the temporal artery (Thomson, 1964), the radial artery (Bucci *et al.*, 1966; Shaw, 1968), or indeed any palpable small artery; specimens can be successfully withdrawn from the dorsalis pedis artery with a syringe, or obtained from the digital arteries by a stab incision into the tip of the finger. The insertion of an indwelling umbilical-artery catheter provides a convenient route for obtaining serial samples as well as monitoring blood pressure and giving intravenous fluids and drugs (Kitterman, Phibbs & Tooley, 1970), and involves minimal disturbance once the catheter has been correctly positioned (Table 8.5, p. 322), but the technique is not without its hazards (Egan & Eitzman, 1971). Because of risk of cardiac arhythmia, haemorrhage, sepsis and thrombosis (Neal *et al.*, 1972), there is now general agreement the catheter should never remain in place longer than is essential. There is some hope that the development of a catheter-tip transducer for measuring oxygen tension continuously *in vivo* will eventually make the management of oxygen therapy more precise (Parker, Key & Davies, 1971), but such a development will not avoid the hazards that are always present whenever an arterial catheter is in use.

Lung damage

A high concentration of oxygen in the inspired air can cause surfactant inactivation, impaired mucous flow and diffuse microhaemorrhages, as well as frank bronchopulmonary fibroplasia, in the lungs of experimental animals, and arterial desaturation has now been shown to provide no significant protection from this toxic effect of oxygen (Miller, Waldhausen & Rashkind, 1970). Significant damage seldom occurs unless the partial pressure of oxygen in the air exceeds about 450 mmHg, but that there is an absolute safe limit is by no means certain. Man is not immune to this hazard, and the suggestion has recently been made that the marked proliferative regeneration and widespread fibrosis sometimes seen in the human infant following prolonged artificial ventilation is due to oxygen toxicity (Northway, Rosan & Porter, 1967). It has also been suggested that pulmonary haemorrhage is sometimes due to oxygen toxicity (Shanklin & Wolfson, 1967), but the evidence for this is equivocal (Fedrick & Butler, 1971a). The histological changes found by Northway *et al.* were very similar to those seen following acute oxygen poisoning in the monkey (Gupta, Winter & Lanphier, 1969), but the

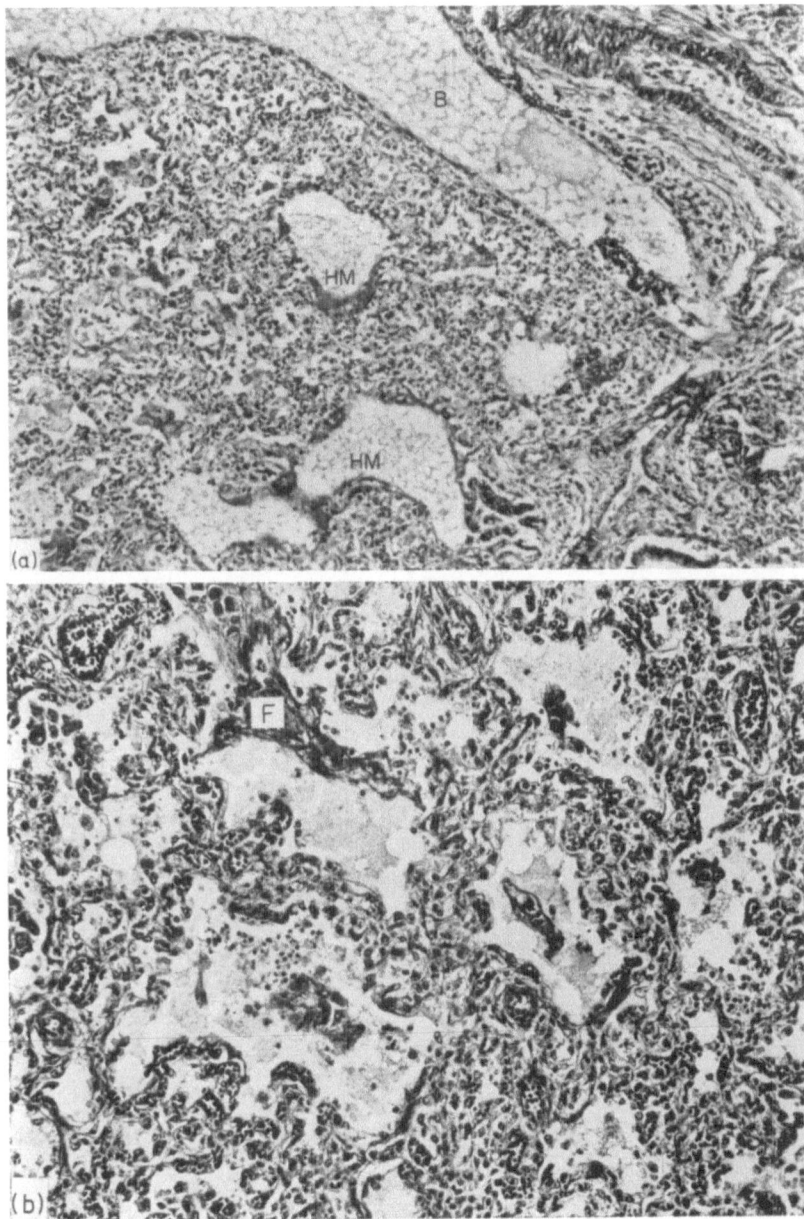

Plate 8.1. Lung damage caused by intubation and artificial positive-pressure ventilation with 70 % oxygen for 6 d in a 2·9-kg baby of 40 weeks' gestation, who had no evidence of respiratory distress at birth but developed congestive cardiac failure as a result of septicaemia at the age of 10 d.

(a) Section of lung showing scattered hyaline membrane (HM) in many of the alveolar ducts, and complete erosion of the epithelium from the terminal bronchiole (B).

(b) Thickening of the alveolar walls due to fibroplasia and numerous dilated and engorged capillaries. There is fibrin deposition (F) in one of the alveolar spaces

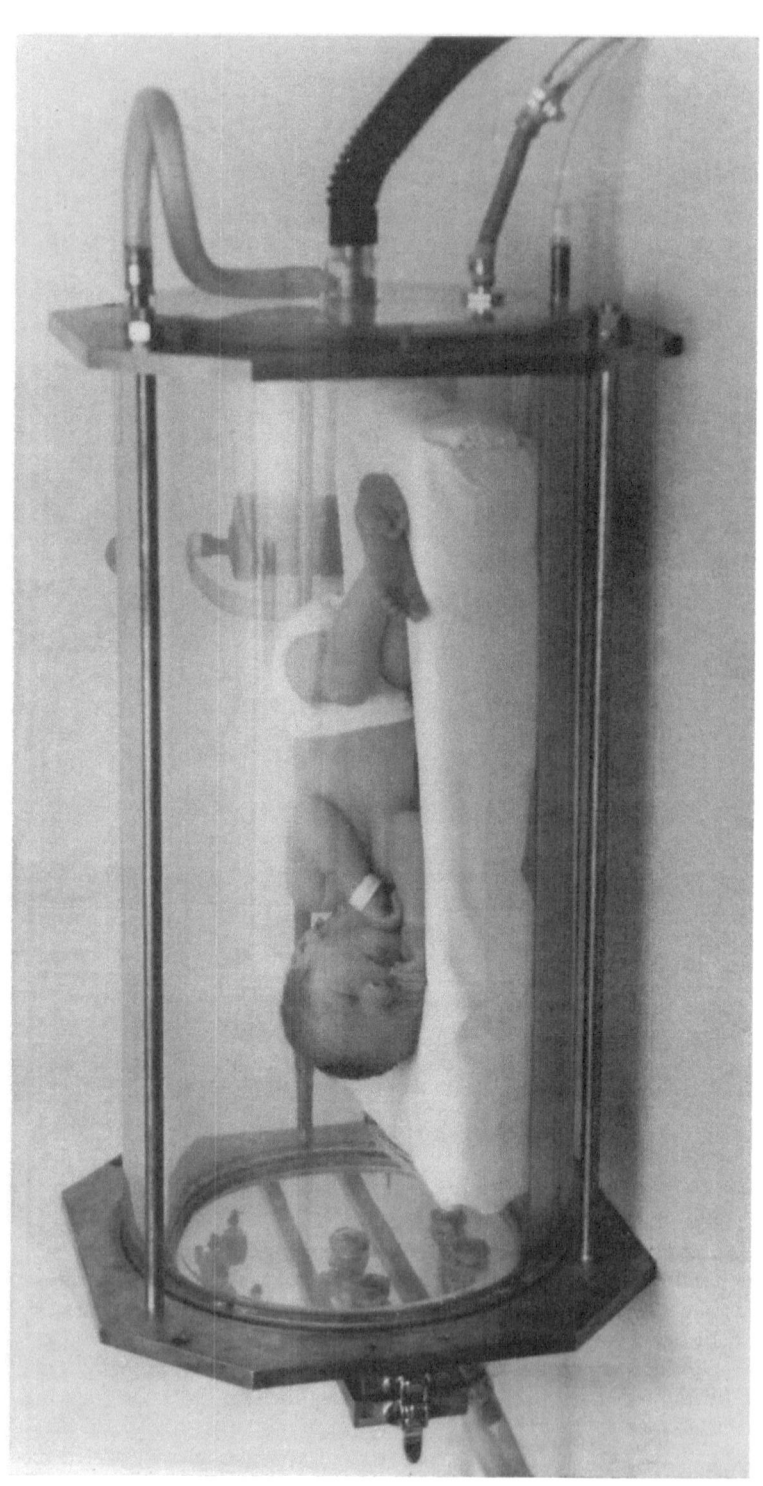

Plate 8.2. A metabolic chamber used to measure heat production and evaporative water loss, and to study thermoregulatory function in the newborn baby

lung displays a fairly stereotyped response to most forms of injury, and it is difficult to be certain that the changes seen after artificial ventilation are not the result of poor humidification, secondary infection and mechanical trauma in a lung already damaged by disease.

All the published cases of severe lung damage have followed intubation and positive pressure ventilation for respiratory distress, and it has been argued that the histological changes are those that would be expected during recovery from hyaline membrane disease of such severity that artificial means were necessary to maintain life. There is no evidence that complications of this nature ever arise in babies with normal lungs subjected to prolonged ventilation because of tetanus neonatorum (Smythe, 1967), but similar changes have been seen in the absence of hyaline membrane disease in infants receiving artificial ventilation with oxygen at high pressure after neonatal surgery (Plate 8.1). The experiments of de Lemos *et al.* (1969) revealed that newborn lambs made to breathe pure oxygen developed signs of pulmonary damage quite rapidly and died after 3 or 4 d, and that artificial ventilation made no difference to the speed with which symptoms developed. However, since unequivocal evidence of pulmonary damage has never been detected in the newborn baby even after more than a week in 80–90% oxygen in the absence of prolonged artificial ventilation, mechanical damage due to prolonged ventilation at high pressure is probably at least as important a factor as oxygen toxicity in the genesis of pulmonary fibroplasia in the human infant (Pusey, Macpherson & Chernick, 1969). This suggests a species difference in the response to oxygen. However, in interpreting the report of de Lemos *et al.* that mechanical ventilation does not increase the rapidity with which signs of oxygen toxicity become apparent in the newborn lamb, it must be remembered that the lungs of these animals were normal at the start of the experiment and that high ventilating pressures were not necessary. The suggestion of Bean (1945) that pre-existing pulmonary damage increases the lung's vulnerability to oxygen has never been thoroughly investigated.

Oxygen poisoning appears to cause generalized exudation, followed later by a proliferative regeneration that culminates in widespread fibrosis, and all these stages have been found in infants artificially ventilated with oxygen in high concentration (Anderson & Strickland, 1971). Rosan has recently discussed what is known about the nature of this toxic process in a stimulating and very personal way (Rosan & Lauweryns, 1971). There is no doubt that the presence of an indwelling endotracheal tube can cause severe local trauma to the trachea (Sinclair,

1969), but for mechanical ventilation to cause generalized progressive pulmonary damage seems unusual if the inspired concentration of oxygen is kept below 60% (Banerjee, Girling & Wigglesworth, 1972). Reports that oxygen toxicity is not encountered despite prolonged exposure to high concentrations of oxygen, as long as endotracheal intubation is avoided and a negative pressure ventilator is used to provide respiratory support (Stern *et al.*, 1970), are difficult to explain. Probably the respiratory tract is better humidified, and mechanical trauma minimized, by this form of ventilation if transthoracic pressure never falls to zero at the end of expiration and repetitive airway collapse and closure are avoided: this form of mechanical support may also fail to ventilate the lungs of the most severely affected infants effectively enough for the children to survive long enough for the complication to develop.

Management

There is no evidence that the continued use of a high concentration of oxygen after the first few hours of life causes a detectable decrease in mortality. The neonatal death rate did not rise when the concentration of oxygen was limited to a maximum of 40% in the three controlled trials already referred to, and the best evidence that a 'liberal' policy has a beneficial effect on mortality remains the uncontrolled retrospective survey undertaken by Avery and Oppenheimer in 1960. They showed that the number of babies dying of hyaline membrane disease at the Johns Hopkins Hospital, Baltimore, increased during the 4-year period in which environmental oxygen was limited to a maximum of 40%, compared to a similar 4-year period a decade earlier when oxygen had been used quite freely. Since, however, mortality from other unrelated conditions also increased during the years in which the use of oxygen was restricted, this study must be regarded as inconclusive. Lubchenco *et al.* (1972) have reported that mortality *decreased* in Denver, Colorado, when environmental oxygen was limited to a maximum of 40%.

Since hypoxaemia is usually caused by direct veno–arterial shunting, rather than impaired pulmonary diffusion or ventilation-perfusion imbalance in the first few days of life, oxygen therapy is least effective where it is most necessary. Where more than 50% of the blood is being shunted across the heart and arterial P_{O_2} falls below 45 mmHg breathing room air, more than 40% oxygen must be given to relieve the incipient hypoxaemia; if the magnitude of the shunt increases a further 10%, arterial P_{O_2} will often be less than 40 mmHg breathing room air, and will only rise about 7 mmHg even in 100% oxygen. While minor improvements in tissue oxygenation may sometimes be critical, an infant who is hypoxic in 60% oxygen is rarely significantly improved by 100% oxygen.

Because of the known risk to the retina and the lung, quite obsessional care is taken over the monitoring of oxygen therapy in many neonatal intensive-care units at the present time. So long as the method used to monitor therapy is not in itself hazardous, it is hard to fault such dedicated attention to detail; the author must, nevertheless, be excused for wondering whether over-concern with this one aspect of therapy may not result in insufficient attention being given to other problems and to other children when staff are scarce. It is important to maintain a sense of proportion and acknowledge that progressive retrolental fibroplasia is rare in babies receiving oxygen for less than 5 d (Fig. 8.2(b)), that pulmonary damage has only been convincingly demonstrated as yet in babies given artificial ventilation, that the hazards associated with prolonged umbilical-artery catheterization are not inconsiderable, and that the risks associated with catheterization are almost certainly inversely related to the experience of the clinician concerned. A comforting thought is that, even if ambient oxygen is restricted to a maximum of 40%, those infants with respiratory distress who survive the initial period of severe hypoxaemia show little evidence of cerebral damage in later life (Ambrus *et al.*, 1970), unless there has been a secondary complication such as intraventricular haemorrhage.

The ambient oxygen concentration should always be monitored, and care should be taken to warm and humidify any supplementary oxygen given into a hood or mask over the face. A small paramagnetic oxygen meter is not expensive, and a continuous-reading servo-regulated device for controlling oxygen concentration is now commercially available (Daily & Kearns, 1970). Paediatricians have often aimed at achieving an oxygen tension of between 75 and 150 mmHg in the descending aorta (Roberton *et al.*, 1968), but this is well above the level of arterial saturation achieved by many healthy babies of short gestation in the first few days of life (Orzalesi *et al.*, 1967), and—since blood samples from the descending aorta will underestimate retinal artery P_{O_2} if there is a significant right-to-left shunt through the ductus arteriosus—it is now considered advisable to keep the P_{O_2} of blood in the descending aorta below 90 mmHg. Retinal-artery vasoconstriction is known to occur in some babies when the arterial oxygen tension is only between 100 and 150 mmHg (Aranda *et al.*, 1971), and many cases of severe retrolental fibroplasia have been documented in babies who never received more

277

than 40% oxygen. There are even a few authenticated reports of severe retrolental fibroplasia developing in babies who never received any supplementary oxygen. Minor degrees of retinal damage leading to myopia and impaired visual acuity are now known to occur quite frequently even when arterial P_{O_2} never exceeds 150 mmHg, and the risk is highest in those babies who receive supplementary oxygen for more than a week; approximately 10% of the babies who weighed less than 1·5 kg at birth had retinal damage of this type at follow up in a recent survey at the Hammersmith Hospital, London (Mushin, 1971). Clearly therefore no safe upper limit for arterial oxygen tension exists and a mistake was made in ever attempting to mimic the arterial tension found during adult life during the treatment of neonatal hypoxaemia. A particularly close watch needs to be kept over arterial oxygen tension during the period of recovery from hyaline membrane disease, since in these circumstances the hypoxaemia is usually due to regional inequalities between ventilation and perfusion rather than to the true shunt that is present in the acute phase of the disease process (Adamson *et al.*, 1969b).

Because it is increasingly realized that oxygen can cause lung damage, attention is now (somewhat belatedly) being given to defining what should be considered the safe *minimum* arterial P_{O_2}. Roberton (1969) and Graziani, Weitzman & Pineda (1972) have shown that few significant changes occur in the encephalogram until arterial P_{O_2} falls below 40 mmHg, and Usher (1970) found no change in overall mortality when he restricted the use of more than 60% oxygen to babies who had an arterial P_{O_2} of less than 40 mmHg or a capillary P_{O_2} of less than 35 mmHg. When Usher compared the results of this conservative policy with one in which as much oxygen was given as was necessary to keep arterial P_{O_2} above 80 mmHg, the babies who were treated liberally had symptoms of respiratory distress for much longer and much more marked X-ray evidence of pulmonary damage (Usher, 1968). There was even a suggestion that mortality might be increased by the injudicious use of oxygen. Paediatricians have been slow to accept this further evidence of oxygen toxicity, partly because Usher used capillary-blood specimens to monitor P_{O_2} in some of his babies; nevertheless this is an important study and stands in urgent need of confirmation from a second centre. Perhaps babies should be nursed in the minimum amount of oxygen necessary to maintain an arterial P_{O_2} of between 40 and 50 mmHg after the first few hours of life, a higher P_{O_2} only called for if neurological deterioration or advancing metabolic acidosis suggest that tissue perfusion and oxygenation are becoming inadequate. To judge clinical cyanosis is not as difficult as is sometimes imagined.

Cardiovascular function

Unlike most of the other major organs of the body, the heart performs a vital function *in utero* from a very early age, and this is reflected in its early structural and chemical maturation. Cardiac output per kilogram body weight in the human fetus reaches a value very similar to that found after delivery at term at quite an early fetal age, and the fetus also responds to asphyxia by redistributing cardiac output and increasing blood flow to the coronary arteries, the brain and the adrenal gland at 10–20 weeks gestation (Rudolph *et al.*, 1971). Active cardiovascular reflexes can be elicited from an early age in experimental animals (Dawes, 1968), and in many respects the cardiovascular system is clearly better adapted to withstand premature delivery than most of the other organ systems of the body.

Fetal circulation

Anatomical closure of the ductus venosus, ductus arteriosus and foramen ovale occurs a variable time after birth; the ductus venosus is nearly always obliterated within a month, but the ductus arteriosus often remains probe patent for many months, and the foramen ovale is often only closed by virtue of the presence of a flap valve even in adult life (Scammon & Morris, 1918). However, functional closure of all these channels nearly always occurs very quickly after birth. Functional closure of the foramen ovale depends on reversal of the inter-atrial pressure gradient, and functional closure of the ductus venosus appears to be a consequence of decreased portal-sinus pressure which seems to cause the opening of the ductus venosus to retract to a mere slit (Meyer & Lind, 1966).

Functional closure of the ductus arteriosus, on the other hand, occurs because of the active contraction of its smooth muscle coat in response to a rise in arterial oxygen tension, and it has long been recognized that closure may be delayed by respiratory difficulty in the first few days of life. A number of these babies even develop signs of heart failure and require treatment with digitalis and mersalyl, but spontaneous closure does occur eventually in the majority. Prolonged patency of the ductus arteriosus appears to be much commoner in babies of short gestation, and something like a quarter of all babies of less than 35 weeks' gestation show evidence of delayed closure at some time during the first three weeks of life (Girling & Hallidie-Smith, 1971); persistent patency coming to the notice of paediatric cardiologists and requiring surgical treatment, on the other hand, does not appear to be particularly associated with a history of premature delivery. Delayed closure in the preterm baby can nevertheless cause severe intractable heart failure and even death on occasion, and persistent gross patency of the duct is occasionally the only abnormal finding at autopsy in a small baby dying after a period of progressive cardiorespiratory difficulty at the age of one or two weeks. Recent studies have shown that the duct of the fetal lamb becomes progressively more responsive to increased arterial oxygen with increasing maturity (McMurphy *et al.*, 1972), and this provides a possible explanation for the high incidence of delayed closure that has been observed in babies of short gestation.

Placental circulation

The circulating blood volume of the newborn infant in the first week of life is influenced by the extent to which the baby receives additional blood from the placenta at birth. A full-term infant may receive as much as 100 ml of blood in this way following delivery, causing the blood volume to rise from 70 to 100 ml/kg and the venous haematocrit to rise by 10% (Moss & Monset-Couchard, 1967). If the baby is laid on the mother's bed after delivery, there is a slow transfer of blood from the placenta to the baby that normally continues for at least 3 min; the rate of transfer is, however, significantly accelerated when the mother is given intravenous methylergometrine to contract the uterus (Yao, Hirvensalo & Lind, 1968). Whether the baby cries or not before the cord is clamped seems to make little difference to the volume of blood transfused but there is some evidence that the transfusion takes place more rapidly in the full-term baby (Saigal *et al.*, 1972).

There is now some evidence of an increased incidence of respiratory distress in babies whose cords are clamped early (Usher *et al.*, 1971) and there are certainly many important physiological consequences of the marked difference in circulating blood volume that results from the variable placental transfusion at birth. Pulmonary, systemic and atrial pressures are higher throughout the first day of life in babies who receive a placental transfusion, and a transfusion also delays the fall in pulmonary pressure that normally occurs after birth. Babies whose umbilical cords are clamped late have a smaller functional residual capacity, a lower compliance and a more rapid respiratory rate with grunting in the first 6 h of life (Oh *et al.*, 1967); they also have a lower arterial oxygen tension, a higher carbon dioxide tension and radiological evidence of a larger heart volume. Renal blood flow, glomerular filtration rate and urine flow are also higher in the first 12 hours of life in babies who receive a large placental transfusion at birth, but these differences diminish with time and are barely detectable at 5 d, even though there is still a large difference in circulating blood volume (Oh, Oh & Lind, 1966).

We do not as yet know what constitutes the optimum circulating blood volume immediately after birth. Such evidence as is currently available would seem to suggest that babies should be placed on the bed alongside their mothers, and that the cord should be clamped about 1 min after delivery; this will allow a modest transfusion of placental blood to occur. It seems likely that a large transfusion could occasionally embarrass cardiorespiratory adaptation to birth (Yao, Lind & Vuorenkoski, 1971) and also cause late hyperbilirubinaemia (Saigal *et al.*, 1972). On the other hand, it is important to remember that gravity has a marked influence on placental transfusion, and that lifting the baby into the air immediately after delivery and before the cord is clamped will not only prevent any placental transfusion taking place but also tend to cause some blood to drain from the baby back into the placenta (Yao & Lind, 1969). Neonatal hypovolaemia is particularly likely to occur following Caesarian section for this reason, and this may be one of the factors responsible for the increased incidence of respiratory distress following delivery by Caesarian section.

Neonatal circulation
Blood pressure in the healthy newborn baby during the first 12 hours of life is related to body weight in the manner shown in Fig. 8.3. During the next five days of life, a progressive rise in blood pressure occurs, so that at the age of 1 week blood pressure is usually some 15% higher than it was on the day of birth. There are no really adequate data for total blood volume in babies of very short gestation, but circulating blood volume per kilogram body weight would seem to be reasonably constant throughout the second half of gestation (85–100 ml/kg), and the documented fall in plasma volume (Cassady, 1966) evidently mirrors the slow rise in haematocrit. Haematologic standards for babies of 20–40 weeks' gestation are summarized in Table 8.3, but the available data are extremely meagre and we have no idea of what constitutes the normal range for haematocrit and haemoglobin concentration in babies of short

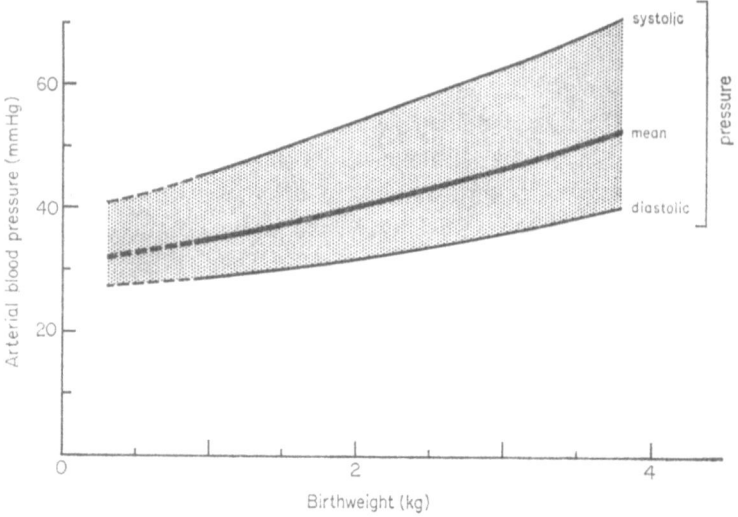

Fig. 8.3. Systolic, diastolic and mean arterial blood pressure during the first 12 h of life in babies of differing weight at birth. Ninety-five per cent of all pressures fall within ±10 mmHg of the values shown. (Taken from the data of Kitterman, Phibbs & Tooley, 1969, with supplementary data on blood pressure in the pre-viable fetus at birth from Rudolph *et al.*, 1971)

gestation; without this information it is impossible to define what constitutes significant anaemia in a pre-term baby. Estimates of haematocrit on capillary blood are higher than estimates obtained from venous blood, but the relation between the two is too variable for a capillary sample to be an adequate substitute for a venous specimen in a sick infant,

Table 8.3. *Mean red-cell values for cord blood. (From the data of Turnbull & Walker, 1955; Marks, Gairdner & Roscoe, 1955; Thomas & Yoffey, 1962 and Lochridge, Pass & Cassady, 1971.)*

		Gestational age (weeks)		
		20	30	40
Haemoglobin (g/100 ml)	{Mean	12·5	14·6	16·8
	{95 % limits	–	–	13·7–20·1
Red-blood-cell count (10^6 [cells/mm^3])		2·8	4·1	5·1
Mean corpuscular volume (μm^3)		135	120	107
Nucleated red cells (cells/mm^3)		8000*	2000	1000
Reticulocyte count (%)		14–20	7–11	4–6
Haematocrit (%)	{Mean	39	45	53
	{95 % limits	–	–	47·3–58·9

* The count is about 2000 if the fetus is delivered by hysterectomy prior to the onset of labour (Playfair, Wolfendale & Kay, 1963).

in the first few days of life. The venous haematocrit and haemoglobin concentration usually rise abruptly by at least 10% in the first few hours of life (Gairdner *et al.*, 1958), and then fall progressively until, by the end of the first week of life, the values are once more much the same as they were in the umbilical cord blood at birth. Mean corpuscular volume and mean corpuscular haemoglobin both fall during the second half of gestation (Table 8.3) as does the amount of fetal haemoglobin, but the mean corpuscular haemoglobin concentration remains nearly constant.

Blood loss
Accidental or iatrogenic blood loss during birth and in the early neonatal period is more common than is usually realized, and such acute blood loss can easily compromise early neonatal adaptation to delivery. Bleeding may have occurred into the mother's circulation, into the uterus, into the placenta or into a twin if there is no evidence of external loss at birth. Internal haemorrhage into the lungs, the abdomen or the skull is an important cause for progressive anaemia after delivery. Loss of 10–15% of the infant's predicted blood volume can cause severe peripheral vasoconstriction throughout the skeletal muscle bed, together with a fall in stroke volume and cardiac output, even though there is little change in heart rate and blood pressure (Wallgren, Hanson & Lind, 1967; Wallgren & Lind, 1967); a loss of this magnitude can also cause a fall in tissue oxygen tension. These babies often become hypothermic and show a very sluggish skin circulation. Larger losses of blood are associated with overt shock, moderate respiratory distress despite good air entry, and significant hypotension. Increased right-to-left shunting may occur, but cyanosis is seldom seen, heart rate only increases slightly and central venous pressure is not always particularly low (Rowe & Arcilla, 1968). A low haemoglobin or haematocrit is strong corroborative evidence (if there is no reason to suspect a haemolytic process), but vasoconstriction may temporarily mask this anaemia making it necessary to take serial specimens.

Clearly any infant who has lost enough blood to become hypotensive has probably lost at least a quarter of his blood volume. Even though the haematocrit may not be particularly low at this stage, the infant is at serious risk of sudden collapse and death, and requires an urgent transfusion with at least 20 ml/kg of group O Rh-negative blood without waiting for the results of a formal crossmatch just as soon as blood has been taken for grouping and other laboratory tests. Plasma, albumin or dextran can be used to expand the circulation temporarily if necessary, but blood must be given just as soon as possible. The infant who has lost 10–15% of his circulating blood volume as a result of an acute haemorrhage may not appear shocked, but he may still have a low cardiac output, marked peripheral vasoconstriction, low tissue oxygen levels and increasing respiratory acidosis. To pick up these milder cases of posthaemorrhagic anaemia is also important, particularly when they are associated with short gestation, respiratory distress, aspiration pneumonia or intrapartum asphyxia. Here the most reliable screening procedure is to monitor the central venous haematocrit or haemoglobin for three or four hours after birth. The first specimen may appear normal but a falling haematocrit in the first four hours of life is very unusual, and a fall of more than 5% strongly suggests recent blood loss.

Chronic blood loss must be handled rather differently. The infant is pale, but cardiac output and peripheral circulation are not usually impaired unless there is evidence of congestive failure with hepatomegaly, cardiomegaly and respiratory distress. The initial haematocrit will be low, but will not alter significantly if monitored for several hours after birth, and the nucleated red blood cell and reticulocyte counts will be high. Here a continuous transfusion with packed cells will help to correct the anaemia, but an over-rapid transfusion could lead to congestive heart failure. An exchange transfusion may occasionally be of help for a severely anaemic baby. Oral supplements of iron are essential, and a further small transfusion may become necessary after a few weeks. The many rarer causes of neonatal anaemia are well discussed by Oski & Naiman (1972).

Polycythaemia

Polycythaemia seldom causes symptoms in the neonatal period, but viscosity increases exponentially when the haematocrit exceeds about 60% (Stone, Thompson & Schmidt-Nielsen, 1968), and decreased blood flow due to high viscosity could well be responsible for the otherwise unexplained cyanosis, respiratory difficulty and congestive heart failure sometimes seen in babies with a central venous haematocrit in excess of 70% (Gatti *et al.*, 1966). Some of these babies exhibit muscular twitching, a few have unexplained convulsions, and significant jaundice is quite common. Renal-vein thrombosis has also been reported. Polycythaemia is particularly common in babies who have suffered severe fetal growth retardation, and is frequently accompanied by hypoglycaemia (Haworth, Dilling & Younaszai, 1967). In babies with symptoms a partial exchange transfusion with fresh plasma in order to bring the haematocrit down to 60% would seem to be the treatment of choice.

Intracranial bleeding

Many babies of short gestation die as a result of intracranial bleeding. Sometimes this is due to damage to the falx, the tentorium or the great vein of Galen during delivery. The tentorium is easily torn by the sudden moulding that occurs during rapid delivery, and breech deliveries are at particular risk; similar damage can often occur when the pre-term baby is delivered by elective Caesarian section before the onset of labour (Fedrick & Butler, 1971b).

Death from intraventricular haemorrhage is an even commoner occurrence in babies of short gestation, particularly where there is a history of intrapartum or early neonatal asphyxia (Fedrick & Butler, 1970b; Harcke *et al.*, 1972). It is not known why bleeding is so common, but it may be that increased cerebral blood flow results in rupture of some of the poorly supported vessels in the area of rapidly growing brain close to the ventricular wall. Towbin (1968), on the other hand, has suggested that intraventricular haemorrhage is frequently a secondary consequence of sub-ependymal venous infarction. Babies of short gestation have decreased levels of the vitamin K-dependent clotting factors (II, VII, IX and X), and Factor XI is also often depressed even in normal full-term babies for several weeks after birth (Cade, Hirsh & Martin, 1969; Hilgartner & Smith, 1965). It is, therefore, somewhat surprising that the pre-term baby's blood often appears hypercoagulable. This is an unsolved paradox.

Gray, Ackerman & Frazer (1968) have argued that, since the asphyxiated low-birth-weight baby is so vulnerable to ventricular haemorrhage, infants with a severe deficiency of factors II, VII and X on Thrombotest screening should be given 10 ml/kg of fresh frozen plasma. While there is no doubt that this improves the coagulation status, it has not yet been shown to produce a statistically significant fall in mortality. There is no doubt that ventricular haemorrhage is commoner in babies with a low 'Thrombotest', and Bryant *et al.* (1970) have also produced evidence that late morbidity is increased in those babies of under 2·5 kg who had a low Thrombotest in the first 12 hours of life. This, however, does not prove that the bleeding was caused by the low factor levels. Furthermore, Appleyard & Cottom (1970) were unable to find any evidence that intrapartum asphyxia affected Thrombotest findings on the day of birth, although asphyxia did seem to impair the response to Vitamin K.

Possibly the low Thrombotest findings on the day of birth were the result of a consumption coagulopathy with disseminated intravascular coagulation; further studies will be necessary to test this possibility. Intracranial bleeding remains a difficult and challenging problem that deserves more attention than it has received in the past.

Thermoregulation

Adult man maintains a constant deep-body temperature with remarkable precision, even in very adverse circumstances, and it is difficult to resist the conclusion that thermal stability is of fundamental importance to the body's welfare. For the fetus to regulate its own body temperature before delivery is almost impossible, and this therefore would seem to be one of the homeostatic mechanisms that are called into play for the first time at birth. Appropriate thermoregulatory responses can, in fact, be demonstrated within a few minutes of birth, even in a baby born 12 weeks before term; there can, therefore, be no doubt that the newborn baby is a true homoeotherm, even though regulatory ability is often extremely limited and easily overwhelmed. The importance of limiting the stress placed on the thermoregulatory mechanism during the first few weeks of life has been underlined by a series of controlled trials in which neonatal mortality in babies weighing less than 2 kg at birth was cut by a quarter when precautions were taken to minimize heat loss and to keep body temperature above 36°C.

Thermal exchange before birth

Fetal metabolism results in the production of heat, and the temperature of the fetus is in consequence higher than that of the surrounding uterine structures. We know nothing about the way in which the temperature gradient varies during development *in utero*, but we do know that the gradient is between 0·6 and 0·8°C at term in man (Mann, 1968) and in sheep. Most of this heat seems to be transferred to the placenta through the umbilical circulation (Abrams *et al.*, 1970), but the capacity for direct heat loss to the surrounding amniotic fluid is larger than is generally appreciated. It can be calculated from a knowledge of tissue insulation that, if the amniotic fluid is 0·3°C cooler than the fetus (Mann, 1968), as much as a quarter of the heat would be dissipated in this way, even if fetal skin were maximally vasoconstricted. Walker, Walker & Wood (1969) showed that the feto-maternal temperature gradient increased when the baby developed a tachycardia during labour, and it is tempting to suggest that both these changes are the result of reduced maternal blood flow to the placenta.

Extremely little is known about metabolic rate during growth *in utero*, and even less about the way in which it can change during periods of stress. Indirect calorimetry is technically difficult since it involves simultaneous measurement of umbilical blood flow and of the arterio-venous oxygen content difference; direct calorimetry provides a largely untried alternative method for monitoring metabolism both *in vivo* and *in vitro*. We do not as yet know whether adipose tissue is ever mobilized during intrauterine life, nor whether the thermogenic brown fat is ever functional at this time. To infuse noradrenaline into a fetal animal and see whether heat production is stimulated might seem a simple matter, but the effect of noradrenaline on the placenta largely vitiates the validity of any such experiment; arterial oxygen saturation is low before delivery, and this is also likely to severely limit any rise in brown-adipose-tissue metabolism.

We know equally little about the way in which skin blood flow varies *in utero*, and nothing at all about the function of the eccrine sweat glands at this time. Since body temperature before birth is higher than the threshold temperature at which babies sweat after birth, it would be interesting to know whether the eccrine sweat glands function before delivery. Their increasing responsiveness in the first week of life indicates that they may have been dormant *in utero*. If this is so the threshold temperature at which sweat-gland stimulation occurs would seem to be 'set' higher than it is after birth, or else some overriding means of suppression operates. Sweat glands produce progressively less sweat when the skin is soaked in water or a dilute salt solution at 36–37°C, and the same process could clearly operate before birth. The decrease in output could be due to swelling of the sweat ducts when the skin is wet; however, visible miliaria usually develops in these circumstances if gland stimulation continues for any length of time. The fact that miliaria is not seen at birth, although a 'sweat rash' often develops soon after birth, may indicate that this thermoregulatory process is normally inactive before birth.

Just how the fetus responds to the diurnal variation in maternal body temperature also remains to be determined. No evidence of any diurnal variation in the set point around which body temperature is regulated has been detected in the newborn baby at birth, and we do not know at what age this rhythm first appears.

Homoeothermy after birth

Although the baby is now generally admitted to be a true homoeotherm, low basal metabolism per unit surface area, small body size and imperfect sweat function are well recognized to severely limit thermoregulatory function after birth. In comparing a baby's ability with that of an adult, the importance of these factors can, however, be exaggerated: the baby's inability to alter, to move away from, or even to complain about an adverse environment in any effective way is in many respects a much more fundamental handicap. Control over body insulation by the use of clothes is man's most effective method of thermal regulation, and dependence on others in this respect is, perhaps, the biggest single handicap of all. Many newborn animals protect themselves from cold stress by snuggling up close to their litter mates or their mothers, and most animals instinctively conserve heat loss by curling up in a ball when cold: all these behavioural responses are usually denied the human infant. Our responsibility for helping a small newborn baby to maintain its body temperature is therefore considerable, and our responsibility increases even further when we deprive the child of clothes and bedding, since a naked baby is much more susceptible to any fluctuation in environmental temperature. Happily, our understanding of thermal exchange and temperature regulation in the newborn baby has been transformed in the last 15 years by studies involving indirect calorimetry under carefully defined environmental conditions within closed metabolic chambers (such as the equipment illustrated in Plate 8.2).

Basal heat production

The resting or basal heat production (basal metabolic rate or BMR) of a healthy baby rises progressively after birth, and similar changes have been documented in a number of animal species. The exact reason for the rise is not known, but it is clearly related to increasing metabolic activity within many of the organs of the body. The idea that the rise is largely due to increased brain activity, or is simply a response to feeding, can be discounted since the increase has been detected in more than one totally unfed anencephalic infant (Cross *et al.*, 1971). Animal experiments have shown that the rise still occurs if the infant is maintained in a neutral thermal environment from the moment of birth, and the magnitude and time course rule out any possibility of the rise being caused merely by the work of breathing.

In the small pre-term baby the rise in basal heat production is a slow progressive change that continues for many weeks after birth (Hill &

Robinson, 1968). The minimum rate is higher than that of an adult, when compared on a weight-for-weight basis, and remains higher throughout childhood. Probably this is at least partly due to the fact that the most metabolically active organs of the body constitute a larger fraction of total body mass at birth than they do in later life. Heat loss, however, is dependent on the total surface area available for thermal exchange, and heat production per unit surface area in the first three months after birth is less than it is during adult life. Indeed, in a baby that weighs under 1·5 kg, basal heat production per unit surface area may be less than half that of an adult throughout at least the first week of life.

Illness can, not surprisingly, affect basal heat production. Babies with poor pulmonary blood flow, and babies with low-output cyanotic congenital heart disease or hypothyroidism, tend to have a low level of heat production and therefore require extra warmth. There is also some evidence that BMR may be marginally depressed in babies with jaundice. In contrast to this, babies with dyspnoea or congestive cardiac failure, and babies who are restless or subconvulsive, often have a high level of heat production and require less warmth. Babies with severe hyaline membrane disease have an unusually low metabolic rate in the first 2 d of life (when the energy cost of breathing is at its highest) and often develop a raised rate for a while during recovery. Attempts to explain the high basal metabolism seen in some infants who are 'light for dates' at birth in terms of an increase in active cell mass have not yet met with success (Bhakoo & Scopes, 1971), since babies who have grown poorly *in utero* have an unusually large amount of extracellular fluid.

Benedict & Talbot were among the first to detect a rise in metabolic rate in the human baby after the ingestion of food, and Mestyán *et al.* (1969) have studied the effect of food intake on metabolic rate in the newborn baby in some detail. It was generally accepted at this time that the 'specific dynamic effect' of a meal was due to its protein content, and was caused by oxidation of amino acids. However, investigations on malnourished children have now shown that a significant postprandial rise in metabolism only occurs during periods of active growth, and appears to reflect the high energy cost of protein synthesis, rather than the energy cost of protein digestion and catabolism within the body (Brooke & Ashworth, 1972). Since the human infant is almost always either actively digesting one meal or actively anticipating the next, a measure of basal metabolic rate uninfluenced by the effect of either physical or metabolic activity is difficult to obtain; estimates of BMR in the newborn almost inevitably contain some allowance for the overall effect of specific dynamic action for this reason. The effect of infusing nutrients intravenously on metabolic rate needs to be investigated; Přibylová and Znamenáček (1970) found that the infusion of 1 g/kg glucose shortly after birth leads to a fall in oxygen consumption. Prolonged starvation is known to cause a fall in BMR in the young infant (Varga, 1959), as in later childhood and in adult life (Ablett & McCance, 1971).

Heat loss

Defective control over skin blood flow was thought to increase the small baby's vulnerability to cold stress, but we now know that even a baby of 28 weeks' gestation can maintain an intense skin vasoconstriction when subjected to cold stress. Nevertheless the ability of the body of a 1-kg baby to conserve heat is only two-thirds of that achieved by a 3-kg baby, and one-third of that achieved in adult life (Hey & Katz, 1970). This reduction in maximum tissue insulation is in part due to small size and to lack of subcutaneous body fat, and in part due to the presence of a relatively high blood flow to the limbs even during maximum vasoconstriction (Kidd *et al.*, 1966; Berg & Celander, 1971).

Heat dissipation through water is rapid and, as a result, the resistance offered by the tissues is the only effective insulation against heat loss from the skin when a fetus is immersed in amniotic fluid. Still air on the other hand is an efficient form of insulation, and the thin boundary layer of still air next to the skin forms a very significant barrier to heat loss from the surface of the body as long as it is not disturbed by draughts or currents of air. Heat loss from the skin by radiation and convection varies inversely with environmental temperature; radiant loss slightly exceeds convective loss in an environment of uniform temperature and high radiant emissivity when air movement is negligible, but draughts cause a rapid rise in convective heat loss. A small child loses heat from the skin slightly faster than an adult under comparable conditions, but the difference is not large and is almost certainly accounted for by increased convective loss from surfaces with a reduced radius of curvature (Hey, Katz & O'Connell, 1970). Direct conductive heat loss to surfaces with which the body is in contact forms a further avenue for heat loss that must never be forgotten, but loss by this route is usually negligible when a baby is lying on a warm mattress of low thermal conductivity.

Evaporation of water from the skin and the respiratory tract causes a further important obligatory loss of heat. Under normal conditions about one-third of this loss occurs through the respiratory tract, but the exact fraction is influenced by respiratory minute volume and rises considerably whenever a baby cries. Loss from the respiratory tract is directly proportional to the difference between ambient vapour pressure and saturation vapour pressure at deep body temperature, and this loss can therefore be controlled by increasing the absolute vapour pressure of the air. Insensible water loss from the skin is a rather more complex process, but is also influenced by ambient vapour pressure (see p. 324). Total evaporative heat loss normally accounts for almost exactly a quarter of basal heat production under conditions of moderate ambient humidity, both immediately after birth and during later life (Hey & Katz, 1969a), but this loss may increase two- to fourfold with the onset of active sweating in the full-term baby.

The relative importance of radiant, convective, conductive and evaporative heat loss in a small baby being nursed in draught-free surroundings of moderate humidity is shown in Fig. 8.6A, p. 301.

Response to cold

Most healthy babies can increase their heat production two-and-a-half times in response to acute cold stress, but the response is always muted on the day of birth, and only increases slowly during the first week of life in babies of low birth-weight (Hey, 1969). Body temperature falls in consequence, and since such a fall often occurs when heat production is well below the maximum that can be achieved, it cannot, by itself, be taken as evidence of an inability to respond. The response to sustained cold stress is frequently less than that seen during brief exposure; indeed in some small babies heat production may actually fall during prolonged cold exposure, and this may be because the fall in body temperature eventually reduces the baby's ability to generate heat.

The response to cold stress is also reduced and sometimes abolished by severe hypoxaemia, but the critical point at which heat production becomes impaired is probably a function of cardiac output and regional blood flow, as well as arterial oxygen saturation, and attempts to define the arterial oxygen tension at which heat production becomes impaired have yielded conflicting data. (Some babies with severe chronic anaemia and many babies with cyanosis as a result of respiratory distress show little metabolic response to cold stress, but babies with cyanotic congenital heart disease usually have a moderately good response as long as cardiac output is not reduced.) Sedation after birth reduces the metabolic response to cold stress, as does maternal sedation with pethidine or diazepam during labour. Severe birth asphyxia also causes a variable reduction in the metabolic response to cold stress that may last many days, and these babies are therefore particularly vulnerable to cold stress.

New-born babies rarely shiver and, while restlessness and muscular activity certainly increase in the cold, most of the observed rise in heat production is now thought to be due to oxidation of fat within brown adipose tissue. This tissue has recently been shown to be responsible for much if not all the non-shivering thermogenesis seen in a wide range of neonatal mammals, and is also now known to be involved in arousal after hibernation as well as in the adaptation of certain small mammals to chronic cold stress (Lindberg, 1970; Jansky, 1971).

Aherne & Hull (1966) described the appearance and distribution of brown fat round the neck and in the axilla, round the kidneys and deep in the thorax of the human infant and reviewed the evidence in favour of its thermogenic role. It was then thought that brown adipose tissue lost its ability to produce heat when depleted of fat, and the discovery that babies dying of hypothermia had fat-depleted brown adipose tissue seemed to be further confirmation of this view, but we now know that brown adipose tissue can produce heat by utilizing fat mobilized from other sites when its own reserves are depleted (Hardman, Hey & Hull, 1969). Most cases of neonatal hypothermia are therefore probably the result of cold stress overwhelming rather than exhausting the thermo-regulatory mechanism. Nevertheless a baby weighing less than 1·5 kg at birth has (weight for weight) only about a fifth as much fat as a normally grown baby at term, and there is some evidence that tiny babies may occasionally develop marasmic hypothermia even in the first two weeks of life, if both brown and white adipose tissue reserves become depleted as a result of undernutrition at this time (Hey & Katz, 1969b).

Response to warmth

Skin blood flow increases in warm surroundings enough to reduce tissue insulation to one-third what it is in the cold, and even very small babies possess a thermoregulatory control over skin blood flow comparable to that achieved in adult life (Hey & Katz, 1970a).

The ability to sweat is more limited, particularly in babies born more than two or three weeks before term, while babies born more than eight weeks before term seem unable to sweat at all. Provocation tests with intradermal acetylcholine indicate that this is almost certainly because of the functional immaturity of the eccrine glands in the skin. It has been possible to show that the glands over the head and upper trunk mature earlier than those elsewhere on the body both structurally and functionally, and that there is, in general, a slow caudal progression of maturation that is marginally hastened by the stimuli resulting from premature delivery (Foster, Hey & Katz, 1969). Because of this immaturity the pre-term baby is particularly vulnerable to the risks associated with hyperthermia if subjected to heat stress.

'Set point' control

Babies only vasodilate in warm surroundings when rectal temperature exceeds 36·6–37·3°C, and never sweat until rectal temperature exceeds 37°C. Threshold temperatures are, if anything, rather higher than this in the first few days of life, and babies of short gestation behave in much the same way as babies born at term. Body temperature is often low in the first few days of life, but there is no evidence that body temperature is usually regulated at this subnormal level, and the low body temperature is certainly not due to lack of vasomotor tone. The 'set point' around which body temperature is regulated seems, therefore, to be very much the same immediately after birth as it is in later life. Indeed, there is some evidence that the 'set point' may be higher than normal at birth, as a result, perhaps, of the high body temperature experienced during fetal life.

It is true, of course, that rectal temperature often deviates from this 'set point' in man, even when the imposed hot or cold stress is sufficiently small to be easily capable of total compensation, particularly in the first few days of life. It could be argued that the temperature-controlling mechanisms might be acting to maintain a constant temperature at some site or sites in the body not adequately reflected by the measurement of rectal or colonic temperature, and temperature sensors have been described in the spinal canal close to the cervical brown adipose tissue in the guinea pig that could function in this way (Brück & Wünnenberg, 1970). This is unlikely to be the complete explanation for the lability of rectal temperature at birth in man because rectal and oesophageal temperature becomes progressively more stable with increasing post-natal age before there is any significant change in body size or over-all thermal conductivity. Conversely, this lability of deep body temperature could be a sign that the regulatory system functions inadequately when first called into play immediately after birth: a limited and controlled change in body temperature during thermal stress would, however, help to conserve water and energy reserves, and this damped response may possibly be a physiological adaptation of some value. This amounts to a suggestion that the newborn baby is behaving in this respect like a number of adult mammals including, for example, the desert camel.

Tiny babies receiving conventional nursing care have often, in the past, had deep-body temperatures of less than 36°C throughout the first week of life, and in some units body temperature was normally allowed to stabilize at 33–35°C if a baby of less than 2 kg was already cold on admission. This hypothermia was often not associated with any sustained increase in heat production (Kintzel, 1966), and in certain circumstances heat production decreased when body temperature was allowed to fall (Silverman & Agate, 1964). Some of these babies probably lacked the ability to increase their heat production in response to cold stress, but most babies show a brisk if poorly maintained metabolic response to a decrease in environmental warmth that increases the gradient between skin and air temperature and threatens a further fall in deep-body temperature. Brück, Parmelee & Brück (1962) concluded that these cold pre-term babies were regulating their deep-body temperature round a low set point, and believed that most babies of less than 1·5 kg behaved similarly. Certainly although most babies maintain an intense skin vasoconstriction when rectal temperature is less than 36·5°C, skin blood flow in some appears to be rather less restricted; nevertheless in most instances there is a further large increase in skin blood flow once rectal temperature exceeds 36·5°C suggesting perhaps not so much a low set point as a broadened thermoregulatory temperature range. Brück *et al.* (1962) believed that these babies had a low set point at birth, but recent observations by the author suggest that this is an adaptation that only develops *after* birth in response to fairly prolonged cold stress, and disappears again as soon as the baby is rewarmed.

It has often been argued that a low body temperature is 'normal' in small babies and that, since heat production is not stimulated, these babies are, by definition, in a neutral environment. Since, however, controlled trials have now shown that mortality rises when deep body temperature is allowed to remain below 36°C (see Table 8.4), it seems wrong to call such an environment 'neutral'.

Thermal neutrality

Between the extremes of environment that provoke either an increase in heat production or an increase in active sweat production there lies a narrow zone within which changes in posture and blood flow are sufficient to control and regulate body temperature in the full-term baby. This neutral zone cannot be defined in terms of temperature alone, since

Table 8.4. *The effect of increasing environmental warmth on the survival of babies weighing less than 1·5 kg nursed naked in incubators from birth.*

Authors	Number of babies studied	Change in environment under test	Resultant decrease in mortality (%)
Silverman, Fertig & Berger, 1958	98	Incubator air temp. maintained at 31·7° instead of 28·9°C	40
Jolly, Molyneux & Newell, 1962	74	Air temp. high enough to keep rectal temp. at 36·5–37·2°C *instead* of a fixed air temp. of 29·4°C	19
Buetow & Klein, 1964	158	Supplementary radiant heat to keep abdominal skin temp. at 36°C *in addition* to an air temp. of 31·5°C	22
Day *et al.*, 1964	81	As for Buetow & Klein (1964) but including babies of up to 1·6 kg in weight	43

thermal balance is also affected by a host of other factors such as the nature of the radiant environment, ambient humidity, mattress conductivity, wind speed, posture and clothing, but neutral conditions can be described in terms of this one variable when the remaining factors are reasonably well defined and relatively constant. As a result, calculation of the air temperature most likely to provide neutral conditions for a full-term baby being nursed in a cot or an incubator has recently become possible (Hey & Katz, 1970b).

The same can be done for a tiny baby weighing only 1 kg at birth, but, since such a baby is unable to sweat and is often only capable of a transient increase in heat production in response to cold stress, the neutral thermal range can only be calculated from a knowledge of basal metabolic rate, evaporative water loss and total thermal insulation. Fig. 8.4 shows the calculated range of thermal neutrality when such a baby is nursed in draught-free surroundings of moderate humidity (∼ 50% RH) and uniform temperature, during the first month of life either naked, or fully clothed and well wrapped under blankets in a cot. (It has been assumed that such a baby will have a 'set point' attempting to maintain deep body temperature at 37°C in the absence of hypothermia due to prior cold stress.) Increasing incubator humidity from

25 to 75% nearly halves the baby's evaporative water loss but this normally only lowers the neutral temperature range by about 0·5°C.

We cannot yet predict the neutral thermal environment for a baby weighing substantially less than 1 kg, because we lack data on basal metabolism and on the rate at which heat is lost by evaporation through

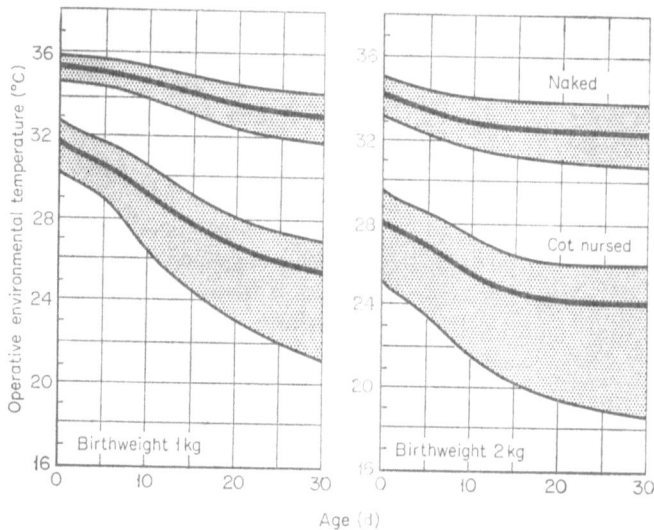

Fig. 8.4. Diagram summarizing the changes that usually occur with age in the mean temperature to provide reasonable warmth for babies weighing 1 and 2 kg at birth, when nursed in draught-free surroundings of uniform temperature and of moderate humidity (50% RH). The thicker lines indicate what is probably the usual 'optimum' temperature (at the lower limit of the neutral environmental temperature range), and the shaded area the range of temperature within which most babies can be expected to maintain a normal body temperature without increasing either their heat production or their evaporative water loss by more than 25%. The higher temperatures are appropriate for a naked baby, and the lower temperatures for a baby that is fully clothed and well wrapped in a cot. (Modified from Hey & Katz (1970b) to make due allowance for the high evaporative heat loss now known to occur in babies of less than 30 weeks' gestation in the first few weeks of life)

the thin, almost transparent, skin. Such an infant would clearly require a higher temperature than one weighing 1 kg. This is a problem that deserves further study.

The provision of optimum warmth

A neutral thermal environment cannot be assumed to provide optimum conditions for intact survival, though there can be little doubt that it approximates quite closely to this ideal. Babies can cope with lowered temperatures up to a point, but four controlled trials have now shown that mortality rises significantly if small babies are nursed in surroundings more than a degree or so below the neutral range. In the studies listed in Table 8.4, mortality in the warmer environment was only 60 to 80% of what it was in colder surroundings. The exact cause of the increased mortality is uncertain: the increased oxygen and calorie requirement may in itself be a hazard, while changes in regional blood flow may compromise other aspects of homoeostasis, but hypothermia is clearly a further danger once the thermoregulatory mechanism is overwhelmed.

Unfortunately, although it is usually easy to detect when environmental temperature exceeds the neutral range because of sweating, restlessness and a rise in body temperature, it is not always easy to detect when environmental temperature falls below this range: there may again be some restlessness and the extremities may become noticeably cool, but the change in rectal temperature is often small at first because of increased heat production. Optimal management is made more difficult by the fact that the neutral range for a naked baby is so narrow. Since apnoeic attacks seem to be more frequent when small babies are nursed in a temperature at the upper end of the neutral range, and since the pre-term baby's ability to regulate body temperature is particularly inadequate when environmental temperature is too high, the lower end of the neutral range would seem to be the most logical temperature to aim for in clinical practice.

Babies grow faster in a neutral environment, although the difference can be offset by increasing the calorie intake of those in a sub-neutral environment (Glass, Silverman & Sinclair, 1968; 1969). Since there is now some evidence that brain growth may be permanently affected if growth is interrupted for even a brief time in the early neonatal period by a combination of low calorie intake and high heat loss due to inadequate environmental warmth, the importance of maintaining optimum growth takes on added relevance. However, one slightly unfortunate consequence of nursing babies naked in a neutral environment is that they fail to develop an increased ability to withstand cold stress with time in the same way as babies nursed under cooler conditions (Glass et al., 1968). Possibly this difference is due to increased thermogenic potential in the brown fat of the cold-exposed baby, analagous to that observed in the newborn guinea-pig (Zeisberger et al., 1967), newborn lamb (Alexander, Bell & Williams, 1970) and newborn rabbit (Hardman & Hull, 1971).

297

Delivery-room care

The human infant is particularly vulnerable at the moment of delivery, and babies subjected to cold stress at this time have both an increased metabolic acidosis (Gandy *et al.*, 1964) and a lower arterial oxygen tension (Stephenson, Du & Oliver, 1970). The metabolic response to cold stress is muted at birth and may be further reduced by sedation or asphyxia. Radiant and convective heat loss will be high in a cool and sometimes draughty delivery room, but just how much heat is lost as a result of water evaporating off the skin is not generally realized. It is not just the easily-removed water on the surface of the skin that evaporates: there is, in addition, a measurable loss as the stratum corneum dries out. In consequence, heat loss due to evaporation can exceed heat production and cause body temperature to fall for more than an hour after birth, even when radiant and convective loss are negligible.

Ingenious plastic devices such as the Silver Swaddler designed by Baum & Scopes (1968) and the transparent Baby Bag of Besch *et al.* (1971) have recently come into use for swaddling babies immediately after birth. These were designed in the first instance to minimize radiant and convective heat loss, respectively, but they may work mainly by reducing evaporative heat loss. Whether these devices are really better than blankets in reducing radiant and convective heat loss remains to be seen. Radiant heat canopies are also in widespread use. These devices serve a useful purpose but it needs to be stressed that the use of a fixed-wattage heater will rarely provide optimum warmth, and their use could do more harm than good if it encouraged staff to believe that transfer to a more adequate environment was no longer urgent. Unfortunately, many commercial heaters have too low a wattage at present. An unfocused 400-watt radiant-heat source suspended 60 cm above a baby delivers enough heat to balance radiant and convective loss from a 2-kg baby in a draught-free room at about 22°C, but it cannot compensate for the variable loss of heat that occurs when the skin is wet, nor for the extra loss caused by draughts.

We do not at the moment know exactly how important it is to prevent excessive heat loss at birth. Analysis of the data published by Buetow and Klein suggests that providing optimum warmth throughout the first two weeks of life is important, but that careful regulation of incubator temperature makes relatively little difference to mortality in the first 24 h after birth (Fig. 8.5). It is however possible that the babies who died towards the end of the first week of life died because they had been cold during the first few critical hours after birth. There are those who argue, on the basis of animal experiments, that deliberate hypothermia is of value in the management of acute birth asphyxia (Dunn & Miller, 1969). Oxygen requirements can be halved by a 10°C fall in deep body temperature, and there is no doubt that hypothermia can prolong survival in small mammals suffering from severe asphyxia; whether body temperature can be lowered fast enough for this to be of any value in an infant weighing more than 1 kg at birth is more debatable, because endotracheal intubation, artificial ventilation and cardiac massage will nearly always restore adequate oxygen uptake within 12 min of birth,

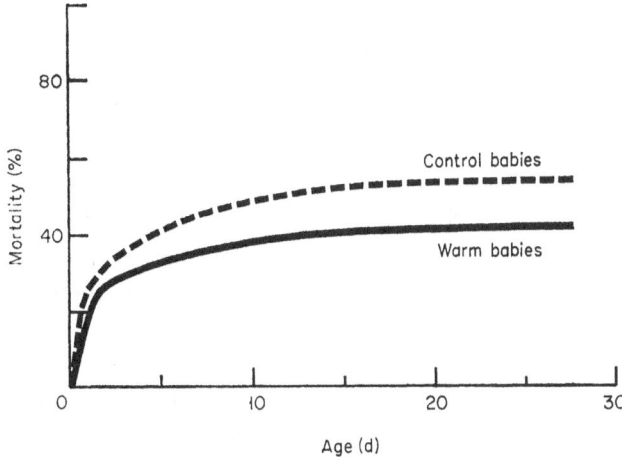

Fig. 8.5. The effect of providing sufficient additional radiant warmth to maintain abdominal skin temperature at 36°C on cumulative neonatal mortality in babies of less than 1·5 kg at birth, nursed in a single-walled commercial incubator with an air temperature of 31·5°C during the first month of life. (From the data of Buetow & Klein, 1964)

even after the most severe acute asphyxia, many minutes before spontaneous respiration is seen (Hey & Kelly, 1968). Hypothermia is less likely to serve any useful function once adequate oxygen uptake is assured and to lower the deep body temperature of a large baby fast is difficult. Whether hypothermia is of any value in the management of the cerebral oedema seen after chronic intra-partum asphyxia is not known. The effect of prolonged partial asphyxia is very different from the effect of acute total asphyxia (Myers, 1972), and the value of hypothermia during recovery from chronic fetal distress merits further study.

The correct way to manage a baby who is already hypothermic is also uncertain. Active rewarming may be dangerous and it may be better to allow body temperature to stabilize at between 32° and 34°C (Kintzel, 1956; Breunung, 1965). This view is difficult to reconcile with the data summarized in Table 8.4 and Fig. 8.5, and most paediatricians believe that these babies should be rewarmed in a cot or warm air incubator at a rate not exceeding 0·5°C per hour. Babies who are hypothermic on arrival in hospital on the first day of life are well known to have an increased mortality (Kunnas, 1968); this does not prove that hypothermia is itself dangerous, since the hypothermia might well be a secondary consequence of some co-existing illness. Further control studies will need to be done to determine whether there is ever any advantage in allowing an *ill* baby to become moderately hypothermic immediately after birth. Further work is also necessary to establish the best way of managing babies who are already cold on admission.

Incubator care

That the small pre-term baby is particularly vulnerable to cold stress has long been known. Swaddling has been employed as one method of minimizing the risk of hypothermia from time immemorial, while specially designed heated cots and incubators have now been in use for nearly a hundred years. The obstetricians and paediatricians who developed the early incubators were convinced that many babies could be saved from death if active steps were taken to prevent hypothermia, and Budin, working in Paris at the turn of the century, became convinced that small babies should be nursed in surroundings maintained at between 25° and 30°C (depending on age and weight), even though every infant was also fully clothed and covered with blankets. Blackfan & Yaglou (1933) also came to the same conclusion 30 years later after four years' experience with the new air-conditioned nurseries at Boston in America, and measurements of thermal exchange have now served to confirm these views (Fig. 8.4).

More recently, as a result of progressive refinements in incubator design, it has gradually become the practice for almost every baby that is cared for in an incubator to be nursed naked, and it is probably true to say that the principal indication for incubator care is now the desire to leave a baby unclothed in order to keep it under constant observation. The problem is that, when a small baby is nursed naked, control and maintenance of environmental temperature within very precise limits is essential. The *minimum* temperature to provide neutral conditions for a 1-kg baby usually exceeds the *maximum* safe temperature for a 2-kg baby during the first two weeks of life. Furthermore, any incubator capable of providing neutral conditions for a 1-kg baby on the day of birth will inevitably expose the same infant, when older, to the risk of hyperthermia, if the control thermostat is turned full on; the concept of a single safe maximum air temperature is therefore illusory.

A further complication of attempting nursing care in a standard single-walled commercial incubator is that the internal radiant environment is very much influenced by room temperature. Fortunately the magnitude of this effect can be predicted fairly precisely. Since the Perspex walls of such an incubator are only heated indirectly, their temperature is influenced by room temperature as much as by incubator temperature; in consequence, while a uniform environment of 34°C will, for example, provide neutral thermal conditions for a healthy four-day-old infant of 1·5 kg, this air temperature will not suffice if the infant is nursed in a single-walled incubator housed in a room at 21°C, because of increased

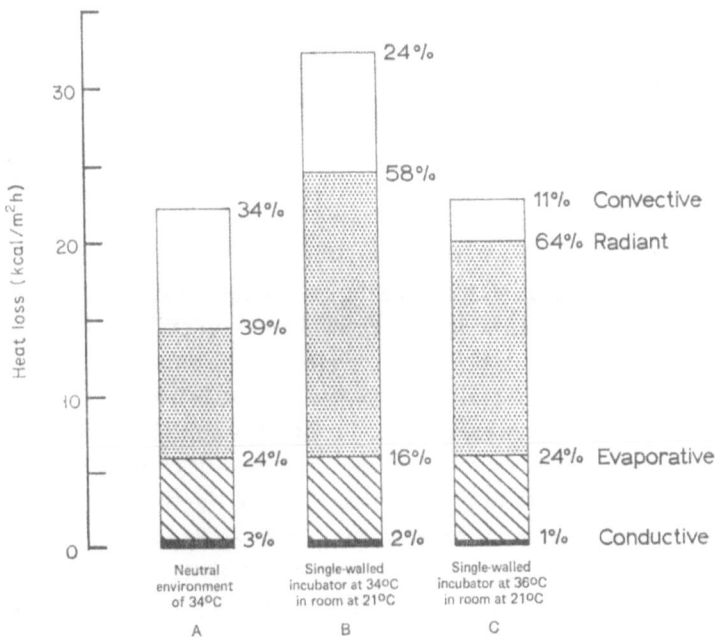

Fig. 8.6. Heat loss from a naked 4-d-old baby of 1·5 kg with a rectal temperature of 37°C in draught-free surroundings of moderate humidity (50% RH).
A. Heat loss by conduction, convection, radiation and evaporation in a neutral environment of uniform temperature (34°C).
B. Heat loss within a single-walled commercial incubator having the same air temperature (34°C) when room temperature is low.
C. Heat loss in a similar incubator after increasing air temperature 2°C

radiant heat loss to the cold Perspex (Fig. 8.6, compare columns A and B). Neutral conditions can of course be achieved either by screening the infant from these cold surfaces with a second free-standing Perspex screen inside the incubator or, more simply, by increasing incubator air temperature about 1°C for every 7°C by which incubator air temperature exceeds room temperature (Fig. 8.6, column C). Unfortunately, this is sometimes impossible without increasing incubator air temperature above 35°C (the highest air temperature allowable under current British Standards), and management is made very much easier if all incubators are kept in a room with a stable temperature of about 30°C wherever possible (Hey & Mount, 1967).

To make any accurate allowance for the influence of localized sources of extraneous radiant heat is more difficult. Perspex, like glass, is opaque to radiant heat from surfaces with a low temperature (like the walls of a room or human skin). Such radiant heat, when it is not reflected, is therefore absorbed by the glass or Perspex which change temperature as a result. Radiation from the sun, on the other hand, is of a sufficiently short wavelength to pass through glass or Perspex, as is a little of the radiation that comes from filament lamps and heaters. A glass greenhouse works because short-wave radiant heat from the sun can enter but long-wave radiant heat from within is unable to escape. In this respect a Perspex incubator and a glass greenhouse are strikingly similar.

Equipment that allows incubator warmth to be servo-controlled by means of a thermistor strapped to the baby's abdomen has become commercially available in the last 10 years. However, there is always some risk of hyperthermia should the sensing thermistor become detached, and the use of a servo-control mechanism also deprives the temperature chart of much of its potential clinical value. It is not always realized just how precise a control over skin temperature has to be maintained if neutral conditions are to be achieved. Some of the thermistors sold for use with such equipment misread the temperature by as much as one degree centigrade and experience has shown that regular checks are necessary to detect and correct drift in the setting of the control mechanism.

There are certain advantages in nursing a very small baby unclothed: the condition of such an infant is easier to keep under continuous observation and any jaundice can be treated with phototherapy. The amount of handling to which the baby is exposed can also be reduced, and this may be more important than is sometimes realized in the first few days of life: babies with respiratory distress often react unfavourably

302

to being handled, while babies who are prone to attacks of apnoea are particularly liable to have a severe episode of apnoea with bradycardia after being disturbed. Babies of very short gestation have a tender and delicate skin that is easily chafed and excoriated. Furthermore there is evidence that moisture around the umbilicus and perineum and in the flexures can facilitate colonization of the skin by yeasts, fungi and potentially pathogenic bacteria.

Nevertheless it needs to be remembered that a naked baby is almost three times as susceptible to any fluctuation in environmental temperature as a baby that is clothed and swaddled in blankets. The face and upper respiratory tract possess important sensory receptors, and there are theoretical reasons for avoiding excessive warmth over the face; clothing certainly helps to minimize the risk that a high and fluctuating air temperature might precipitate temporary apnoea. Gentle swaddling has a subtle soothing effect that should not be overlooked merely because it is poorly understood. There is certainly a case for providing some babies in incubators with light clothing; indeed whether incubator care is ever really necessary for a healthy baby of more than about 1·2 kg after the first few days of life is questionable. Care in a cot in a well-designed temperature-controlled room is easy and inexpensive; access is easy and the anxiety caused by maternal separation can be minimized. Clothes and blankets provide a warm and stable environment that is almost entirely free of the twin perils of serious cold stress and accidental hyperthermia, and the ability to withstand cold stress seems to develop rather better under conditions of thermal neutrality, when small babies are nursed in cots instead of incubators (Glass, 1971). While the explanation for this observation is not as yet certain, the brief periods of mild cold stress that occur each time the cot-nursed baby is fed or toileted probably serve to enhance thermoregulatory ability.

Liver function

The liver is one of the largest organs of the body at birth, being responsible for more than 4% of the body's entire weight, and at mid-gestation its relative weight is even greater (Tanimura et al., 1971). It serves as an important storage organ in fetal life, and glycogen is accumulated in high concentration during the last third of gestation before being rapidly liberated following delivery. There is some evidence that a similar reserve of readily mobilizable fat is also built up in the human liver during fetal life (Aherne, 1965), but this phenomenon does not appear to have been studied in detail. Similar stores have been documented in a number of animal species, but in man the amount of fat in the liver is only a small fraction of the total reserve at term.

The liver performs an important haemopoietic function during fetal life: it is the chief organ of erythropoiesis from the third to the sixth month of gestation, and red-cell precursors account for nearly half the nucleated cells within the organ at this time. Marrow cellularity becomes maximal by about the thirtieth week of gestation, but erythropoiesis usually continues within the liver until shortly after term. Synthesis of the many important proteins present in plasma is another of the many functions assumed by the liver at an extremely early gestational age. The various proteins first appear at differing times, the magnitude of the various fractions varies during gestation, and a number of protein fractions such as α fetoprotein are only found during fetal life (Gittin & Boesman, 1966; Gittin & Biasucci, 1969). Liver synthesis of the vitamin K-dependent clotting factors also increases slowly during fetal life, and these factors only reach adult levels after birth. There is no evidence that this is due to any lack of vitamin K.

The activity of many liver enzymes increases rapidly at the time of birth, and a number of enzymes only show detectable activity after delivery. The change seems to be triggered off by the process of birth itself, irrespective of gestational age at the time of delivery. Some of the enzymes associated with amino-acid degradation behave in this way. Thus tyrosine aminotransferase, p-hydroxyphenyl pyruvate hydroxylase and phenylalanine 4-hydroxylase appear to remain relatively inactive within the human liver until after birth, but show a rapid increase in their activity after delivery, and this probably accounts for the significant loss of tyrosine that occurs in the urine in the first few days of life, and for the high plasma phenylalanine and tyrosine levels found in the pre-term baby in the immediate post-natal period (Menkes & Avery, 1963). Increased adrenocortical activity probably accounts for at least part of the observed post-natal activation of tyrosine aminotransferase, since the increase can be prevented by adrenalectomy; however, an additional inhibitory factor appears also to be operating during fetal life (Sereni & Principi, 1971). High tyrosine levels can persist in the blood of the pre-term baby for several weeks, particularly if the infant is on a high-protein diet (Snyderman, 1971), and there is now some evidence that this can have a detrimental effect on intellectual performance in later life (Menkes *et al.*, 1972).

Bilirubin metabolism

Bilirubin is the final breakdown product that results from the catabolism of haeme within the reticuloendothelial system. The details of the metabolic pathway are not known, but an oxidative process is clearly involved in opening the alpha methene bridge, with the loss of one carbon atom and the formation of one molecule of carbon monoxide for each molecule of haeme so metabolized. One gram of haemoglobin yields 34 mg of bilirubin and about three-quarters of the bilirubin produced by the newborn infant normally comes from the destruction of circulating red cells (Vest, Strebel & Hauenstein, 1965). Once bilirubin leaves the reticuloendothelial system, it is transported to the parenchymal cells of the liver where a complex enzyme system controls the conjugation of bilirubin to form bilirubin glucuronide. The key enzyme in this process is glucoronyl transferase, but bilirubin conjugation appears to be a complex process resulting in the formation of sulphates as well as mono- and di-glucuronides. Serum containing water-soluble conjugated bilirubin gives an immediate red colour when added to diazotized sulphanilic acid in the Van den Bergh reaction, and this is often termed direct-reacting bilirubin; unconjugated (or indirect-reacting) bilirubin, on the other hand, is insoluble in aqueous solutions, and only gives a diazo reaction after the addition of alcohol. Conjugation is essential for effective biliary excretion of bilirubin; significant levels of conjugated (direct-reacting) bilirubin in the serum are usually only seen in patients with obstructive jaundice.

Little is known about the rate at which bilirubin is produced in the fetus, but there are reasons for believing that production is at least as rapid as it is in adult life (weight for weight), and recent studies on the fetal monkey have confirmed that a good deal of unconjugated bilirubin crosses the placenta (Bernstein et al., 1969). Conjugated bilirubin on the other hand does not cross the placenta, and the fetal primate liver has little ability to conjugate or excrete bilirubin. This has frequently been regarded as evidence of enzymatic 'immaturity', but, since lack of conjugating ability allows most of the bilirubin produced to cross the placenta and be excreted by the mother, the lack of conjugating ability may more appropriately be regarded as an example of integrated adjustment to intrauterine life.

However, as a consequence of this, liver function has to undergo a rapid and fundamental change immediately after birth. Almost every baby develops some degree of 'physiological' jaundice in the first few days of life and, whereas serum unconjugated bilirubin is seldom more

305

than 2 mg% in adult life, it is frequently much higher than this immediately after birth. While serum bilirubin seldom exceeds 12 mg% in babies of more than 36 weeks' gestation, it exceeds 15 mgm% in about 10% of babies of less than 35 weeks' gestation, usually reaching a maximum between 4 and 6 d after birth (Barton, Wilson & Walker, 1962).

This hyperbilirubinaemia is probably the result of a number of interrelated factors. In the first place, measurements of carbon-monoxide production show that the normal newborn baby produces twice as much bilirubin per day as an adult per kilogram body weight (Maisels *et al.*, 1971), and this in turn is probably due to the fact that the baby has a higher circulating red-cell volume, and a shorter mean red-cell life span (Garby, Sjölin & Vuille, 1964). Profound changes in hepatic blood supply occur at birth; the liver receives most of its blood supply from the well-oxygenated umbilical vein before birth, but this supply is abruptly cut off at delivery, leaving the liver to be supplied by the poorly oxygenated portal vein. Blood flow to the portal system may well remain small until the gastro-intestinal tract becomes active, and there may, in addition, be a significant 'shunt' through the ductus venosus.

Reabsorption of unconjugated bilirubin from the gut and enterohepatic recirculation will further increase the load on the neonatal liver. Recirculation is normally minimal in the adult because conjugated bilirubin is reduced to urobilin by the bacterial flora within the gut. Lack of urobilin production causes increased recirculation immediately after birth and, in the presence of significant beta-glucuronidase activity within the gut, deconjugation of bilirubin also occurs (Brodersen & Hermann, 1963). The faeces of the newborn baby thus contain unconjugated bilirubin, which is free to recirculate through the liver. Poland & Odell (1971) were recently able to demonstrate the significance of this enterohepatic circulation by feeding babies with agar, which bound the bilirubin present in the gut, prevented recirculation, lowered the serum bilirubin level and increased bilirubin excretion. Serum bilirubin levels tend to be lower in babies who are allowed to start feeding immediately after birth (Wennberg, Schwartz & Sweet, 1966; Wu *et al.*, 1967), and earlier bacterial colonization of the gut, together with earlier meconium elimination (Smallpeice & Davies, 1964), may be contributory factors. In addition, there is also some evidence that starvation enhances the conversion of haeme to bilirubin at least in the neonatal rat (Thaler, Gemes & Bakken, 1972).

Immaturity of the conjugating system of the liver has usually been considered to be the major cause of physiological jaundice immediately after birth, and glucuronyl-transferase activity is low in the newborn pre-term infant (Lathe & Walker, 1958), as it is in a number of other neonatal mammals. However Di Toro, Lupi & Ansanelli (1968) took liver-biopsy samples from pre-term babies and studied the ability of these samples to conjugate 4-methyl-umbelliferone: they found no significant conjugating capacity until the fourth week of life, although jaundice had vanished within two weeks of birth. This suggests that the resolution of jaundice seen soon after birth is not directly related to maturation of the conjugating system (assuming, of course, that the transferase system responsible for glucuronide conjugation is the same for both these substrates). Furthermore, although glucuronyl-transferase activity may be low at birth, there is no doubt that the human fetus can conjugate bilirubin to a limited extent (Brodersen et al., 1967; Bakken, 1970).

Recent studies suggest that the presence and activity of two recently isolated intracellular hepatic receptor proteins may be rather more closely correlated with the incidence and time-course of neonatal jaundice in the new-born monkey (Levi, Gatmaitan & Arias, 1970). These two proteins (designated y and z) seem to be responsible for the subcellular localization of bilirubin within the liver cell; the concentration of y is low in the liver of monkeys at birth, but reaches an adult level within five days of birth, at very much the same time as the serum bilirubin level starts to fall. Further work may well show that a similar relative deficiency of y is implicated in the pathogenesis of neonatal jaundice in man.

Kernicterus

The association between severe unconjugated hyperbilirubinaemia and neonatal kernicterus was first appreciated by Mollison and Cutbush in 1949, and the identification of unconjugated bilirubin within the yellow-stained brain-stem nuclei of babies dying with kernicterus later gave added substance to the belief that bilirubin is itself directly responsible for the clinical and histological evidence of central nervous damage (Claireaux, Cole & Lathe, 1953). This damage may result in early neonatal death, or leave the baby handicapped by gross athetosis and deafness; hyperbilirubinaemia can probably also cause more subtle forms of neurological handicap (Hyman et al., 1969).

Bilirubin is normally transported in the plasma firmly bound to albumin, and bound bilirubin appears to be non-toxic; free, unbound and unconjugated bilirubin, on the other hand, is able to cross the

THE MAMMALIAN FETUS *IN VITRO*

blood–brain barrier and damage the cells of the central nervous system. Guinea-pigs given an infusion of unbound bilirubin developed severe neurotoxicity, but toxicity and mortality were greatly reduced if albumin was administered soon after bilirubin first entered the CNS, in spite of the fact that albumin administration caused a marked increase in serum bilirubin as pigment that had entered the tissues returned to the circulation (Diamond & Schmid, 1966).

Kernicterus is rare except in the newborn baby, and this may indicate that the blood–brain barrier of the infant is 'immature' in some way. However, no such variation in vulnerability with age has been detected in animals, and kernicterus has recently been seen to develop in a 16-year-old boy with Crigler-Najjar syndrome as a result of hyperbiliru-binaemia (Blumenschein *et al.*, 1968). Severe unconjugated hyperbiliru-binaemia is, of course, rare outside the neonatal period, and the low plasma albumin level of the newborn and pre-term infant seems likely to increase the risk that significant quantities of unbound bilirubin will accumulate at this age. Kernicterus seldom occurs before the serum bilirubin level exceeds 20 mg%, and clinical management is at the present time guided by this generalization. Some anxiety has recently been expressed that more subtle forms of brain damage may occur before jaundice becomes as severe as this (Boggs, Hardy & Frazier, 1967), but the conclusions of this study have been criticized because of the short period of follow up and the failure to make any adjustment for gestational age at the time of reassessment. Other long-term follow-up studies agree in concluding that there is little correlation between neo-natal hyperbilirubinaemia and neurological handicap, as long as the serum bilirubin level is kept below 20 mg% (Culley *et al.*, 1970; Crichton *et al.*, 1972).

There have nevertheless been a number of reports of kernicterus occurring in small pre-term infants when the serum bilirubin level was well below 20 mg%, and in one instance kernicterus occurred when the serum bilirubin level was only 9·4 mg% (Gartner *et al.*, 1970; Ackerman, Dyer & Leydorf, 1960). Plasma albumin levels will often be particularly low in such babies, and this may partly explain the increased risk. Many of these babies were ill in the neonatal period, and there is now a good deal of evidence that acidosis seriously diminishes plasma-albumin binding capacity. Acidosis and asphyxia are also known to increase the vulnerability of the brain to subsequent hyperbilirubinaemia, since the metabolically active brain-stem nuclei seem to be particularly vulnerable to both hyperbilirubinaemia and asphyxia (Lucey *et al.*, 1964; Lending,

Slobody & Mestern, 1967). In addition to this, the clinician needs to remember that many other ions compete with bilirubin for the various binding sites available, and that drugs such as sodium benzoate, which is often used as a preservative in the preparation of drugs for intravenous infusion (Schiff, Chan & Stern, 1971), also behave in this way. The same is true for certain antibiotics such as the sulphonamides, cephalo-thin and novobiocin (Malaka-Zafiriu & Strates, 1969). A rise in the plasma non-esterified-fatty-acid level as a result of starvation or cold stress will also cause increased competition for binding sites.

For these reasons a number of attempts have been made to assess protein-binding capacity more directly. The measurement of plasma-albumin concentration will give an approximate index of capacity, but it gives no indication of the extent to which binding sites are being monopolized by other competitive anions. Many of the methods so far proposed have ultimately been found to give misleading results; a recently described technique for measuring the amount of bilirubin that can be displaced from albumin by the addition of salicylate shows greater promise (Odell, Storey & Rosenberg, 1970) and warrants further assessment. A method for measuring the amount of free bilirubin by adsorbing all unbound bilirubin onto a Sephadex column has also shown great promise (Blondheim et al., 1972). However, it is doubtful whether very much free bilirubin ever accumulates in the blood stream, since bilirubin will tend to enter the tissues of the body just as soon as the most avid binding sites in the blood stream are saturated. An estimate of *reserve* binding capacity for bilirubin is what is really needed (Chan, Schiff & Stern, 1971).

Clinical management

The first aim of management should be to reduce the amount of haemolysis to a minimum, and the second aim to reduce the competition for plasma binding of bilirubin to a minimum. Where these prophylactic measures fail, three methods of aiding the disposal of excess bilirubin are now available. The first involves the mechanical removal of bilirubin by means of an exchange transfusion with fresh heparinized or acid-citrate-dextrose (ACD) stored blood. The second involves stimulation of the normal metabolic pathway for bilirubin excretion with an enzyme inducer such as phenobarbitone. The third involves the use of alternative pathways that normally only play a minor role in bilirubin excretion: the use of phototherapy to increase photo-oxidation of bilirubin within the skin is an example of this latter approach. A recent symposium on bilirubin metabolism in the newborn concentrated largely on these aspects of clinical management (Bergsma, 1970).

Exchange transfusion is a rapid and effective way of lowering the serum bilirubin level, but, in the absence of any active haemolytic process, this is only the treatment of choice if the bilirubin level is already high. Heparinized blood is probably ideal when a very sick baby requires exchange, but protamine should be given to counteract the effects of the heparin, unless there is evidence of disseminated intravascular coagulation. Since it is difficult to judge just how quickly the infused heparin will be metabolized, enough protamine should probably be given to neutralize about half the infused heparin. Where fresh ACD blood is used, enough of the supernatant plasma should be removed to reduce the citrate load and increase the packed-cell volume to 40%; the use of ACD blood presents the infant with a large load of metabolizable acid, and buffering the blood used during the early part of the first exchange is probably right when treatment is contemplated on an ill or only recently delivered baby. Usually, 180 ml of blood per kilogram body weight is exchanged over a period of about 1 h.

Treatment with phenobarbitone is only effective if continued for at least 3 d, but in countries such as Greece and Hong Kong, where severe hyperbilirubinaemia is extremely common in babies of low birth weight, prophylaxis with phenobarbitone may well prove to be the most appropriate method of management. Phenobarbitone can be given to the mother before delivery, but a prophylactic approach of this nature may not be appropriate except in communities where the incidence of severe neonatal hyperbilirubinaemia is high. Phenobarbitone seems to increase both hepatic uptake and hepatic excretion of bromsulphalein (Yeung & Yu, 1971); it has been shown to induce glucuronyl transferase enzyme activity, increase the concentration of the receptor protein Y within the liver and stimulate bile flow (Crigler & Gold, 1969; Thaler & Schmid, 1971). However, it also increases the activity of many other microsomal enzymes and may cause significant depression of the vitamin K-dependent clotting factors.

Cremer, Perryman & Richards (1958) reported that exposure to sunlight caused a fall in the serum bilirubin level of pre-term babies with 'physiological' jaundice in the first week of life; they also showed that this effect could be mimicked by fluorescent lighting. Many controlled trials have since been conducted into this, and there is no doubt that phototherapy started shortly after birth can greatly decrease the incidence of hyperbilirubinaemia, while therapy started later is also effective in the management of existing jaundice. Phototherapy has now been used to treat many thousand babies in Europe and the Americas without apparent ill-effect; unfortunately very few controlled studies have yet been undertaken. The products of photodecomposition themselves were once thought to be toxic, but there is as yet no evidence of this. Phototherapy was at one time thought to cause growth retardation in the Gunn rat, but this has not been confirmed. Continuous bright light has been found to influence puberty and other biological rhythms in a number of laboratory animals, but these effects seem to be mediated through the retina and there is no reason to believe that phototherapy would have any such effect in man, as long as the eyes are shielded during treatment.

The real danger inherent in the use of phototherapy would appear to be the extent to which its liberal use tends to result in illness being masked. Jaundice in the first week of life may be the first indication of liver disease, urinary infection, generalized sepsis or mild haemolytic anaemia. Use of phototherapy on a prophylactic basis may remove this valuable sign. There is also always a real risk that signs of jaundice will be treated symptomatically, while their underlying cause is ignored. Serious progressive anaemia may pass undetected for all too long, when the continuing jaundice associated with severe haemolytic disease is masked by phototherapy. The degree of jaundice may also be underestimated during phototherapy because the exposed skin often loses its yellow appearance before the underlying hyperbilirubinaemia is controlled. Blue light with a wavelength of 450–460 nm is particularly effective (Ente et al., 1972), but white fluorescent light with a spectrum similar to that of sunlight is also quite efficient, and cyanosis is much easier to assess when this sort of lamp is used (Medical Research Council, 1965).

A cradle of lamps suspended above the incubator and producing about 5000 lux have been used in most of the studies published to date, but Giunta & Rath (1969) have shown that a very modest increase in general room illumination is enough to secure a significant decrease in the serum bilirubin level. They secured their results by providing approximately 1000 lux of light within each incubator; an infant exposed to direct sunlight will experience more like 10,000 lux of light. Too little is known for any general recommendation to be possible as yet, but the intensity of illumination at present employed in some nurseries is less than 300 lux, particularly at night, and this must be well below the optimum for detecting early signs of change in a baby's clinical condition.

Optimum results may well be achieved by arranging for a cautious increase in general ward lighting, while reserving intensive photo-therapy with a cradle of lights above an incubator for the few babies at greatest risk from hyperbilirubinaemia (Giunta, 1971). It has been customary to cover the eyes of babies treated by artificial phototherapy, and we shall need to know much more about the effect of bright light on the pre-term baby's eyes before this precaution is abandoned when the level of illumination exceeds 1000 lux (Sisson *et al.*, 1970).

Detoxication

The liver performs a vital function in securing the detoxication of many foreign compounds and drugs prior to their excretion in the urine or faeces. Detoxication is sometimes secured by a process of oxidation, reduction or hydrolysis, but most characteristically by conjugation. The mechanism by which bilirubin undergoes glucuronide conjugation has already been discussed; many other organic compounds such as steroids, and drugs such as sulphonamides, salicylates and chloramphenicol are conjugated in a similar manner. Studies have shown that elimination of these drugs is delayed in the neonatal period; naladixic acid (Rohwedder, Simon, Kubler & Hoheraner, 1970) and certain of the sulphonamides (Krauer, Spring & Dettli, 1968; Sereni, Perletti, Marubini & Mars, 1968) are among the more frequently used antibiotics where there is evidence of a prolonged half-life in the neonatal period and an age-dependent rate of elimination.

Many tragic deaths occurred in the late 1950s as a result of injudicious treatment with chloramphenicol (Burns, Hodgman & Cass, 1959). Newborn babies treated with 100 mg/kg per day for more than 3 d frequently developed abdominal distension, soon followed by progressive pallid cyanosis and profound circulatory collapse. Many of the infants with the signs of this 'gray baby' syndrome experienced troublesome vomiting; mortality appeared to be highest in babies of short gestation, but even in the more mature babies treatment with chloramphenicol seemed to cause a tenfold increase in mortality. Investigation revealed that these babies had grossly elevated blood chloramphenicol levels, apparently as a result of delayed conjugation and excretion (Weiss, Glazko & Weston, 1960). Chloramphenicol inhibits protein synthesis, and high blood levels may have caused toxicity by inhibiting enzyme synthesis.

Renal function

While the kidney is now known to be normally very active during fetal life, homoeostasis is clearly not dependent on adequate renal function, since babies with bilateral renal agenesis appear biochemically normal at birth. Fetal urine is one of the main sources of amniotic fluid in the second half of pregnancy, once the fetal skin has become keratinized, and many babies with renal agenesis show evidence of arthrogryposis at birth as a result of oligohydramnios during development *in utero*. These infants also suffer from pulmonary hypoplasia and this usually compromises their survival after birth, before the other consequences of renal agenesis have time to become apparent; growth is nevertheless normal in most respects until the time of birth, and the cause of the associated pulmonary growth retardation is unknown. It has been suggested that pulmonary hypoplasia occurs as a result of oligohydramnios, but it may be that hormonal factors are involved.

The mesonephros is an almost rudimentary organ in the human infant and the loop of Henle, which is characteristic of the metanephros, appears to become functional about the fourteenth week of gestation. Coincident with this there is a gradual decrease in urine flow and an increase in the percentage of glomerular filtrate that is reabsorbed, in those species where fetal kidney function has been studied in any detail. Glomerular development starts in the juxtamedullary region and progresses towards the capsule and, as a result, there are relatively more mature nephrons in the juxtamedullary region than there are in the outer cortex throughout fetal life. The number of nephrons increases progressively between the sixth and thirty-sixth week of gestation, by which time there are usually some two million nephrons. No new nephrons are thought to develop after this (MacDonald & Emery, 1959), all subsequent kidney growth being due to a small increase in mean cell size together with a sevenfold increase in total cell number (Widdowson, Crabb & Milner, 1972).

Glomerular function

Renal blood flow and glomerular filtration are probably low *in utero* when the kidney does not need to function as an excretory or regulatory organ; blood flow and glomerular filtration certainly increase rapidly during the first few weeks of life as the kidney assumes its homoeostatic role, and recent longitudinal studies have now served to demonstrate for the first time just how rapid the increase in glomerular filtration usually

313

is in the first week of life (Fig. 8.7). Barnett *et al.* (1948) concluded from a study of pre-term infants of differing post-natal age but similar weight that post-natal age might be a more important factor in determining renal function than gestational age or body size; Fig. 8.7 reveals that this is an understatement and that glomerular filtration is very largely a function of post-natal age almost uninfluenced by body size or gestational age, in babies of more than 30 weeks' gestation at birth. Surprisingly little is known at the present time about renal function immediately after birth in babies of less than 30 weeks' gestation, and this is a deficiency that needs to be put right.

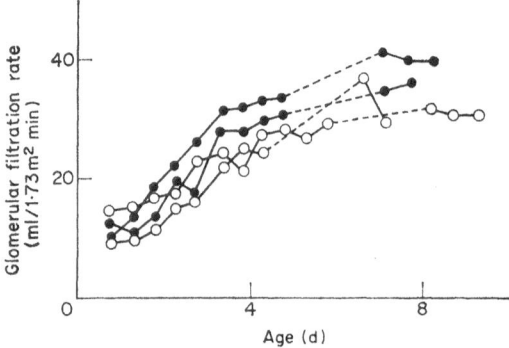

Fig. 8.7. Serial studies of glomerular filtration rate during the first 10 d of life. Results from two babies of 39–40 weeks' gestation (●), and from two babies of 31–32 weeks' gestation (o). (Author's unpublished data.)

Glomerular function remains low in relation to kidney weight but continues to increase slowly during the remainder of the first year of life (West, Smith & Chasis, 1948) until it reaches the adult rate of filtration of 125 ml/1·73 m² min. Increasing renal blood flow would seem responsible for most of the early increase in filtration in man, but other factors such as low filtration pressure and low glomerular-membrane permeability may be partially responsible for poor excretory function in the first year of life. Renal vascular resistance is high at birth in several species including the lamb, the pig and the guinea-pig and, in the latter, decreasing resistance in the afferent arteriole (proximal to the glomerulus) has been found responsible for much of the progressive increase in glomerular filtration that occurs after birth. In the puppy, on the other hand, no significant fall in renal vascular resistance has been found, and the rise in glomerular filtration that occurs during the first month of life is largely a function of increased arterial blood pressure (Kleinman &

Lubbe, 1972a). Whether changes in blood pressure have a similar effect on glomerular filtration in the human infant is unfortunately not known, and a detailed analysis of the factors affecting glomerular filtration is unfortunately bedevilled by the difficulty of obtaining an accurate measure of renal blood flow. Excretion of para-aminohippurate provides a reasonable measure of effective renal blood flow in adult man, but extraction is very incomplete for some time after birth both in man (Calcagno & Rubin, 1963) and in a number of animals, and the reason for this has not yet been established (Kleinman & Lubbe, 1972b).

The impediment to filtration offered by the cuboidal epithelium of the fetal and neonatal glomerular membrane was for many years considered to be the main reason for glomerular filtration being low, and Gruenwal & Popper (1940) believed that rupture of this epithelial cell layer at birth contributed to the post-natal rise in filtration. Evidence that fetal urine formation was more rapid in the lamb at mid-gestation (when the glomerulus still had a columnar epithelium) than it was at term (when the epithelial cells had become cuboidal) was at one time held to rule out this possibility, but there is now some preliminary evidence from a study of dextran clearance that the pore size of the glomerular membrane may increase from 20 to 40 Å during post-natal growth in man. Whether this has a major effect on glomerular filtration may, however, be doubted, as the proportion of blood flowing through the kidney that is 'cleared' by glomerular filtration is much the same at birth as it is in adult life.

There is some suggestion that glomerular filtration is more labile in the neonatal period than it is in adult life, and capable of increasing as much as 50% during an osmotic diuresis (Calcagno *et al.*, 1954). Furthermore, the nature of the diet also appears to have an effect on the rate at which renal function matures, for Edelmann & Wolfish (1968) were able to show that pre-term infants fed from birth on a high-protein high-solute diet for 4 weeks were able to clear inulin and para-aminohippuric acid, and to excrete hydrogen ion, nearly twice as fast as control infants. Changes in blood flow to the outer cortical glomeruli may be responsible for some of this lability.

Excretion
The newborn baby is well known to have a strictly limited capacity for handling a sudden water load. A baby given a water load equivalent to 3% of its body weight on the day of birth is only capable of excreting 10% of this load in the first 3 h, whereas an adult given a similar water load would have excreted all the water in this time. The ability to

handle a slow continuous intravenous infusion has frequently been underestimated however: the pre-term baby is as capable of excreting a dilute urine (50 mosmol of solute per litre) on the day after birth as an adult man. The ability to withstand water deprivation, on the other hand, is more limited (maximum urine concentration being about 700 mosmol/l) largely because the low urea load results in a low urea concentration in the medulla. Low solute load and a reduced filtration rate are the main factors limiting water excretion immediately after birth; Both minimum and maximum safe limits for water intake in the neonatal period are therefore easy to calculate from a knowledge of these factors, once due allowance has been made for insensible loss through the lungs and skin (p. 324). The availability or efficacy of antidiuretic hormone immediately after birth is still doubted by some, but there is very little evidence that this factor limits concentrating capacity. Remarkably little is known about the effect of 'weightlessness' *in utero* (Avery, 1965; Wood, 1970), or the effect of gravity on renal function after delivery.

The kidneys' limited excretory capacity immediately after birth has also to be born in mind when prescribing drugs, because a number of important antibiotics, including gentamycin, naladixic acid and most of the penicillins, are excreted largely unchanged in the urine.

Tubular function

The studies of Fetterman *et al.* (1965) have shown that the ratio of glomerular surface area to proximal tubular volume is higher at birth than in adult life; it is not surprising therefore that the capacity of the proximal tubule to absorb glucose and secrete para-aminohippurate increases threefold between birth and adult life. Increased urinary loss of a number of other nutrients including aminoacids, folate and riboflavin in the neonatal period point to a generalized mild tubular inadequacy. Whether these differences are due to uneven and sometimes excessive blood flow to the proximal tubule, to sluggish tubular function or to postglomerular blood bypassing the proximal tubule is not known. These limitations are probably of little significance, but the same cannot be said of the kidney's extremely limited ability to regulate or compensate for acidaemia in the first week of life.

Acid-base balance

Renal tubular capacity for ammonium and hydrogen-ion excretion is extremely limited in the first week of life, particularly in infants of short

316

gestation, and Allen & Usher (1971) have shown that the kidneys of pre-term infants with respiratory distress fail to compensate either for the increased bicarbonate excretion or for the accumulation of hydrogen ion caused by respiratory acidaemia in the first few days of life. These findings suggest that there may be a particularly marked immaturity of the carbonic anhydrase and glutaminase enzyme systems in the kidneys of infants who are born before term.

Maximum excretory capacity for hydrogen ion in the pre-term infant rises rapidly in the first 2 weeks of life, and then rather more slowly during the next 4 weeks (Sulyok & Heim, 1971), and limited excretory capacity probably accounts for the late metabolic acidosis seen in many small infants during the second and third week of life (Kerpel-Fronius, Heim & Sulyok, 1970). Sulyok (1971) has pointed to the close parallelism between the post-natal fall in serum sodium and the development of progressive metabolic acidosis in these babies, and has argued that both may be due to immaturity of adrenocortical hormone control. Late metabolic acidosis is seen to be a consequence of dietary intake exceeding excretory capacity, and renal maturation almost certainly explains the self-limited nature of the problem.

Acid production is clearly related to nitrogen catabolism and is significantly higher in babies fed on cows' milk than it is in babies fed on breast milk. Metabolic acid production also varies with the rate of growth, and anabolism, as emphasized by McCance and Widdowson, greatly limits the number of molecules originating from the food that require to be excreted; the incorporation of calcium into bone is the only important exception to this generalization, because this involves the liberation of hydrogen ions that have to be excreted by the kidney. Prolonged late metabolic acidosis is particularly likely to occur, therefore, in infants who are failing to grow, and conversely acidosis can itself also inhibit growth. A reduction in dietary protein and a single dose of intravenous bicarbonate is sufficient to correct this problem in many instances.

Nutrition

Babies of very low birth weight have a poorly developed suck reflex, and there is some evidence that oesophageal peristalsis is also relatively inefficient at this age (Gryboski, 1969). These babies, therefore, require gavage feeding. When adequate staff are available, it is probably best to pass a wide-calibre tube through the mouth into the oesophagus each time the baby requires food. Where this is not possible, a fine tube of non-irritant plastic can be passed through the nose into the stomach, and left in place for several days in order to facilitate feeding; however, this inevitably causes some respiratory obstruction. There is no place for routine gastrostomy in the management of even the smallest baby (Vengusamy et al., 1969). One or more 'clear feeds' have traditionally been given before offering milk to a newborn baby, and water is better than dextrose for this as it leaves the stomach more rapidly (Husband & Husband, 1969), and causes less damage to the lung if aspirated (Olson, 1970). There is, however, little need to give a first 'clear feed' if gavage feeding is employed.

Even extremely immature babies digest the protein and carbohydrate provided by a milk diet very competently. The ability to absorb fat is rather less than that of a full-term baby, especially in the first week or two after birth, and only 50% of the fat in cows milk may be absorbed; however, more like 85% of the high fat content of breast milk is satis-factorily absorbed, even in babies of very short gestation. The small baby's biggest handicap is an inability to cope with the *volume* of milk required for optimum fluid and calorie intake in the first few weeks of life. These babies are therefore traditionally offered regular small feeds eight or twelve times a day, and the volume of the feed increased pro-gressively just as soon as there is clearly no residue still present when the next feed is due. These babies are best nursed prone, with the head of the mattress slightly elevated (Hood, 1964).

It was for many years considered wisest not to attempt oral feeding for at least 24 h after birth, and to increase the volume of the feed gradually during the first week of life, because of the risk of aspiration. Indeed 15 years ago many hospitals were withholding all food from the smallest babies for a full three days after birth (Beard et al., 1963). However, we now realize that the disadvantages of this precautionary starvation far outweigh the potential advantages (Smallpeice & Davies, 1964), and that early feeding with milk almost eradicates the risk of significant hypoglycaemia (Wu et al., 1967). Milk also lowers the incidence of hyperbilirubinaemia, as does water (Wennberg, Schwartz & Sweet, 1966) and intravenous glucose (Bucci et al., 1971). Glial-cell growth is now also recognized to occur with particular rapidity during the last 10 weeks before and first six months after birth (Dobbing, 1970; Winick, 1970); brain growth can be permanently damaged by a period of undernutrition at this time (Platt & Stewart, 1971). Davies & Davis

(1970) have produced retrospective evidence that subsequent head growth is permanently affected by a brief period of undernutrition in babies weighing 1500 g or less at birth. They believed that the increased calorific demand caused by inadequate warmth potentiated the effects of a low-calorie intake at this time, and Glass, Silverman & Sinclair (1971) reported a series of observations that tend to corroborate this view.

Milk

There are many reasons for favouring the use of human breast milk. The fine protein curd of human milk leaves the stomach sooner than that of cow's milk (Silverio, 1964), the low mineral content reduces the risk of hypertonic dehydration, and the high calcium : phosphate ratio reduces the incidence of late neonatal tetany (Oppé & Redstone, 1968; Barltrop & Oppé, 1970). The breast-fed baby is also less susceptible to infection (Wimberg & Wessner, 1971); immune globulins present in human milk probably inhibit viral and bacterial multiplication in the gastro-intestinal tract (Kenny, Boesman & Michaels, 1967), while the high lactose, low protein and low phosphate content, poor buffering capacity, small bulk and low residue modify the lactobacillary flora and cause a lower faecal pH (Bullen & Willis, 1971). Sudden unexpected 'cot death' is probably less common in breast-fed babies, and the potential hazard of the high strontium content is eliminated.

Human milk contains relatively little protein, and there is some evidence that weight gain is more rapid with a milk that contains rather more protein, especially in babies of less than average weight for their gestational age at birth (Rezza et al., 1971). Unfortunately the high mineral content of most high-protein diets can sometimes cause spurious weight gain due to fluid retention, and measurements of linear growth have not been made in many of the studies reported to date. There is, however, some evidence that the rate of growth can be influenced by the protein content as distinct from the mineral content of the diet (Babson & Bramhall, 1969). Two grams of protein per kilogram of body weight per day would seem to be close to the minimum requirement for the full-term baby in the first 8 weeks of life (Fomon et al., 1973), but the fetus usually accumulates protein rather more rapidly than this during the last trimester of pregnancy, and a protein intake of 2 g/kg a day is almost certainly less than the optimum intake for the pre-term baby (Snyderman, 1971).

Large increases in protein intake make remarkably little difference to weight gain, as long as calorie intake is adequate, but balance studies suggest that nitrogen retention increases greatly. The nitrogen content of the body rises rapidly during fetal and early post-natal life (Widdowson & Spray, 1951), and these balance studies would *seem* to indicate that an increased protein intake accelerates the chemical maturation of the body. Whether this is advantageous is not known; indeed, animal studies have failed to reveal any evidence of accelerated maturation over and above the effects directly related to increasing body size (Weil & Miller, 1971), and have thrown some doubt on the accuracy of these long-term metabolic balance studies. It is certainly unwise to offer more than 6 grams of protein per day per kilogram body weight because enzyme immaturity (p. 304) causes an abnormally high plasma level of tyrosine, phenylalanine and methionine to develop in babies of short gestation (Valman *et al.*, 1971).

The rapid growth of the skeleton in the last three months before term causes the fetus to accumulate more than 100 mg of calcium per kilogram body weight per day. To maintain growth at anything like this rate after birth is difficult, and both the calcium content and the calcium : nitrogen ratio of fat-free bone fall rapidly after birth and remain low throughout the first year of life even in the full-term baby (Dickerson, 1962). A similar fall in the degree of calcification has been observed in a number of other mammals, and matrix deposition would appear to outstrip the supply of calcium at this time. The magnitude of the calcium deficit is even more striking following premature delivery, and the calcium content of human milk is so low that such a deficit is unavoidable. Cow's milk contains nearly four times as much calcium, but poor fat absorption interferes with the absorption of the calcium that is available, and net uptake is only marginally better than it is for breast milk (J. C. Shaw, personal communication).

On present evidence human milk would seem to provide the best available diet for the small pre-term baby except, perhaps, in respect of its extremely low calcium content, and its low and rather variable protein content. 'Humanized' milk is probably the best alternative food in the first two weeks of life, though there may be a case for supplementing protein intake if it is not possible to increase the volume of milk sufficiently to achieve a protein intake of about 2·5 g/kg per day. Unfortunately, while the mineral content of breast milk can be mimicked reasonably easily, matching the digestive property of the fat provided by human breast milk with an artificial milk has as yet proved much more difficult, and we must rely either on a half-skim cow's milk with a high protein and carbohydrate content, or on a 'filled' milk, in which the animal fat has been replaced by vegetable oil (Davidson et al., 1967).

Intravenous nutrition

Small babies experience so much respiratory difficulty in the first few days of life that there may be doubts about attempting oral feeding: indeed some babies with respiratory distress develop a transient ileus, and in these circumstances it is clearly wise to employ an intravenous infusion to provide the baby with calories and water. Cornblath et al. (1966) found that an infusion of 10% dextrose marginally decreased over-all mortality in babies of less than 1250 g at birth, but Mamunes et al. (1969), Mendicini et al. (1971) and Kitchen & Campbell (1971) were unable to confirm this. Nevertheless there is no doubt that the intense catabolism that occurs during starvation can be greatly reduced by intravenous glucose (Auld et al., 1966), and that hypoglycaemia can be almost abolished. Starved babies develop a high serum sodium and potassium level (Beard et al., 1963); they may also develop bradycardia, a 2:1 atrio-ventricular block and a prolonged electrocardiographic QT interval in association with the hypokalaemia (Bucci et al., 1971). These abnormalities are not seen when babies are given a minimum of 65 ml/kg of 10% dextrose per day intravenously.

This rate of infusion is probably not enough to cover the baby's fluid need, and this volume of 10% dextrose will certainly not provide enough calories to cover basal metabolism, let alone the additional calories necessary for growth. That enough nutrients can be provided intravenously to support growth when oral feeding is impossible has, however, been amply demonstrated. Progress has been rapid in the last 5 years, as is clear from the reviews of Harries (1971), Heird et al. (1972), Wretlind (1972) and Shaw (1973). While relying *entirely* on intravenous feeding in babies of low birth weight is seldom necessary for more than 2 or 3 d, there is no doubt that a supplementary infusion can help to minimize the inevitable interruption to growth. An infusion of 10% or 15% glucose may be all that is necessary when the baby is also receiving milk, and glucose is certainly preferable to fructose, because of the risk of metabolic acidosis. However, there may be a case for adding a protein hydrolysate or an amino-acid mixture to the intravenous fluid, if the intake of milk remains low for more than 4 or 5 d. Unfortunately, the available protein hydrolysates contain a high concentration of ammonia, and there is a real risk that hyperammonaemia may also lead to metabolic acidosis (Ghadimi et al., 1971). Infusions of emulsified fat appear to be well tolerated by babies of short gestation (Gustafson *et al.*, 1972), but their use in the newborn is still largely experimental, and an adequate calorie intake can usually be achieved without infusing fat. So little is known about the possible long-term effect of this form of treatment on the growing child that caution must be counselled (American Academy of Pediatrics, 1972). Electrolytes are frequently added to the solution, but the need for this has not been established in babies also receiving milk. Hypocalcaemia is relatively common in babies of low birth weight (Tsang & Oh, 1970), and symptoms of tetany merit correction with intravenous calcium.

Table 8.5. *Optimum nasotracheal tube and umbilical catheter lengths.*

Body weight (kg)	Crown-heel length (cm)	Nasotracheal tube length (cm)	Umbilical vein catheter (cm)	Umbilical artery catheter (cm)
0·75	32	6	5	8·5
1·0	36	7	6	10
1·5	40	8	7	11·5
2·0	44	9	8	13

A nasotracheal tube should normally lie with its tip half way between the vocal cords and the carina if inserted down the nose the distance shown (Coldiron, 1968); an umbilical-vein catheter should lie with its tip in the inferior vena cava just above the diaphragm, and an umbilical-artery catheter should lie in the ascending aorta with its tip at the level of the diaphragm, if inserted the distance shown (Dunn, 1966).

The umbilical artery or umbilical vein are frequently utilized when infusing glucose and other nutrients into a small baby, but there are dangers inherent in their use. When they are used, it is probably wise to try and position the tip of the catheter in a large vessel away from the origin of any important tributary; this can most readily be achieved by placing the tip of the catheter at the level of the diaphragm (Table 8.5). The risk of portal-vein thrombosis and subsequent portal hypertension is considerable if a venous catheter is left *in situ* for more than about 36 h; the risk of arterial catheterization is also considerable, and prolonged arterial catheterization is probably only justified when there are important reasons for monitoring the arterial blood gases regularly (p. 274). The risk of sepsis is also much greater with an indwelling catheter than with a fine needle placed in a peripheral vein (Peter, Lloyd-Still & Lovejoy, 1972), and the latter is, therefore, nearly always the route of choice when a small baby requires intravenous nutrition.

Requirements

Calories

Growth provides the most sensitive index as to the adequacy of the calorie intake in a small baby of short gestation. Weight is increasing *in utero* by nearly 3% every day during the early part of the third trimester of pregnancy, and to aim for a weight gain comparable to that usually achieved *in utero* seems reasonable if a baby is born before term.

The calorie requirement for basal metabolism in a 1-kg baby is about 40 kcal/d in the first week of life, and intermittent physical activity accounts for a further 5 to 10 kcal/d. The calorie requirement for growth is more difficult to assess with accuracy. The body of the small pre-term baby only contains about 2% of fat, and growth is largely a matter of assimilating and incorporating water, minerals and protein. The calorie cost of growth may, therefore, be as low as 2 kcal/g; however, even on this conservative estimate growth requires the provision of at least a further 50 kcal/d. If due allowance is made for the significant if rather variable loss that occurs into the stool, we are left with a need for a total intake of at least 120 kcal/kg d. (1 kcal $\approx$ 4·18 kJ.)

Basal metabolic rate increases by 50% during the first six weeks of life (p. 288), and the calorie cost of growth almost certainly rises a good deal when fat is laid down (Close & Mount, 1971). Indeed the calorie requirement for optimum growth probably exceeds the requirement for basal metabolism at this age. In these circumstances, optimum growth is only likely to be achieved by a total calorie intake at least 150 kcal/d per kg body weight; to achieve this a breast-fed baby would need to ingest about 250 ml of milk per day per kilogram body weight. An intake as large as this considerably exceeds the intake recommended by Gordon *et al.* (1940), but it is not impracticable, and it produces a very satisfactory weight gain. Some nurseries utilize intermittent gavage feeding, while others employ a continuous nasogastric drip in order to achieve this intake (Valman *et al.*, 1971).

Water

The minimum obligatory urinary water loss required to maintain homoeostasis on a low-protein low-electrolyte diet is about 30 ml/kg d, when urine osmolality is in the range 200–300 mosmol/l; obligatory water loss doubles when there is significant endogenous catabolism, and can be higher still on a cow's milk diet. Insensible water loss also usually accounts for about 30 ml/kg of water per day in an environment of moderate humidity; however, this loss may increase threefold in babies of less than about 30 weeks' gestation, because the skin is so much more permeable to water (p. 327). The minimum safe requirement almost certainly exceeds 100 ml/d in a 1-kg baby, and when milk feeding is introduced the optimum intake is probably between 1 and 1·5 ml/kcal. The ability to excrete excess water in the first few days of life is much higher than is generally realized (p. 315), and intravenous fluid intake is often needlessly restricted.

Vitamins and Minerals

Milk contains little iron and babies of low birth weight rapidly deplete their available reserves. Ferrous sulphate drops providing between 10 and 15 mg of elemental iron per day are the usual form of supplementation employed.

Vitamin C is not stored in the body and is not available in cow's milk; it is therefore recommended that all babies should receive 20 mg of supplemental ascorbic acid each day after the first week of life, unless they are being breast fed or fed on a proprietary milk already fortified with vitamin C. Stores of vitamin A and vitamin D are small in the pre-term baby. Milk contains a good deal of vitamin A, but relatively little vitamin D and rickets is now the commonest vitamin-deficiency disease in Great Britain. All proprietary infant milks contain added vitamin D, but the total fluid intake of many small infants is so small that to rely on this source of vitamin D in the first month of life is unwise (Lewin *et al.*, 1971). The optimum requirement is probably about 400 international units a day (FAO/WHO Expert Group, 1970). Moderation is necessary in prescribing supplementary vitamin D, because overdosage is not without its risks. There is fortunately no significant risk of overdosage with vitamin A.

Environmental influences

Humidity

Ambient humidity has an effect on evaporative heat loss, and hence on temperature regulation, and the homoeostatic consequences of this have already been discussed (p. 290). Silverman and Blanc showed in 1957 that high humidity significantly reduced the mortality of small pre-term babies nursed in incubators, with an air temperature of 29°C during the first week of life. They noticed, however, that the axillary temperature of the infants in an environment of 80–90% humidity was, on average, half a degree higher than that of infants in an environment of 45–55% humidity. They speculated that this might be the cause of the decreased mortality and confirmed this belief by a further controlled trial the following year (Silverman et al., 1958). There remained the possibility that a high relative humidity had a favourable effect on mortality, independent of and additional to its effect on body temperature, but Silverman, Agate & Fertig reported in 1963 that they were unable to detect any such effect. Earlier trials had shown that there was equally little advantage to be gained from employing a nebulizer in order to achieve a supersaturated mist around the infant (Silverman & Andersen, 1956).

The fashion for nursing small babies in a really humid environment has accordingly receded, and most babies are now nursed in warm-air incubators with a relative humidity of about 50–70% (that is, in an environment with an ambient water-vapour pressure of between 18 and 28 mmHg). Whether it is either safe or reasonable to nurse a baby in a much drier environment than this has never been tested. Blackfan & Yaglou reported in 1933 that when they cot-nursed babies weighing less than 2 kg in a nursery with an ambient humidity of 7·5 mmHg (30% RH) instead of 15 mmHg (65% RH) the incidence of diarrhoea increased, body temperature became less stable, average weight gain was reduced and more babies died. It is doubtful whether these differences can be explained by differences in heat loss or body temperature, because Blackfan & Yaglou increased air temperature in order to compensate for the estimated increase in evaporative heat loss.

The climate of the British Isles is mild and relatively uniform; the average water-vapour pressure of the atmosphere is about 8 mmHg and the average seasonal variation between 6 and 11 mmHg. The absolute humidity of hospital wards will only exceed that of the atmosphere by a small and variable amount, unless expensive air conditioning equipment is installed. It follows that many small babies lying clothed in a cot or naked under an open heat cradle are currently being nursed in an environment with a relative and absolute humidity very considerably below the minimum recommended by Blackfan & Yaglou. There is an urgent need for further work on these lines, in order to define what constitutes the minimum safe ambient humidity for small babies with greater precision, since, if the need for an artificially raised ambient humidity could be avoided, then the expense and the attendant hazards of humidifying equipment could also be avoided.

Evaporative water loss from the body is dependent not on the relative but on the absolute humidity of the environment. Expired air leaves the body nearly 95% saturated, however dry the inspired air, and, although respiratory water loss normally accounts for between a quarter and a third of all evaporative water loss, the amount of water lost is very much influenced by ambient humidity (Hey & Katz, 1969a). Recent measurements of evaporative water loss in the newborn baby show that, as in

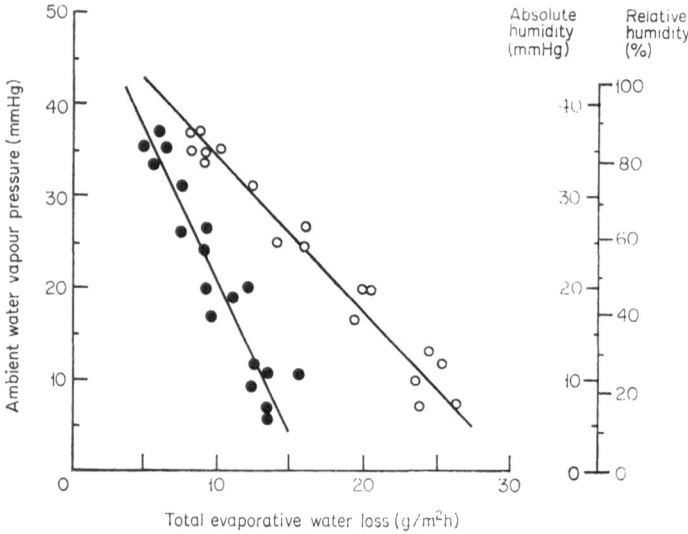

Fig. 8.8. The relation between absolute humidity and total evaporative water loss from a sleeping baby in a neutral thermal environment in the second week of life. Data from a baby of 1·7 kg of 32 weeks' gestation (●) and from a baby of 1·1 kg and 28 weeks' gestation (o). The immature baby has a significantly higher water loss. The relation between absolute and relative humidity shown is at environmental temperature 35°C. (Author's unpublished data).

the adult, the relation between absolute humidity and total evaporative water loss is effectively linear, and that a considerable change in absolute humidity is necessary to produce a significant reduction in total evaporative water loss (Fig. 8.8). Evaporative loss per unit surface area is lower in the baby than it is in adult man, mean heat loss due to evaporation accounting for approximately a quarter of basal heat production in both cases. There is no evidence that clothing normally has any effect on the equilibrium relationship between ambient humidity and evaporative water loss, and a change in absolute humidity of the magnitude studied by Blackfan & Yaglou can be calculated to result in a difference in evaporative water loss of between 5 and 10 g a day, an amount that, although small, would be enough to account for the observed difference in weight gain.

Babies of less than 30 weeks' gestation
The skin of the small pre-term baby of less than about 30 weeks' gestation is particularly thin and delicate. Indeed we now realize that the skin only starts to become keratinized at about 20–26 weeks' gestation, that permeability is extremely high during the earlier stages of development *in utero* (Parmley & Seeds, 1970; Lind, Kendall & Hytten, 1972), and that the epidermal cell surface is rich in microvilli and cytoplasmic granules during the third and fourth month of gestation (Breathnach & Wyllie, 1965). These latter features suggest that the skin may perform an active secretory function during early gestation.

Further evidence that the fine, almost transparent skin of the very small pre-term infant is particularly permeable to water was obtained by Fanaroff *et al.* (1972). These authors showed that babies of less than about 30 weeks' gestation frequently had an unexpectedly high evaporative heat loss, when nursed in a standard commercial warm-air incubator. Somewhat surprisingly they found that the use of a radiant-heat shield within the incubator caused a decrease in evaporative water loss as well as an increase in body temperature; it is difficult to explain how the presence of the baffle could affect evaporative water loss, unless it caused an increase in the ambient humidity around the child by modifying the circulation of air within the incubator. Detailed experiments into the relationship between ambient humidity and evaporative water loss carried out within the metabolic chamber illustrated in Plate 8.2 have confirmed that many small babies of less than about 30 weeks' gestation lose excessive quantities of water through their skin (Fig. 8.8). Basal evaporative water loss is almost constant, irrespective of birth weight or gestational age, in babies of more than 30 weeks' gestation, and amounts to approximately 1 g/h per kilogram body weight when ambient humidity is between about 10 and 15 mmHg. Babies of less than 30 weeks' gestation almost always show an increased loss; the magnitude is variable but some babies certainly lose more than 2 g/h per kilogram body weight and allowing for this high obligatory loss is important when calculating the fluid and nutrient requirements (p. 323). Evaporative water loss frequently increases by as much as 50% when a baby becomes active and restless, and insensible water loss per day will therefore be significantly higher than these estimates of basal loss.

Bacterial and viral infection

Bacterial infection is still the root cause of many neonatal deaths every year. Generalized septicaemia is common at this age and is sometimes accompanied by meningitis, while gastroenteritis, pneumonia and infection of the urinary tract are also common, if somewhat less lethal, conditions. Infection can occur *in utero* by transplacental passage from the mother's blood stream, or secondarily following infection of the amniotic sac, usually as a result of prolonged rupture of membranes. Following delivery, the baby is at the mercy of the environment. Dust-born staphylococci can cause infection, but most babies probably acquire their staphylococci from other infants via the hands of ward staff. Enterobacteria may be transmitted in the same way, but a number of hygrophilic pathogens, of which *Pseudomonas aeruginosa* is the most notorious, lurk in the humidifying units of incubators and respirators, and in hospital suction and resuscitation equipment, ready to colonize and invade any sick or debilitated baby. The β haemolytic streptococcus was once the commonest pathogen, but it was then replaced by the ubiquitous hospital staphylococcus; gram-negative bacilli including *Escherichia coli* and *Pseudomonas aeruginosa* are now becoming responsible for an increasing proportion of all neonatal infections.

The newborn baby shows a normal if slightly delayed and poorly localized inflammatory response. There is good evidence that all the immunoglobulins except 1gD and 1gA are synthesized *in utero*, but maternally derived 1gG crosses the placenta, particularly during the last 3 months of gestation, and constitutes the bulk of the healthy infant's serum immunoglobulin at birth. The relative paucity of 1gM has been held to be responsible for the known susceptibility to gram-negative infection, but the level rises rapidly in response to any challenge. Severe staphylococcal sepsis is not infrequent in mature babies, despite the high 1gG level present at birth, and the ability to synthesize antibodies is probably of more value than the presence of passively acquired antibody. Viewed in this light, it is wrong to consider the fetus and newborn baby as immunologically incompetent. There is, however, evidence that complement levels are low in the pre-term baby, and that cell-mediated immunity is only poorly developed at birth. This latter finding may well explain the high risk of disseminated disease with *Listeria monocytogenes*, tubercle bacilli and *Toxoplasma gondii*, and with cytomegalovirus and the rubella, varicella and herpes hominis viruses. Lack of primary 1gM opsonins as a result of poor placental transference almost certainly contributes to the baby's increased susceptibility to gram-negative infection;

polymorphonuclear neutrophils from the newborn are only able to phagocytose and kill bacteria as effectively as in adult life in the presence of adult opsonin. Neutrophils from the newborn also show a poor migratory response to chemotactic factors. Some of the reasons for the newborn baby's vulnerability to infection have recently been reviewed by Dossett (1972) and Wood (1972).

Gershon (1971) has discussed the role of virology in the diagnosis of neonatal infection, and Davies (1971) has reviewed the diagnosis and management of bacterial infection in the fetus and newborn, and also stressed the need to take conscientious yet realistic steps to minimize the risk of infection. Prevention of infection is better than its cure, and this involves maintaining a scrupulously clean environment and guarding against cross infection.

Human contact

It would be wrong to conclude this chapter without stressing the importance of nurturing the vital but fragile ties that constitute the bond between a child and its parents. Worry, anxiety and separation in the period immediately after childbirth can have a serious and enduring effect on the forging of a normal human relationship between mother and child. In our concern for the baby's physical well-being we have in the past often ignored the equally important emotional needs of the family. Psychologists have speculated that some of the behaviour problems commonly seen in pre-term infants during later childhood might be due to a period of isolation within an incubator after birth (Rothschild, 1967), and it is even more important to meet the emotional needs of the mother at this time (Klaus & Kennell, 1970).

The delivery of a small pre-term baby must inevitably cause worry and anxiety for the whole family; physical separation adds to the strain and may prevent the mother from developing any emotional attachment for her offspring. These ties have, of course, already started to form before birth; physical separation after birth will, however, prevent the maturation and reinforcement of these bonds during the key period of heightened sensitivity immediately after delivery. Concern for the baby's health and fear that it could die may even precipitate anticipatory mourning and severe emotional withdrawal. The long-term consequence of such a disturbance can take many different forms. Some children react to overprotection by developing a severe emotional disturbance (Green & Solnit, 1964). Others react to their mother's emotional withdrawal by failing to grow normally. In a few cases the relationship between the child and its parents is so disturbed that there is a real risk of the baby being injured or even killed in a moment of parental stress (Klein & Stern, 1971).

The successful medical management of a severely ill newborn baby requires diagnostic skill and manual dexterity, but it calls for more than this since, unless the parents are handled with tact, sensitivity and insight, the bonds that unite the family are likely to be seriously damaged. The more sensitive and discerning members of the profession recognized this many years ago. Budin noted as long ago as 1900 that 'mothers separated from their young soon lost all interest in those whom they were unable to nurse or cherish,' and Sir James Spence expanded on this same theme in his Charles West lecture to the Royal College of Physicians in London in 1946, when he suggested that many maternity hospitals were failing to fulfil their essential functions adequately:

'I take it that the function of a maternity hospital is to deliver a woman safely of her child, and afterwards to care for them in such a manner as to ensure their health, to establish their intimate and interdependent relationship, and finally to leave the woman free from the fears of having another child. Our maternity hospitals are ensuring the safe delivery of a woman to a greater and greater degree, but are they fulfilling their other two functions ?

'I know maternity hospitals which are the hygienist's dream of perfection. The women lie for their 10 days in immaculate beds placed equidistantly along sterile walls. Their ward is a picture of calm repose and passive immobility. You ask what has happened to create this atmosphere of silence and subdued conversation and fail to get an answer; but the truth is that the mothers are mystified by an arrangement under which their babies have been taken away from them at a time when, at the end of nine months' waiting, they had expected to possess them. The babies are away from them congregated in a room along the corridor beyond their earshot, out of sight but not out of mind. At regular intervals of the day they are placed on a trolley, wheeled along the corridor, and with a ringing of a bell which announces that milking-time is at hand, they are delivered by one masked woman to another masked woman at the door of the ward in which the mothers wait. Milking-time over, the babies are re-embarked for their nursery where they are solaced with sugar and water while the mothers wait again with empty arms and quiet resignation.'

Many paediatricians and obstetricians have long recognized the unnatural nature of such a system, and relaxed the rules that caused such unnecessary maternal separation, but tradition dies hard and segregation still persists in many hospitals both in America and in Europe. The development of this tradition appears to have been influenced by the psychological needs, the inevitable bias and the convenience of the nurses, doctors and administrators who form the dominant groups within the hospital community. Careful documentation of the psychological damage that such an approach can engender would appear to be the only way of convincing those who still doubt the need for change that these arrangements are inhuman.

If these arguments hold good for the mother of a healthy newborn child, they must surely hold with added force for the mother of an ill baby. If this is true then doctors and nurses will need to devote more time to fostering a sense of attachment, to keeping each mother informed about progress, and to keeping the flame of hope alive without becoming falsely over optimistic. There is a growing belief that mothers should always be encouraged to share in their baby's care and that it is vital that every mother should have the chance to see and touch her child.

Accessibility to the special-care nursery has usually been carefully restricted in the past, and the few visitors allowed in have often been required to wear caps, masks, aprons and overshoes. However, the value of these precautions is very difficult to demonstrate (Williams & Oliver, 1969) and they may well do more harm than good. Special precautions may be necessary for some particularly frail and tiny babies, but only bad nursery design, inflexible routine and lack of insight and sensitivity can explain why these precautions are so frequently applied indiscriminately to all—to large and small, young and old, sick and healthy alike.

There is, of course, a very special need to prepare a newly delivered mother for her first sight of a sick child, particularly if it is small. There is also a need to prepare her for the sight of the accumulation of equipment that may surround such an infant. Unless this is done the mother's first visit may be a frightening and damaging experience: however, if she is not allowed to visit, her imagination will run riot, which is just as bad. To ban access at such a time is therefore manifestly unwise, and to procrastinate dangerous; parents need properly prepared guidance together with opportunities for regular contact and continuing support following discharge home.

Prognosis

It is important that the problems associated with very low birth weight and premature delivery should be seen in perspective. Only a third of

1% of all the babies born alive during the British Perinatal Mortality Survey in 1958 had a gestational age of less than 30 weeks. A similar small proportion weighed less than 1 kg at birth, but approximately 1% of all the babies weighed less than 1·5 kg at birth. Babies weighing between 1 and 1·5 kg at birth have at least an even chance of surviving, and in many centres three-quarters of these babies now survive the neonatal period. Survival is unusual if a baby weighs less than 1 kg at birth, but 20% of those who reach the nursery alive may expect to survive the neonatal period with expert care, and there is no doubt that prognosis has improved considerably in the last 10 years. While babies of low birth-weight only form a small proportion of all deliveries, the numbers involved are considerable.

Unfortunately, the outlook for babies less than 1·5 kg at birth has in the past been bleak, and several centres found that more than half their surviving babies were handicapped (Lubchenco *et al.*, 1972). Mental retardation, learning disorder, behavioural disturbance and deafness are among the problems most frequently encountered. Myopia and blindness due to oxygen toxicity have been common in the past. There is some evidence that babies of less than 28 weeks' gestation have a particularly gloomy prognosis (Lubchenco, Delivora-Papadopoulos & Searls, 1972), but the more mature ('light-for-dates') baby may be the one at particular risk (Drillien, 1971).

It was once thought that any decline in mortality would be accompanied by a rise in the number of survivors with handicap. Had this prediction been fulfilled, there would have been grounds for the view that modern methods of intensive care were unjustifiable 'meddlesome medicine' (Zuelzer, 1971). Happily there are signs that this assumption was wrong, and that with skilled nursing care less than 20% of the survivors will be seriously handicapped (Davies & Russell, 1968; Rawlings *et al.*, 1971); follow-up studies will need to be continued for several more years, however, before a completely accurate estimate of handicap is possible. These results were obtained in centres where care was taken to anticipate and prevent complications such as birth trauma, infection, haemostatic failure, hypoxaemia, hypothermia, hypoglycaemia and hyperbilirubin aemia that might be expected to damage the brain. Early and generous feeding with undiluted human breast milk was also employed in these two centres, in order to maintain growth and minimize the severe biochemical disturbance that is associated with starvation after delivery. Babies requiring artificial ventilation because of severe respiratory distress at birth sometimes have evidence of disturbed lung function for many months after discharge home from hospital, and an increased incidence of respiratory-tract infection in the first few years of life, but the long-term prognosis seems to be reasonably encouraging (Shephard *et al.*, 1968; Outerbridge *et al.*, 1972).

There are few indications for inducing labour or securing delivery by Caesarian section before 37 weeks' gestation at the present time. Indeed many of the babies now delivered prematurely following the spontaneous onset of labour would probably have been better off growing undisturbed *in utero*. However, there are pregnancies complicated by uncontrollable pre-eclamptic toxaemia, severe intrauterine growth retardation (associated with signs of what has been termed 'placental insufficiency'), and sudden accidental haemorrhage due to placental bleeding, where there are strong indications for securing delivery many weeks before term. Intrauterine transfusion is currently employed in the treatment of haemolytic disease due to severe rhesus incompatability, but no controlled trial has ever been done to show that this form of management is preferable to premature delivery, and, as methods of post-natal management continue to improve, the relative advantage of attempting treatment by blind intraperitoneal transfusion *in utero* might reasonably be expected to decline. Current attitudes to the premature induction of labour are the result of much accumulated experience. If the recently reported fall in mortality, and in subsequent handicap, can be maintained and surpassed, then the attitude of obstetricians to premature induction will begin to change. However, no radical change in policy is likely to emerge until techniques are available for studying placental function, for assessing fetal growth and maturity, for monitoring fetal well being before and during labour, and for determining the risk of respiratory distress after birth.

References

ABLETT, J. G. & McCANCE, R. A. (1971) Energy expenditure of children with kwashiorkor. *Lancet*, **2**, 517–519.

ABRAMS, R., CATON, D., CLAPP, J. & BARRON, D. H. (1970) Thermal and metabolic features of life in utero. *Clinical Obstetrics and Gynaecology*, **13**, 549–564.

ACKERMAN, B. D., DYER, G. Y. & LEYDORF, M. M. (1970) Hyperbilirubinemia and kernicterus in small premature infants. *Pediatrics, Springfield*, **45**, 918–925.

ADAMSON, T. M., BOYD, R. D. H., PLATT, H. S. & STRANG, L. B. (1969a)

Composition of alveolar liquid in the fetal lamb. *Journal of Physiology*, **204**, 159–168.

ADAMSON, T. M., HAWKER, J. M., REYNOLDS, E. O. R. & SHAW, J. L. (1969b) Hypoxaemia during recovery from severe hyaline membrane disease. *Pediatrics, Springfield*, **44**, 168–178.

AHERNE, W. (1965) Fat infiltration in the tissues of the newborn infant. *Archives of Disease in Childhood*, **40**, 406–410.

AHERNE, W. A., CROSS, K. W., HEY, E. N. & LEWIS, S. R. (1967) Lung function and pathology in a premature infant with chronic pulmonary insufficiency (Wilson–Mikity Syndrome). *Pediatrics, Springfield*, **40**, 962–974.

AHERNE, W. & HULL, D. (1966) Brown adipose tissue and heat production in the newborn infant. *Journal of Pathology and Bacteriology*, **91**, 223–234.

ALEXANDER, G., BELL, A. W. & WILLIAMS, D. (1970) Metabolic response of lambs to cold. Effects of prolonged treatment with thyroxine and of acclimatization to low temperatures. *Biologia Neonatorum*, **15**, 198–210.

ALLEN, A. C. & USHER, R. (1971) Renal acid excretion in infants with the respiratory distress syndrome. *Pediatric Research*, **5**, 345–355.

AMBRUS, C. M., WEINTRAUB, D. H., NISWANDER, K. R., FISCHER, J., BROSS, I. D. J. & AMBRUS, J. L. (1970) Evaluation of survivors of respiratory distress syndrome at 4 years of age. *American Journal of Diseases in Children*, **120**, 292–302.

AMERICAN ACADEMY OF PEDIATRICS COMMITTEE ON FETUS AND NEWBORN (1967) Nomenclature for duration of gestation, birth weight and intrauterine growth. *Pediatrics, Springfield*, **39**, 935–939.

AMERICAN ACADEMY OF PEDIATRICS COMMITTEE ON NUTRITION (1972) Parenteral feeding—a note of caution. *Pediatrics, Springfield*, **49**, 776–779.

ANDERSON, W. R. & STRICKLAND, M. B. (1971) Pulmonary complications of oxygen therapy in the neonate. *Archives of Pathology*, **91**, 506–514.

APPLEYARD, W. J. & COTTON, D. G. (1970) Effect of asphyxia on thrombotest value in low birthweight infants. *Archives of Disease in Childhood*, **45**, 705–707.

ARANDA, J. V., SAHEB, N., STERN, L. & AVERY, M. E. (1971) Arterial oxygen tension and retinal vasoconstriction in newborn infants. *American Journal of Diseases of Children*, **122**, 189–201.

ASHTON, N., GARNER, A. & KNIGHT, G. (1971) Intermittent oxygen in retrolental fibroplasia. *American Journal of Ophthalmology*, **71**, 153–160.

ASHTON, N., WARD, B. & SERPELL, G. (1954) Effect of oxygen on developing retinal vessels with particular reference to the problem of retrolental fibroplasia. *British Journal of Ophthalmology*, **38**, 397–432.

AULD, P. A. M., BHANGANANDA, P. & MEHTA, S. (1966) The influence of an early calorie intake with I-V glucose on catabolism of premature infants. *Pediatrics, Springfield*, **37**, 592–596.

AVERY, M. E. (1965) Some effects of altered environments. Relationships between space medicine and adaptation at birth. *Pediatrics, Springfield*, **35**, 345–354.

AVERY, M. E. (1968) *The Lung and Its Disorders in the Newborn Infant.* (2nd Edition). Philadelphia: Saunders.

AVERY, M. E. & MEAD, J. (1959) Surface properties in relation to atelectasis and hyaline membrane disease. *American Journal of Diseases of Childhood*, **97**, 517–523.

AVERY, M. E. & OPPENHEIMER, E. H. (1960) Recent increase in mortality from hyaline membrane disease. *Journal of Pediatrics*, **57**, 553–559.

BABSON, S. G. & BRAMHALL, J. L. (1969) Diet and growth in the premature infant. The effect of different dietary intakes of ash-electrolyte and protein on weight gain and linear growth. *Journal of Pediatrics*, **74**, 890–900.

BACOLA, E., BEHRLE, F. C., DE SCHWEINITZ, L., MILLER, H. C. & MIRA, M. (1966) Perinatal and environmental factors in late neurogenic sequelae. *American Journal of Diseases in Children*, **112**, 359–368.

BADEN, M., BAUER, C. R., COLLE, E., KLEIN, G., TAEUSCH, H. W., Jr. & STERN, L. (1972) A controlled trial of hydrocortisone therapy in infants with respiratory distress syndrome. *Pediatrics, Springfield*, **50**, 526–534.

BAKKEN, A. F. (1970) Bilirubin excretion in newborn infants. II. Conjugated bilirubin as a possible trigger for bilirubin excretion. *Acta Paediatrica Scandinavica*, **59**, 153–156.

BANERJEE, C. K., GIRLING, D. J. & WIGGLESWORTH, J. S. (1972) Pulmonary fibroplasia in newborn babies treated with oxygen and artificial ventilation. *Archives of Disease in Childhood*, **47**, 509–518.

BARLTROP, D. & OPPÉ, T. E. (1970) Dietary factors in neonatal calcium homoeostasis. *Lancet*, **2**, 1333–1335.

BARNETT, H. L., HARE, W. K., McNAMARA, H. & HARE, R. S. (1948) Influence of postnatal age on kidney function of premature infants. *Proceedings of the Society for Experimental Biology and Medicine*, **69**, 55–57.

BARTON, M. E., WILSON, J. & WALKER, W. (1962) Idiopathic jaundice in premature infants. *Lancet*, **2**, 847–851.

BAUM, J. D. & SCOPES, J. W. (1968) The silver swaddler. Device for preventing hypothermia in the newborn. *Lancet*, **1**, 672–673.

BEAN, J. W. (1945) Effects of oxygen at increased pressure. *Physiological Reviews*, **25**, 1–147.

BEARD, A. G., PANOS, T. C., BURROUGHS, J. C., MARASIGAN, B. V. & ÖZTALAY, A. G. (1963) Perinatal stress and the premature neonate. I. Effect of fluid and calorie deprivation. *Journal of Pediatrics*, **63**, 361–385.

BELGAUMKAR, T. K. & SCOTT, K. E. (1972) The possible role of low environmental humidity in exacerbating apnoeic spells in prematures. (Abstract), *Pediatric Research*, **6**, 407.

BERG, K. & CELANDER, O. (1971) Circulatory adaptation in the thermoregulation of fullterm and premature newborn infants. *Acta Paediatrica Scandinavica*, **60**, 278–284.

BERGSMA, D. (Ed.) (1970) Bilirubin metabolism in the newborn. *Birth defects: original article series*, **6**, no. 2.

BERNSTEIN, R. B., NOVY, M. J., PIASECKI, G. J., LESTER, R. & JACKSON, B. T. (1969) Bilirubin metabolism in the fetus. *Journal of Clinical Investigation*, **48**, 1678–1688.

BESCH, N. J., PERLSTEIN, P. H., EDWARDS, N. K., KEENAN, W. J. & SUTHERLAND, J. M. (1971) The transparent baby bag. A shield against heat loss. *New England Journal of Medicine*, **284**, 121–124.

BHAGWANANI, S. G., FAHMY, D. & TURNBULL, A. C. (1972) Prediction of neonatal respiratory distress by estimation of amniotic fluid lecithin. *Lancet*, **1**, 159–162.

BHAKOO, O. N. & SCOPES, J. W. (1971) Weight minus extracellular fluid as metabolic reference standard in newborn baby. *Archives of Disease in Childhood*, **46**, 483–489.

BLACKFAN, K. D. & YAGLOU, C. P. (1933) The premature infant. A study on the effects of atmospheric conditions on growth and on development. *American Journal of Diseases of Children*, **46**, 1176–1236.

BLAKE, A. M., COLLINS, L. M., LANGHAM, J. & REYNOLDS, E. D. R. (1970) Clinical assessment of apnoea—alarm mattress for newborn infants. *Lancet*, **2**, 183–185.

BLOICE, J. A. (1972) Contactless apnoea detector using low energy radar. *Journal of Physiology*, **223**, 3–4P.

BLONDHEIM, S. H., KAPITULNIK, J., VALAES, T. & KAUFMANN, N. A. (1972) Use of a sephadox column to evaluate the bilirubin-binding capacity of the serum of infants with neonatal jaundice. *Israel Journal of Medical Science*, **8**, 22–28.

BLUMENSCHEIN, S. D., KALLEN, R. J., STOREY, B., NATZSCHKA, C., ODELL, G. B. & CHILDS, B. (1968) Familial nonhaemolytic jaundice with late onset of neurological damage. *Pediatrics, Springfield*, **42**, 786–792.

BODA, D., MARÁNYI, L., ALTORIAY, I. & VERESS, I. (1971) Peritoneal dialysis in the treatment of hyaline membrane disease of newborn premature infants. *Acta Paediatrica Scandinavica*, **60**, 90–92.

BOGGS, T. R., HARDY, J. B. & FRAZIER, T. M. (1967) Correlations of neonatal serum total bilirubin concentrations and developmental status at age eight months. *Journal of Pediatrics*, **71**, 553–560.

BOSTON, R. W., HUMPHREYS, P. W., NORMAND, I. C. S., REYNOLDS, E. O. R. & STRANG, L. B. (1968) Formation of liquid in the lungs of the foetal lamb. *Biologia Neonatorum*, **12**, 306–315.

BOUGHTON, K., GANDY, G. & GAIRDNER, D. (1970) Hyaline membrane disease. II. Lung lecithin. *Archives of Disease in Childhood*, **45**, 311–327.

BOYDEN, E. A. & TOMPSETT, D. H. (1965) The changing patterns in the developing lungs of infants. *Acta Anatomica* (Basel), **61**, 164–192.

BRADY, J. P. & CERUTI, E. (1966) Chemoreceptor reflexes in the newborn infant: Effects of varying degrees of hypoxia on heart rate and ventilation in a warm environment. *Journal of Physiology*, **184**, 631–645.

BREATHNACH, A. S. & WYLLIE, L. M. (1965) Fine structures of cells forming the surface layer of the epidermis in human fetuses at fourteen and twelve weeks. *Journal of Investigative Dermatology*, **45**, 179–189.

BREUNUNG, M. (1965) Physiologische und pathologische Hypothermie bei Frühgeborenen. *Kinderarztliche Praxis*, **33**, 1–5.

BRODERSEN, R. & HERMANN, L. S. (1963) Intestinal reabsorption of unconjugated bilirubin: A possible contributing factor in neonatal jaundice. *Lancet*, **1**, 1242.

BRODERSEN, R., JACOBSEN, J., HERTZ, H., REBBE, H. & SPRENSEN, B. (1967) Bilirubin conjugation in the human fetus. *Scandinavian Journal of Clinical and Laboratory Investigation*, **20**, 41–48.

BROOK, D. G. & ASHWORTH, A. (1972) The influence of malnutrition on the postprandial metabolic rate and respiratory quotient. *British Journal of Nutrition*, **27**, 407–415.

BRÜCK, K., PARMELEE, A. H. & BRÜCK, M. (1962) Neutral temperature range and range of 'thermal comfort' in premature infants. *Biologia Neonatorum*, **4**, 32–51.

BRÜCK, K. & WÜNNENBERG, W. (1970) Meshed control of two effector systems. In *Physiological and Behavioural Temperature Regulation*, ed. Hardy, J. D., Gagge, A. P. & Stolwijk, J. A. J. Springfield: Thomas.

BRYANT, G. M., GRAY, O. P., FRASER, A. J. & ACKERMAN, A. (1970) Fate of surviving low birthweight infants with coagulation deficiencies on the first day of life. *British Medical Journal*, **4**, 707–709.

BUCCI, G., MENDICINI, M., SCALAMANDRÈ, A., ANNIBALDI, L., SAVIGNONI, P. G. & NODARI, S. (1971) A controlled trial on therapy for newborns weighing 750–1250 g. II. Blood chemistry and electrocardiographic observations in the newborn period. *Acta Paediatrica Scandinavica*, **60**, 417–427.

BUCCI, G., SCALAMANDRÈ, A., SAVIGNONI, P. G., ORZALERI, M. & MENDICINI, M. (1966) Crib-side sampling of blood from the radial artery. *Pediatrics, Springfield*, **37**, 497–498.

BUETOW, K. C. & KLEIN, S. W. (1964) Effect of maintenance of 'normal' skin temperature on survival of infants of low birth weight. *Pediatrics, Springfield*, **34**, 163–170.

BULLEN, C. L. & WILLIS, A. T. (1971) Resistance of the breast-fed infant to gastroenteritis. *British Medical Journal*, **3**, 338–343.

BURNARD, E. D., GRATTAN-SMITH, P., PICTON-WARLOW, C. G. & GRAUAUG, A. (1965) Pulmonary insufficiency in prematurity. *Australian Paediatric Journal*, **1**, 12–38.

BURNARD, E. D. & GRAUAUG, A. (1965) Dyspnoea and apnoea in the newborn; some results of investigation. *Medical Journal of Australia*, **1**, 445–455.

BURNS, L. E., HODGMAN, J. E. & CASS, A. B. (1959) Fatal circulatory collapse in premature infants receiving chloramphenicol. *New England Journal of Medicine*, **261**, 1318–1321.

CADE, J. F., HIRSH, J. & MARTIN, M. (1969) Placental barrier to coagulation factors: its relevance to the coagulation defect at birth and to haemorrhage in the newborn. *British Medical Journal*, **2**, 281–283.

CALCAGNO, P. L. & RUBIN, M. I. (1963) Renal extraction of para-amino hypurate in infants and children. *Journal of Clinical Investigation*, **42**, 1632–1639.

CALCAGNO, P. L., RUBIN, M. I. & WEINSTRAUB, D. H. (1954) Studies on the renal concentrating and diluting mechanisms of the premature infant. *Journal of Clinical Investigation*, **33**, 91–96.

CARO, C. G. & BLOICE, J. A. (1971) Contactless apnoea detector based on radar. *Lancet*, **2**, 959–961.

CASSADY, G. (1966) Plasma volume studies in low birth weight infants. *Pediatrics, Springfield*, **38**, 1020–1027.

CERUTI, E. (1966) Chemoreceptor reflexes in the newborn infant: effect of cooling on the response to hypoxia. *Pediatrics, Springfield*, **37**, 556–564.

CHAN, G., SCHIFF, D. & STERN, L. (1971). Competitive binding of free fatty acids and bilirubin to albumin: differences in HBABA dye versus Sephadex G-25. Interpretation of results. *Clinical Biochemistry*, **4**, 308–214.

CHERNICK, V. & VIDYASAGAR, D. (1972) Continuous negative chest wall pressure in hyaline membrane disease: one year experience. *Pediatrics, Springfield*, **49**, 753–760.

CHU, J., CLEMENTS, J. A., COTTON, E. K., KLAUS, M. H., SWEET, A. Y. & TOOLEY, W. H. (1967) Neonatal pulmonary ischaemia. Part 1: Clinical and Physiological Studies. *Pediatrics, Springfield*, **40**, 709–782.

CHU, J. S., DAWSON, P., KLAUS, M. & SWEET, A. Y. (1964) Lung compliance and lung volume measured concurrently in normal full-term and premature infants. *Pediatrics, Springfield*, **34**, 525–532.

CLAIREAUX, A. E., COLE, P. G. & LATHE, G. H. (1953) Icterus of the brain in the newborn. *Lancet*, **2**, 1226–1230.

CLOSE, W. H. & MOUNT, L. E. (1971) Energy retention in the pig at several environmental temperatures and levels of feeding. *Proceedings of the Nutrition Society*, **30**, 33a–34a.

CORNBLATH, M., FORBES, A. E., PILDES, R. S., LUEBBEN, G. & GREENGARD, J. (1966) A controlled study of early fluid administration on survival of low birth weight infants. *Pediatrics, Springfield*, **38**, 547–554.

COLDIRON, J. S. (1968) Estimation of nasotracheal tube length in neonates. *Pediatrics, Springfield*, **41**, 823–828.

CREMER, R. J., PERRYMAN, P. W. & RICHARDS, D. H. (1958) Influence of light on the hyperbilirubinaemia of infants. *Lancet*, **1**, 1094–1097.

CRICHTON, J. U., DUNN, H. G., McBURNEY, A. K., ROBERTSON, A. B. & TREDGER, E. (1972) Long term effects of neonatal jaundice on brain function in children of low birth weight. *Pediatrics, Springfield*, **49**, 656–670.

CRIGLER, J. F. & GOLD, N. I. (1969) Effect of sodium phenobarbital on bilirubin metabolism in an infant with congenital, non-hemolytic, unconjugated hyperbilirubinemia, and kernicterus. *Journal of Clinical Investigation*, **48**, 42–55.

CROSS, K. W., HEY, E. N., KENNAIRD, D. L., LEWIS, S. R. & URICH, H. (1971) Lack of temperature control in infants with abnormalities of the central nervous system. *Archives of Disease in Childhood*, **46**, 437–443.

CULLEY, P., POWELL, J., WATERHOUSE, J. & WOOD, B. (1970) Sequelae of neonatal jaundice. *British Medical Journal*, **3**, 383–386.

DAILY, W. J. R. & KEARNS, K. L. (1970) Servo regulation of environmental oxygen concentration in neonatal period. *Lancet*, **1**, 123–124.

DAILY, W. J. R., KLAUS, M. & MEYER, H. B. P. (1969) Apnoea in premature infants: monitoring, incidence, heart rate changes, and an effect of environmental temperature. *Pediatrics, Springfield*, **43**, 510–518.

DAVIDSON, M., LEVINE, S. Z., BAUER, C. H. & DANN, M. (1967) Feeding studies in low birth weight infants. I. Relationships of dietary protein, fat, and electrolyte to rates of weight gain, clinical courses, and serum chemical concentrations. *Journal of Pediatrics*, **70**, 695–713.

DAVIES, P. A. (1971) Bacterial infection in the fetus and newborn. *Archives of Disease in Childhood*, **46**, 1–27.

DAVIES, P. A. & DAVIS, J. P. (1970) Very low birthweight and subsequent head growth. *Lancet*, **2**, 1216–1219.

DAVIES, P. A. & RUSSELL, H. (1968) Later progress of 100 infants weighing 1,000 to 2,000 g. at birth fed immediately with breast milk. *Developmental Medicine and Child Neurology*, **10**, 725–735.

DAWES, G. S. (1968) *Foetal and Neonatal Physiology. A Comparative Study of the Changes at Birth.* Year Book Medical Publisher: Chicago.

DAY, R. L., CALIGUIRI, L., KAMENSKI, C. & EHRLICH, F. (1964) Body temperature and survival of premature infants. *Pediatrics, Springfield*, **34**, 171–181.

DE LEMOS, R., WOLFSDORF, J., NACHMAN, R., BLOCK, A. J., LEIBY, G., WILKINSON, H. A., ALLEN, T., HALLER, J. A., MORGAN, W. & AVERY, M. E. (1969) Lung injury from oxygen in lambs. The role of artificial ventilation. *Anaesthesiology*, **30**, 609–618.

DELIVORA-PAPADOPOULOS, M., RONCEVIC, N. P. & OSKI, F. A. (1971) Postnatal changes in oxygen transport of term premature and sick infants: the role of red cell 2,3-diphosphoglycerate and adult haemoglobin. *Pediatric Research*, **5**, 235–245.

DE SA, D. J. (1968) Respiratory distress accompanied by changes in the pulmonary water content and alveolar lung layer in newborn rabbits recovering from oxygen lack. *Biologia Neonatorum*, **13**, 271–280.

DE SA, D. J. (1969) Pulmonary fluid content in infants with respiratory distress. *Journal of Pathology*, **97**, 469–479.

DIAMOND, I. & SCHMID, R. (1966) Experimental bilirubin encephalopathy. The mode of entry of bilirubin-^{14}C into the central nervous system. *Journal of Clinical Investigation*, **45**, 678–689.

DICKERSON, J. W. T. (1962) Changes in the composition of the human femur during growth. *Biochemical Journal*, **82**, 56–61.

DI TORO, R., LUPI, L. & ANSANELLI, V. (1968) Glucuronation of the liver in premature babies. *Nature*, **219**, 265–267.

DOBBING, J. (1970) Undernutrition and the developing brain. The relevance of animal models to the human problem. *American Journal of Diseases of Children*, **120**, 411–415.

DOSSETT, J. H. (1972) Microbial defences of the child and man. *Pediatric Clinics of North America*, **19**, 355–372.

DRILLIEN, C. M. (1971) Prognosis of infants of very low birth-weight (letter). *Lancet*, **1**, 697.

DUNN, J. M. & MILLER, J. A. (1969) Hypothermia combined with positive pressure ventilation in resuscitation of the asphyxiated neonate. *American Journal of Obstetrics and Gynaecology*, **104**, 58–67.

DUNN, P. M. (1966) Localisation of the umbilical catheter by post-mortem measurement. *Archives of Disease in Childhood*, **41**, 69–75.

EDELMANN, C. M. & WOLFISH, N. M. (1968) Dietary influence on renal maturation in premature infants (Abstract). *Pediatric Research*, 2, 421–422.

EGAN, E. A. & EITZMAN, D. V. (1971) Umbilical vessel catheterization. *American Journal of Diseases of Children*, 121, 213–218.

EMERY, J. (1969) *The Anatomy of the Developing Lung*. London: Heinemann.

ENTE, G., LANNING, E. W., CUKOR, P. & KLEIN, R. M. (1972) Chemical variables and new lamps in phototherapy. *Pediatric Research*, 6, 246–251.

EUROPEAN ASSOCIATION OF PERINATAL MEDICINE (1970) Working party to discuss nomenclature based on gestational age and birthweight. *Archives of Disease in Childhood*, 45, 730.

EVANS, R. S., OLVER, R. E., APPLEYARD, W. J. & NEWMAN, C. G. H. (1970) Effects of intragastric and intravenous sodium bicarbonate on rate of recovery from post-asphyxial acidosis in the neonate. *Archives of Disease in Childhood*, 45, 321–324.

FANAROFF, A. A., WALD, M., GRUBER, H. S. & KLAUS, M. H. (1972) Insensible water loss in low birth weight infants. *Pediatrics, Springfield*, 50, 236–244.

FAO/WHO EXPERT GROUP (1970) Requirements of ascorbic acid, vitamin D, vitamin B_{12}, folate and iron. World Health Organisation Technical Report Series No. 452, Geneva.

FARBER, S. & SWEET, L. K. (1931) Amniotic sac contents in the lungs of infants. *American Journal of Diseases of Children*, 42, 1372–1383.

FEDRICK, J. & BUTLER, N. R. (1970a) Certain causes of neonatal death. I. Hyaline membranes. *Biology of the Neonate*, 15, 229–255.

FEDRICK, J. & BUTLER, N. R. (1970b) Certain causes of neonatal death. II. Intraventricular haemorrhage. *Biology of the Neonate*, 15, 257–290.

FEDRICK, J. & BUTLER, N. R. (1971a) Certain causes of neonatal death. IV. Massive pulmonary haemorrhage. *Biology of the Neonate*, 18, 243–262.

FEDRICK, J. & BUTLER, N. R. (1971b) Certain causes of neonatal death. V. Cerebral birth trauma. *Biology of the Neonate*, 18, 321–329.

FEDRICK, J. & BUTLER, N. R. (1972) Hyaline-Membrane disease. *Lancet*, 2, 768–769.

FETTERMAN, G. H., SHUPLOCK, N. A., PHILIPP, F. J. & GREGG, H. S. (1965) The growth and maturation of human glomeruli and proximal convolutions from term to adulthood. *Pediatrics, Springfield*, 35, 601–619.

FOMON, S. J., THOMAS, L. N., FILER, L. J., ANDERSON, T. A., & BERGMANN, K. E. (1973) Requirements for protein and essential amino acids in early infancy. *Acta Paediatrica Scandinavica*, 62, 33–45.

FOSTER, K. G., HEY, E. N. & KATZ, G. (1969) The response of the sweat glands of the new-born baby to thermal stimuli and to intra-dermal acetylcholine. *Journal of Physiology*, 203, 13–29.

GAIRDNER, D., MARKS, J., ROSCOE, J. D. & BRETTELL, R. O. (1958) The fluid shift from the vascular compartment immediately after birth. *Archives of Disease in Childhood*, 33, 489–498.

GANDY, G. M., ADAMSONS, K., CUNNINGHAM, N., SILVERMAN, W. A. & JAMES, L. S. (1964) Thermal environment and acid-base homeostasis in human infants during the first few hours of life. *Journal of Clinical Investigation*, 43, 751–758.

GANDY, G., JACOBSON, W. & GAIRDNER, D. (1970) Hyaline membrane disease. I: Cellular changes. *Archives of Diseases in Childhood*, **45**, 289–310.

GARBY, L., SJÖLIN, S. & VUILLE, J. C. (1964) Studies on erythro-kinetics in infancy. V. Estimations of the life span of red cells in the newborn. *Acta Paediatrica Scandinavica*, **53**, 165–171.

GARTNER, L. M., SNYDER, R. N., CHABON, R. S. & BERNSTEIN, J. (1970) Kernicterus: High incidence in premature infants with low serum bilirubin concentrations. *Pediatrics, Springfield*, **45**, 906–917.

GATTI, R. A., MUSTER, A. J., COLE, R. B. & PAUL, M. H. (1966) Neonatal polycythaemia with transient cyanosis and cardiorespiratory abnormalities. *Journal of Pediatrics*, **69**, 1063–1072.

GERSHON, A. A. (1971) Diagnostic Virology. *Pediatric Clinics of North America*, **18**, 73–86.

GHADIMI, H., ABACI, F., KUMAR, S. & RATHI, M. (1971) Biochemical aspects of intravenous alimentation. *Pediatrics, Springfield*, **48**, 955–965.

GIRLING, D. J. & HALLIDIE-SMITH, K. A. (1971) Persistent ductus arteriosus in ill and premature babies. *Archives of Disease in Childhood*, **46**, 177–181.

GITLIN, D. & BIASUCCI, A. (1969) Development of γG, γM, γM, β_{1C}/β_{1A}, C'1 esterase inhibitor, ceruloplasmin, transferrin, hemopexin, haptoglobin, fibrinogen, plasminogen, α_1-antitrypsin, orosmomucoid, β-lipoprotein, α_2-macroglobulin, and prealbumin in the human conceptus. *Journal of Clinical Investigation*, **48**, 1433–1446.

GITLIN, D. & BOESMAN, M. (1966) Serum α-fetoprotein, albumin, and γG-globulin in the human conceptus. *Journal of Clinical Investigation*, **45**, 1826–1838.

GIUNTA, F. (1971) A one year experience with phototherapy for jaundice of prematurity. *Pediatrics, Springfield*, **47**, 123–125.

GIUNTA, F. & RATH, J. (1969) Effect of environmental illumination in prevention of hyperbilirubinaemia of prematurity. *Pediatrics, Springfield*, **44**, 162–167.

GLASS, L. (1971) Personal communication.

GLASS, L., SILVERMAN, W. A. & SINCLAIR, J. C. (1968) Effect of the thermal environment on cold resistance and growth of small infants after the first week of life. *Pediatrics, Springfield*, **41**, 1033–1046.

GLASS, L., SILVERMAN, W. A. & SINCLAIR, J. C. (1969) Relationship of thermal environment and calorie intake to growth and resting metabolism in the late neonatal period. *Biologia Neonatorum*, **14**, 324–340.

GLASS, L., SILVERMAN, W. A. & SINCLAIR, J. C. (1971) Food, temperature and head growth in neonates. *Lancet*, **1**, 1186–1187.

GLUCK, L., KULOVICH, M. V., BORER, R. C., BRENNER, P. H., ANDERSON, G. G. & SPELLACY, W. N. (1971) The diagnosis of the respiratory distress syndrome (RDS) by amniocentesis. *American Journal of Obstetrics and Gynecology*, **109**, 440–445.

GLUCK, L., KULOVICH, M. V., EIDELMAN, A., CORDERO, L. & KHAZIN, A. F. (1972) Biochemical development of surface activity in mammalian lung. IV. Pulmonary lecithin synthesis in the human fetus and newborn and etiology of the respiratory distress syndrome. *Pediatric Research*, **6**, 81–99.

GORDON, H. H., LEVINE, S. Z., DEAMER, W. C. & McNAMARA, H. (1940) Respiratory metabolism in infancy and in childhood. XXIII. Daily energy requirements of premature infants. *American Journal of Diseases of Children*, **59**, 1185–1202.

GORDON, H. H., LUBCHENICO, L. & HIX, I. (1954) Observations on the etiology of retrolental fibroplasia. *Bulletin of the Johns Hopkins Hospital*, **94**, 34–44.

GRAHAM, B. D., REARDON, H. S., WILSON, J. L., MAKEPEACE, U. T. & BAUMANN, M. L. (1950) Physiologic and chemical responses of premature infants to oxygen-enriched atmosphere. *Pediatrics, Springfield*, **6**, 55–71.

GRAY, O. P., ACKERMAN, A. & FRAZER, A. J. (1968) Intracranial haemorrhage and clotting defects in low birthweight infants. *Lancet*, **1**, 545–548.

GRAZIANI, L. J., WEITZMAN, E. D. & PINEDA, G. (1972) Visual evoked responses during neonatal respiratory disorders in low birth weight infants. *Pediatric Research*, **6**, 203–210.

GREEN, M. & SOLNIT, A. (1964) Reactions to the threatened loss of a child: a vulnerable child syndrome. Pediatric Management of the Dying Child, Part III. *Pediatrics, Springfield*, **34**, 58–66.

GREGORY, G. A. (1972) Respiratory care of newborn infants. *Pediatric Clinics of North America*, **19**, 311–324.

GREGORY, G. A., KITTERMAN, J. A., PHIBBS, R. H., TODEY, W. H. & HAMILTON, W. K. (1971) Treatment of the idiopathic respiratory distress syndrome with continuous positive airway pressure. *New England Journal of Medicine*, **284**, 1333–1340.

GRUENWALD, P. & POPPER, H. (1940) The histogenesis and physiology of the renal glomerulus in early postnatal life: histological examinations. *Journal of Urology*, **43**, 452–458.

GRYBOSKI, J. D. (1969) Suck and swallow in the premature infant. *Pediatrics, Springfield*, **43**, 96–102.

GUPTA, J. M., DAHLENBURG, G. W. & DAVIS, J. A. (1967) Changes in blood gas tensions following administration of amine buffer THAM to infants with respiratory distress. *Archives of Diseases in Childhood*, **42**, 416–427.

GUPTA, R. K., WINTER, P. M. & LANPHIER, E. H. (1969) Histochemical studies in pulmonary oxygen toxicity. *Aerospace Medicine*, **40**, 500–504.

GUSDON, J. P. & WAITE, B. M. (1972) A colorimetric method for amniotic fluid phospholipids and their relationship to the respiratory distress syndrome. *American Journal of Obstetrics and Gynaecology*, **112**, 62–71.

GUSTAFSON, A., KJELLMER, I., OLEGÅRD, R. & VICTORIN, L. (1972) Nutrition in low birth weight infants. I. Intravenous injection of fat emulsions. *Acta Paediatrica Scandinavica*, **61**, 149–158.

HANSON, J. S. & SHINOZAKI, T. (1970) Hybrid computer studies of ventilatory distribution and lung volume. *Pediatrics, Springfield*, **46**, 900–914.

HARCKE, H. T., NAEYE, R. L., STORCH, A. & BLANC, W. A. (1972) Perinatal cerebral ventricular haemorrhage. *Journal of Pediatrics*, **80**, 37–42.

HARDMAN, M. J., HEY, E. N. & HULL, D. (1969) Fat metabolism and heat production in young rabbits. *Journal of Physiology*, **205**, 51–59.

HARDMAN, M. J. & HULL, D. (1971) The effect of environmental conditions on the growth and function of brown adipose tissue. *Journal of Physiology*, **214**, 191–199.

HARRIES, J. K. (1971) Intravenous feeding in infants. *Archives of Disease in Childhood*, **46**, 855–863.

HARRISON, V. C., HEESE, H. DE V. & KLEIN, M. (1968) The significance of grunting in hyaline membrane disease. *Pediatrics, Springfield*, **41**, 549–559.

HAUGAARD, N. (1968) Cellular mechanisms of oxygen toxicity. *Physiological Reviews*, **48**, 311–373.

HAWORTH, J. C., DILLING, L. & YOUNASZAI, M. K. (1967) Relation of blood-glucose to haematocrit, birthweight, and other body measurements in normal and growth-retarded newborn infants. *Lancet*, **2**, 901–905.

HEIRD, W. C., DRISCOLL, J. M., SCHULLINGER, J. N., GREBIN, B. & WINTER, R. W. (1972) Intravenous alimentation in pediatric patients. *Journal of Pediatrics*, **80**, 351–372.

HEY, E. N. (1969) The relation between environmental temperature and oxygen consumption in the newborn baby. *Journal of Physiology*, **200**, 589–603.

HEY, E. & HULL, D. (1971) Lung function at birth in babies developing respiratory distress. *Journal of Obstetrics and Gynaecology of the British Commonwealth*, **78**, 1137–1146.

HEY, E. N. & KATZ, G. (1969a) Evaporative water loss in the newborn baby. *Journal of Physiology*, **200**, 605–619.

HEY, E. N. & KATZ, G. (1969b) Temporary loss of a metabolic response to cold stress in infants of low birthweight. *Archives of Disease in Childhood*, **44**, 323–330.

HEY, E. N. & KATZ, G. (1970a) The range of thermal insulation in the tissues of the newborn baby. *Journal of Physiology*, **207**, 667–681.

HEY, E. N. & KATZ, G. (1970b) The optimum thermal environment for naked babies. *Archives of Disease in Childhood*, **45**, 328–334.

HEY, E. N., KATZ, G. & O'CONNELL, B. (1970) The total thermal insulation of the newborn baby. *Journal of Physiology*, **207**, 683–698.

HEY, E. N. & KELLY, J. (1968) Gaseous exchange during endotracheal ventilation for asphyxia at birth. *Journal of Obstetrics and Gynaecology of the British Commonwealth*, **75**, 414–424.

HEY, E. N. & MOUNT, L. E. (1967) Heat losses from babies in incubators. *Archives of Disease in Childhood*, **42**, 75–84.

HILGARTNER, M. W. & SMITH, C. H. (1965) Plasmathromboplastin antecedent (factor XI) in the neonate. *Journal of Pediatrics*, **66**, 747–752.

HILL, J. R. & ROBINSON, D. C. (1968) Oxygen consumption in normally grown, small-for-dates and large-for-dates newborn infants. *Journal of Physiology*, **199**, 685–703.

HOBEL, C. J., HYVARINEN, M. A., ERENBERG, A., EMMANOUILIDES, G. C. & OH, W. (1971) Early treatment of neonatal acidosis in low birthweight infants in relation to respiratory distress syndrome (Abstract). *Pediatric Research*, **5**, 415.

HODGMAN, J. E., MIKITY, V. G., TATTER, D. & CLELAND, R. S. (1969)

Chronic respiratory distress in the premature infant. Wilson–Mikity syndrome. *Pediatrics, Springfield,* **44,** 179–195.

HOOD, J. H. (1964) Effect of posture on the amount and distribution of gas in the intestinal tract of infants and young children. *Lancet,* **2,** 107–110.

HUMPHREYS, P. W. & STRANG, L. B. (1967) Effects of gestation and prenatal asphyxia on pulmonary surface properties of the foetal rabbit. *Journal of Physiology,* **192,** 53–62.

HUSBAND, J. & HUSBAND, P. (1969) Gastric emptying of water and glucose solutions in the newborn. *Lancet,* **2,** 409–411.

HYMAN, C. B., KEASTER, J., HANSON, V., HARRIS, I., SEDGWICK, R., WURSTEN, H. & WRIGHT, A. R. (1969) CNS abnormalities after neonatal hemolytic disease or hyperbilirubinemia. *American Journal of Diseases of Children,* **117,** 395–405.

JANSKY, L. (1971) *Nonshivering Thermogenesis.* Proceedings of the symposium held in Prague, April 1–2, 1970. Prague: Academia.

JOHNSON, J. W. C. (1962) A study of fetal intrathoracic pressures during labour and delivery. *American Journal of Obstetrics and Gynecology,* **84,** 15–19.

JOHNSON, K. G. & BABSON, S. G. (1967) Resuscitation of the apnoeic premature infant. *Pediatrics, Springfield,* **40,** 99–100.

JOHNSON, P., DAWES, G. S. & ROBINSON, J. S. (1972) Maintenance of breathing in newborn lamb (Abst.). *Archives of Disease in Childhood,* **47,** 151.

JOLLY, H., MOLYNEUX, P. & NEWELL, D. J. (1962) A controlled study of the effect of temperature on premature babies. *Journal of Pediatrics,* **60,** 889–894.

KENNY, J. F., BOESMAN, M. I. & MICHAELS, R. H. (1967) Bacterial and viral coproantibodies in breast-fed infants. *Pediatrics, Springfield,* **39,** 202–213.

KERPEL-FRONIUS, E., HEIM, T. & SULYOK, E. (1970) The development of the renal acidifying processes and their relation to acidosis in low-birthweight infants. *Biology of the Neonate,* **15,** 156–168.

KIDD, B. S. L., LEVISON, H., GEMMEL, P., AHARON, A. & SWYER, P. R. (1966) Limb blood flow in the normal and sick newborn. *American Journal of Diseases of Childhood,* **112,** 402–407.

KINTZEL, H.-W. (1956) Zur Frage der Warm- oder Kalthaltung der Frühgeborenen. *Archiv fur Kinderheilkunde,* **154,** 238–247.

KINTZEL, H.-W. (1966) Der Energiestoffwechsel Frühgeborenen bei langanhaltender Hypothermie. *Monatsschrift für Kinderheilkunde,* **114,** 544–550.

KITCHEN, W. H. & CAMPBELL, D. G. (1971) Controlled trial of intensive care for very low birth weight infants. *Pediatrics, Springfield,* **48,** 711–714.

KITTERMAN, J. A., PHIBBS, R. H. & TOOLEY, W. H. (1969) Aortic blood pressure in normal newborn infants during the first 12 hours of life. *Pediatrics, Springfield,* **44,** 959–968.

KITTERMAN, J. A., PHIBBS, R. H. & TOOLEY, W. H. (1970) Catheterization of umbilical vessels in newborn infants. *Pediatric Clinics of North America,* **17,** 895–912.

KLAUS, M. H. & KENNELL, J. H. (1970) Mothers separated from their newborn infants. *Pediatric Clinics of North America,* **17,** 1015–1037.

KLEIN, M. & STERN, L. (1971) Low birth weight and the battered child syndrome. *American Journal of Diseases of Children,* **122,** 15–18.

KLEINMAN, L. I. & LUBBE, R. J. (1972a) Factors affecting the maturation of glomerular filtration rate and renal plasma flow in the newborn dog. *Journal of Physiology*, **223**, 395–409.

KLEINMAN, L. I. & LUBBE, R. J. (1972b) Factors affecting the maturation of renal PAH extraction in the newborn dog. *Journal of Physiology*, **223**, 411–418.

KOTAS, R. V. & AVERY, M. E. (1971) Accelerated appearance of pulmonary surfactant in the fetal rabbit. *Journal of Applied Physiology*, **30**, 358–361.

KOTAS, R. V., FLETCHER, B. D., TORDAY, J. & AVERY, M. E. (1971) Evidence of independent regulators of organ maturation in fetal rabbits. *Pediatrics, Springfield*, **47**, 57–64.

KRAUER, B., SPRING, P. & DETTLI, L. (1968) Zur Pharmakokinetik der Sulfonamide im ersten Lebensjahr. *Pharmacology Clinics*, **1**, 47–53.

KRAUSS, A. N. & AULD, P. A. M. (1970) Measurement of functional residual capacity in distressed neonates by helium rebreathing. *Journal of Pediatrics*, **77**, 228–232.

KRAUSS, A. N. & AULD, P. A. M. (1971) Pulmonary gas trapping in premature infants. *Pediatric Research*, **5**, 10–16.

KRAUSS, A. N., LEVIN, A. R., GROSSMAN, H. & AULD, P. A. M. (1970) Physiologic studies on infants with Wilson–Mikity syndrome. *Journal of Pediatrics*, **77**, 27–36.

KRAUSS, A. N., SOODALTER, J. A. & AULD, P. A. M. (1971) Adjustment of ventilation and perfusion in the full term normal and distressed neonate as determined by urinary alveolar nitrogen gradients. *Pediatrics, Springfield*, **47**, 865–869.

KRAVATH, R. E., AHARON, A. S., ABAL, G. & FINBERG, L. (1970) Clinically significant physiological changes from rapidly administered hypertonic solutions: Acute osmol poisoning. *Pediatrics, Springfield*, **46**, 267–275.

KUNNAS, M. (1968) Mortality of premature infants according to the temperature on admission to hospital. *Annales Paediatriae Fenniae*, **14**, 98–101.

LANMAN, J. T., GUY, L. P. & DANCIS, J. (1954) Retrolental fibroplasia and oxygen therapy. *Journal of the American Medical Association*, **155**, 223–226.

LATHE, G. H. & WALKER, M. (1958) The synthesis of bilirubin glucoronide in animal and human liver. *Biochemical Journal*, **70**, 705–712.

LENDING, M., SLOBODY, L. B. & MESTERN, J. (1967) The relationship of hypercapnia to the production of kernicterus. *Developmental Medicine and Child Neurology*, **9**, 145–151.

LEVI, A. J., GATMAITAN, Z. & ARIAS, I. M. (1970) Deficiency of hepatic organic anion-binding protein, impaired organic anion uptake by liver and 'physiologic' jaundice in newborn monkeys. *New England Journal of Medicine*, **283**, 1136–1139.

LEWIN, J. E. (1969) An apnoea-alarm mattress. *Lancet*, **2**, 667–668.

LEWIS, P. K., REID, M., REILLY, B. J., SWYER, P. R. & FRASER, D. (1971) Iatrogenic rickets in low birthweight infants. *Journal of Pediatrics*, **78**, 207–210.

LIGGINS, G. C. (1969) Premature delivery of foetal lambs infused with glucocorticoids. *Journal of Endocrinology*, **45**, 515–523.

LIGGIN, G. C. & HOWIE R. N. (1972). A controlled trial of antepartum glucocorticoid treatment for prevention of the respiratory distress syndrome in premature infants. *Pediatrics, Springfield*, **50**, 515–525.

LIND, T. & BILLEWICZ, W. Z. (1971) A point scoring system for estimating gestational age from examination of amniotic fluid. *British Journal of Hospital Medicine*, **5**, 681–685.

LIND, T., KENDALL, A. & HYTTEN, F. E. (1972) The role of the fetus in the formation of amniotic fluid. *Journal of Obstetrics and Gynaecology of the British Commonwealth*, **79**, 289–298.

LINDBERG, O. (1970) *Brown Adipose Tissue*. New York: Elsevier.

LOCHRIDGE, S., PASS, R. & CASSADY, G. (1971) Reticulocyte counts in intrauterine growth retardation. *Pediatrics, Springfield*, **47**, 919–923.

LUBCHENCO, L. O., DELIVORA-PAPADOPOULOS, M., BUTTERFIELD, L. J., FRENCH, J. H., MEDCALF, D., HIX, I. E., DANICK, J., DODDS, J., DOWNS, M. & FREELAW, E. (1972) Long term follow-up studies of pre-term newborn infants. I. Relationship of handicaps to nursery routines. *Journal of Pediatrics*, **80**, 501–508.

LUBCHENCO, L. O., DELIVORA-PAPADOPOULOS, M. & SEARLS, D. (1972) Long term follow-up studies of pre-term newborn infants. II. Influence of birth weight and gestational age on sequelae. *Journal of Pediatrics*, **80**, 509–512.

LUCEY, J. F., HIBBARD, E., BEHRMAN, R. E., ESQUIVAL DE GALLARDO, F. O. & WINDLE, W. F. (1964) Kerniterus in asphyxiated newborn monkeys. *Experimental Neurology*, **9**, 43–58.

MACDONALD, M. S. & EMERY, J. L. (1959) The late intrauterine and post-natal development of human renal glomeruli. *Journal of Anatomy*, **93**, 331–340.

McMURPHY, D. M., HEYMANN, M. A., RUDOLPH, A. M. & MELMON, K. L. (1972) Developmental changes in constriction of the ductus arteriosus: response to oxygen and vasoactive agents in the isolated ductus arteriosus of the fetal sheep. *Pediatric Research*, **6**, 231–238.

MAISELS, M. J., PATHAK, A., NELSON, N. M., NATHAN, D. G. & SMITH, C. A. (1971) Endogenous production of carbon monoxide in normal and erythroblastotic newborn infants. *Journal of Clinical Investigation*, **50**, 1–8.

MALAKA-ZAFIRIU, K. & STRATES, B. S. (1969) The effect of antimicrobial agents on the binding of bilirubin by albumin. *Acta Paediatrica Scandinavica*, **58**, 281–286.

MAMUNES, P., BADEN, M., BASS, J. W. & NELSON, J. (1969) Early intravenous feeding of the low birthweight neonate. *Pediatrics, Springfield*, **43**, 241–250.

MANN, T. P. (1968) Observations on temperatures of mothers and babies in the perinatal period. *Journal of Obstetrics and Gynaecology of the British Commonwealth*, **75**, 316–321.

MARINI, A., BARBARANI, V., BONCOMPAGNI, P. & CATTANEO, F. (1970) Alkalinising agents in the delivery room. *Lancet*, **2**, 265.

MARKS, J., GAIRDNER, D. & ROSCOE, J. D. (1955) Blood formation in infancy. Part III. Cord Blood. *Archives of Disease in Childhood*, **30**, 117–120.

MEDICAL RESEARCH COUNCIL (1955) Retrolental fibroplasia in the United

Kingdom. A report to the Medical Research Council by their conference on retrolental fibroplasia. *British Medical Journal*, **2**, 78–82.

MEDICAL RESEARCH COUNCIL (1965) Spectral requirements of light sources for clinical purpose. Joint committee on lighting and vision. MRC Memorandum No. 43. London: HMSO.

MENDICINI, M., SCALAMANDRÈ, A., SAVIGNONI, P. G., PICECE-BUCCI, S., ESUPERANZI, R. & BUCCI, G. (1971) A controlled trial on therapy for newborns weighing 750–1250 g. I. Clinical findings and mortality in the newborn period. *Acta Paediatrica Scandinavica*, **60**, 407–416.

MENKES, J. H. & AVERY, M. E. (1963) The metabolism of phenylalanine and tyrosine in the premature infant. *Bulletin of the Johns Hopkins Hospital*, **113**, 301–319.

MENKES, J. H., WELCHER, D. W., LEVI, H. S., DALLAS, J. & GRETSKY, N. E. (1972) Relationship of elevated blood tyrosine to the ultimate intellectual performance of premature infants. *Pediatrics, Springfield*, **49**, 218–224.

MESTYÁN, S., JÁRAI, I., KEKETE, M. & SOLTÉSZ, GY. (1969) Specific dynamic action in premature infants kept at and below the neutral temperature. *Pediatric Research*, **3**, 41–50.

MEYER, W. W. & LIND, J. (1966) The ductus venosus and the mechanism of its closure. *Archives of Disease in Childhood*, **41**, 597–605.

MILLER, H. C., BEHRLE, F. C. & SMULL, N. W. (1959) Severe apnoea and irregular respiratory rhythms among premature infants. A clinical and laboratory study. *Pediatrics, Springfield*, **23**, 676–685.

MILLER, H. C. & HAMILTON, T. R. (1949) The pathogenesis of the 'Vernix Membrane'. Relation to aspiration pneumonia in stillborn and newborn infants. *Pediatrics, Springfield*, **3**, 735–748.

MILLER, W. W., WALDHAUSEN, J. A. & RASHKIND, W. J. (1970) Comparison of oxygen poisoning of the lungs in cyanotic and acyanotic dogs. *New England Journal of Medicine*, **282**, 943–947.

MINKOWSKI, A., MONSET-COUCHARD, M. & AMIEL-TISON, C. (1970) (Editors) Symposium on Artificial Ventilation. *Biology of the Neonate*, **16**, 1–196.

MOLLISON, P. L. & CUTBUSH, M. (1949) Haemolytic disease of the newborn: criterion of severity. *British Medical Journal*, **1**, 123–130.

MOSS, A. J. & MONSET-COUCHARD, M. (1967) Placental transfusion: early versus late clamping of the umbilical cord. *Pediatrics, Springfield*, **40**, 109–126.

MOTOYAMA, E. K., ORZALESI, M. M., KIKKAWA, Y., KAIBARA, M., WU, B., ZIGAS, C. G. & COOK, C. D. (1971) Effect of cortisol on the maturation of fetal rabbit lungs. *Pediatrics, Springfield*, **48**, 547–555.

MURDOCK, A. I., KIDD, B. S. I., LLEWELLYN, M. A., REID, M. McC. & SWYER, P. R. (1970) Intrapulmonary venous admixture in the respiratory distress syndrome. *Biology of the Neonate*, **15**, 1–7.

MUSHIN, A. (1971) Ocular changes in premature babies receiving controlled oxygen therapy in the neonatal period. *Proceedings of the Royal Society of Medicine*, **64**, 779–780.

MYERS, R. E. (1972) Two patterns of perinatal brain damage and their condi-

347

tions of occurrence. *American Journal of Obstetrics and Gynaecology,* **112,** 246–276.

NEAL, W. A., REYNOLDS, J. W., JARVIS, C. W. & WILLIAMS, H. J. (1972) Umbilical artery catheterization: demonstration of arterial thrombosis by aortography. *Pediatrics, Springfield,* **50,** 6–13.

NELIGAN, G. A., OMER, M. & MARR, J. (1972) Personal communication.

NELIGAN, G., ROBSON, E. & HEY, E. (1969) Hyaline membrane disease in twins. *Pediatrics, Springfield,* **43,** 143.

NELSON, G. H. (1972) Relationship between amniotic fluid lecithin concentration and respiratory distress syndrome. *American Journal of Obstetrics and Gynaecology,* **112,** 827–833.

NELSON, N. M. (1966) Neonatal Pulmonary Function. *Pediatric Clinics of North America,* **13,** 769–799.

NELSON, N. M. (1970) On the etiology of hyaline membrane disease. *Pediatric Clinics of North America,* **17,** 943–965.

NELSON, N. M., PROD'HOM, L. S., CHERRY, R. B., LIPSITZ, P. J. & SMITH, C. A. (1962) Pulmonary function in the newborn infant. II. Perfusion-estimation by analysis of the arterial-alveolar carbon dioxide difference. *Pediatrics, Springfield,* **30,** 975–989.

NELSON, N. M., PROD'HOM, L. S., CHERRY, R. B., LIPSITZ, P. J. & SMITH, C. A. (1963a) Pulmonary function in the newborn infant. V. Trapped gas in the normal infant's lung. *Journal of Clinical Investigation,* **42,** 1850–1857.

NELSON, N. M., PROD'HOM, L. S., CHERRY, R. B., LIPSITZ, P. J. & SMITH, C. A. (1963b) Pulmonary function in the newborn infant: the alveolar-arterial oxygen gradient. *Journal of Applied Physiology,* **18,** 534–538.

NORMAND, I. C. S., OLVER, R. E., REYNOLDS, E. O. R. & STRANG, L. B. (1971) Permeability of lung capillaries and alveoli to non-electrolytes in the fetal lamb. *Journal of Physiology,* **219,** 303–330.

NORMAND, I. C. S., REYNOLDS, E. O. R. & STRANG, L. B. (1970) Passage of macromolecules between alveolar and intestinal spaces in foetal and newly ventilated lungs of the lamb. *Journal of Physiology,* **210,** 151–164.

NORTHWAY, W. H., ROSAN, R. C. & PORTER, D. Y. (1967) Pulmonary disease following respirator therapy of hyaline-membrane disease. Bronchopulmonary dysplasia. *New England Journal of Medicine,* **276,** 357–368.

NOURSE, C. H. & NELSON, N. M. (1969) Uniformity of ventilation in the newborn infant: direct assessment of the arterial alveolar N_2 difference. *Pediatrics, Springfield,* **43,** 226–232.

ODELL, G. B., STOREY, G. N. B. & ROSENBERG, L. A. (1970) Studies in kernicterus. III. The saturation of serum proteins with bilirubin during neonatal life and its relationship to brain damage at five years. *Journal of Pediatrics,* **76,** 12–21.

OH, W., OH, M. A. & LIND, J. (1966) Renal function and blood volume in newborn infant related to placental transfusion. *Acta Paediatrica Scandinavica,* **56,** 197–210.

OH, W., WALLGREN, G., HANSON, J. S. & LIND, J. (1967) The effects of placental transfusion on respiratory mechanics of normal term newborn infants. *Pediatrics, Springfield,* **40,** 6–12.

OLSON, M. (1970) The benign effects on rabbits' lungs of the aspiration of water compared with 5 per cent glucose or milk. *Pediatrics, Springfield*, **46**, 538–547.

OPPÉ, T. E. & REDSTONE, D. (1968) Calcium and phosphorus levels in healthy newborn infants given various types of milk. *Lancet*, **1**, 1045–1048.

ORZALESI, M. M., MENDICINI, M., BUCCI, G., SCALAMANDRÈ, A. & SAVIGNONI, P. G. (1967) Arterial oxygen studies in premature newborns with and without respiratory distress. *Archives of Disease in Childhood*, **42**, 174–180.

OSKI, F. A. & NAIMAN, J. L. (1972) *Hematological Problems in The Newborn*. 2nd Edition. Chapter 3. W. B. Saunders: Philadelphia.

OSTREA, E. M. & ODELL, G. B. (1972) The influence of bicarbonate administration on blood pH in a 'closed system': clinical implications. *Journal of Pediatrics*, **80**, 671–680.

OUTERBRIDGE, E. W., NOGRAL, M. B., BEANDRY, P. H. & STERN, L. (1972) Idiopathic respiratory distress syndrome. Recurrent respiratory illness in survivors. *American Journal of Diseases of Children*, **123**, 99–104.

PARKER, D., KEY, A. & DAVIES, R. S. (1971) Catheter-tip transducer for continuous in-vivo measurement of oxygen tension. *Lancet*, **1**, 952–953.

PARMLEY, T. H. & SEEDS, A. E. (1970) Fetal skin permeability to isotopic water (THO) in early pregnancy. *American Journal of Obstetrics and Gynecology*, **108**, 128–131.

PATZ, A., HOECK, L. E. & DE LA CRUZ, E. (1952) Studies on the effect of high oxygen administration in retrolental fibroplasia. I. Nursery observations. *American Journal of Ophthalmology*, **35**, 1248–1253.

PERLSTEIN, P. H., EDWARDS, N. K. & SUTHERLAND, J. M. (1970) Apnoea in premature infants and incubator-air-temperature changes. *New England Journal of Medicine*, **282**, 461–466.

PETER, G., LLOYD-STILL, J. D. & LOVEJOY, F. H. (1972) Local infection and bacteremia from scalp vein needles and polyethylene catheter in children. *Journal of Pediatrics*, **80**, 78–83.

PLAYFAIR, J. H. L., WOLFENDALE, M. R. & KAY, H. E. M. (1963) The leucocytes of peripheral blood in the human fetus. *British Journal of Haematology*, **9**, 336–344.

PLATT, B. S. & STEWART, R. J. C. (1971) Reversible and irreversible effects of protein-calorie deficiency on the central nervous system of animals and man. *World Review of Nutrition and Diatetics*, **13**, 43–85.

POLAND, R. D. & ODELL, G. B. (1971) Physiological jaundice: the enterohepatic circulation of bilirubin. *New England Journal of Medicine*, **284**, 1–6.

PŘIBYLOVÁ, H. & ZNAMENÁČEK, K. (1970) Influence of glucose infusion on oxygen and carbohydrate metabolism during adaptation of the newborn. *Biology of the Neonate*, **15**, 368–374.

PUSEY, V. A., MACPHERSON, R. I. & CHERNICK, V. (1969) Pulmonary fibroplasia following prolonged artificial ventilation of newborn infants. *Canadian Medical Association Journal*, **100**, 451–457.

RAWLINGS, G., REYNOLDS, E. O. R., STEWART, A. & STRANG, L. B. (1971) Changing prognosis for infants of very low birth weight. *Lancet*, **1**, 516–519.

REYNOLDS, E. O. R. (1970) Hyaline membrane disease. *American Journal of Obstetrics and Gynaecology*, **106**, 780–797.

REYNOLDS, E. O. R., ROBERTON, N. R. C. & WIGGLESWORTH, J. S. (1968) Hyaline membrane disease, respiratory distress and surfactant deficiency. *Pediatrics, Springfield*, **42**, 758–768.

REZZA, E., COLOMBO, U., BUCCI, G., MENICINI, M. & UNGARI, S. (1971) Early postnatal weight gain of low-weight newborns: relationships with various diets and with intrauterine growth. *Helvetica Paediatrica Acta*, **26**, 340–352.

RIGATTO, H. & BRADY, J. P. (1972). Periodic breathing and apnea in preterm infants. II. Hypoxia as a primary event. *Pediatrics, Springfield*, **50**, 219–228.

ROBERTON, N. R. C. (1969) Effect of acute hypoxia on blood pressure and encephalogram of newborn babies. *Archives of Disease in Childhood*, **44**, 719–725.

ROBERTON, N. R. C. (1970) Apnoea after THAM administration in the newborn. *Archives of Disease in Childhood*, **45**, 206–214.

ROBERTON, N. R. C., GUPTA, J. M., DAHLENBURG, G. W. & TIZARD, J. P. M. (1968) Oxygen therapy in the newborn. *Lancet*, **1**, 1323–1328.

ROHWEDDER, H. J., SIMON, C., KUBLER, W. & HOHERANER, M. (1970) Untersuchungen über die Pharmakokinetik von Nalidixinsäure bei Kindern verschiedenen Alten. *Zeitschrift für Kinderheilkunde*, **109**, 124–134.

ROSAN, R. C. & LAUWERYNS, J. M. (1971) Oxygen pollution: its effect on the ecology of lung cells. *Maandscrift voor Kindergen*, **39**, 143–161.

ROTHSCHILD, B. F. (1967) Incubator isolation as a possible contributing factor to the high incidence of emotional disturbance among prematurely born persons. *Journal of Genetical Psychology*, **110**, 287–304.

ROWE, M. I. & ARCILLA, R. A. (1968) Haemodynamic adaptation of the newborn to haemorrhage. *Journal of Pediatric Surgery*, **3**, 278–285.

RUDOLPH, A. M., HEYMANN, M. A., TERAMO, K. A. W., BARRETT, C. T. & RÄIHÄ, N. C. R. (1971) Studies on the circulation of the previable human fetus. *Pediatric Research*, **5**, 452–465.

RUSSELL, G. & COTTON, E. K. (1968) Effects of sodium bicarbonate by rapid injection and of oxygen in high concentration in respiratory distress syndrome of the newborn. *Pediatrics, Springfield*, **41**, 1063–1073.

SAIGAL, S., O'NEILL, A., SURAINDER, Y., CHUA, L. B. & USHER, R. (1972) Placental transfusion and hyperbilirubinemia in the premature. *Pediatrics, Springfield*, **49**, 406–419.

SAVIGNONI, P. G., BUCCI, G., CECCAMEA, A., MENDICINI, M., SCALA-MANDRÈ, A. & ORZALESI, M. M. (1969) Intravenous infusion of glucose and sodium bicarbonate in hyaline membrane disease. A controlled trial. *Acta Paediatrica Scandinavica*, **58**, 1–9.

SCAMMON, R. E. & MORRIS, E. H. (1918) On the time of the post-natal obliteration of the fetal blood-passages (foramen ovale, ductus arteriosus, ductus venosus). *Anatomical Record*, **15**, 165–180.

SCHIFF, D., CHAN, G. & STERN, L. (1971) Fixed drug combinations and the displacement of bilirubin from albumin. *Pediatrics, Springfield*, **48**, 139–141.

SCOPES, J. (1971) Respiratory Distress Syndrome. In *Recent Advances in*

Paediatrics, ed. Gairdner, D. & Hull, D. 4th Edition. London: Churchill.

SERENI, F., PERLETTI, L., MARUBINI, E. & MARS, G. (1968) Pharmacokinetic studies with a long-acting sulfonamide in subjects of different ages. *Pediatric Research*, **2**, 29–37.

SERENI, F. & PRINCIPI, N. (1971) The regulation of liver function during development. In *The Biochemistry of Development*, ed. Benson, P. Clinics in Developmental Medicine, No. 37. London: Heinemann.

SHANKLIN, D. R. & WOLFSON, S. L. (1967) Therapeutic oxygen as a possible cause of pulmonary haemorrhage in premature infants. *New England Journal of Medicine*, **277**, 833–837.

SHAW, J. C. L. (1968) Arterial sampling from the radial artery in premature and full-term infants. *Lancet*, **2**, 389–390.

SHAW, J. C. L. (1973). Parenteral nutrition in the management of sick low birthweight infants. *Pediatric Clinics of North America*, **20**, (in press).

SHEPHARD, F. M., JOHNSON, R. B., KLATTE, E. C., BURKO, H. & STAHLMAN, M. (1968) Residual pulmonary findings in clinical hyaline-membrane disease. *New England Journal of Medicine*, **279**, 1063–1071.

SIASSI, B., McDONALD, J. S., HON, E. & HODGMAN, J. E. (1972) Cardiovascular effects of apnoea in premature infants. (Abstract), *Pediatric Research*, **6**, 405.

SILVERIO, J. (1964) Gastric emptying times in the newborn and nursling. *American Journal of Medical Science*, **247**, 732–738.

SILVERMAN, W. A. & AGATE, F. J. (1964) Variation in cold resistance among small newborn infants. *Biologia Neonatorum*, **6**, 113–127.

SILVERMAN, W. A., AGATE, F. J. & FERTIG, J. W. (1963) A sequential trial of the nonthermal effect of atmospheric humidity on survival of newborn infants of low birth weight. *Pediatrics, Springfield*, **31**, 719–724.

SILVERMAN, W. A. & ANDERSEN, D. H. (1956) A controlled clinical trial of effects of water mist on obstructive respiratory signs, death rate and necropsy findings among premature infants. *Pediatrics, Springfield*, **17**, 1–9.

SILVERMAN, W. A. & BLANC, W. A. (1957) The effect of humidity on survival of newly born premature infants. *Pediatrics, Springfield*, **20**, 477–486.

SILVERMAN, W. A., BLODI, F. C., LOCKE, J. C., DAY, R. L. & REESE, A. B. (1952) Incidence of retrolental fibroplasia in a New York nursery. *Archives of Ophthalmology*, **48**, 698–711.

SILVERMAN, W. A., FERTIG, J. W. & BERGER, A. P. (1958) The influence of the thermal environment upon the survival of newly born premature infants. *Paediatrics, Springfield*, **22**, 876–885.

SINCLAIR, J. C. (1966) Prevention and treatment of the respiratory distress syndrome. *Pediatric Clinics of North America*, **13**, 711–730.

SINCLAIR, J. C. (1969) Problems associated with prolonged nasotracheal intubation. In *Problems of Neonatal Intensive Care Units*. Fifty-ninth Ross Conference on Pediatric Research, ed. Lucey, J. F. Columbus: Ross Laboratories.

SINCLAIR, J. C., ENGEL, K. & SILVERMAN, W. A. (1968) Early correction of hypoxaemia and acidaemia in infants of low birth weight: a controlled trial of oxygen breathing, rapid alkali infusion and assisted ventilation. *Pediatrics, Springfield*, **42**, 565–589.

SISSON, T. R. C., GLANSER, S. C., GLANSER, E. M., TASMAS, W. & KUWABARA, T. (1970) Retinal changes produced by phototherapy. *Journal of Pediatrics*, **77**, 221–227.

SMALLPEICE, V. & DAVIES, P. A. (1964) Immediate feeding of premature infants with undiluted breast milk. *Lancet*, **2**, 1249–1352.

SMYTHE, P. M. (1967) The treatment of tetanus in neonates. *Lancet*, **1**, 335.

SNYDERMAN, S. E. (1971) The protein and aminoacid requirements of the premature infant. In *Nutricia Symposium on the Metabolic Processes in the Foetus and Newborn Infant*, ed. Jonxis, J. H. P., Visser, H. K. A. & Troelstra, J. A. Leiden: H. E. Stenfert Kroese, N.V.

STEPHENSON, J. M., DU, J. N. & OLIVER, T. K. (1970) The effect of cooling on blood gas tensions in newborn infants. *Journal of Pediatrics*, **76**, 848–852.

STERN, L., RAMOS, A. D., OUTERBRIDGE, E. W. & BEANDRY, P. H. (1970) Negative pressure artificial respiration. Use in treatment of respiratory failure of the newborn. *Canadian Medical Association Journal*, **102**, 595–601.

STONE, H. O., THOMPSON, H. K. & SCHMIDT-NIELSEN, K. (1968) Influence of erythrocytes on blood viscosity. *American Journal of Physiology*, **214**, 913–918.

STRANG, L. B. & McGRATH, M. W. (1962) Alveolar ventilation in normal infants studied by air wash-in after oxygen. *Clinical Science*, **23**, 129–139.

STRAUSS, J. (1968) *Tris*(hydroxymethyl)amino-methane (THAM): a pediatric evaluation. *Pediatrics, Springfield*, **41**, 667–689.

SULYOK, E. (1971) The relationship between electrolyte and acid-base balance in the premature infant during early postnatal life. *Biology of the Neonate*, **17**, 227–237.

SULYOK, E. & HEIM, T. (1971) Assessment of maximal urinary acidification in premature infants. *Biology of the Neonate*, **19**, 200–210.

TANESCH, H. W., AVERY, M. E. & SUGG, J. (1972) Premature delivery without accelerated lung development in fetal lambs treated with long acting methylprednisolone. *Biology of the Neonate*, **20**, 85–92.

TANIMURA, T., NELSON, T., HOLLINGSWORTH, R. R. & SHEPARD, T. H. (1971) Weight standards for organs from early human fetuses. *Anatomical Record*, **171**, 227–236.

THALER, M. M., GEMES, D. L. & BAKKEN, A. F. (1972) Enzymatic conversion of heme to bilirubin in normal and starved fetuses and newborn rats. *Pediatric Research*, **6**, 197–201.

THALER, M. & SCHMID, R. (1971) Drugs and bilirubin. *Pediatrics, Springfield*, **47**, 807–810.

THIBEAULT, D. W., POBLETE, E. & AULD, P. A. M. (1967) Alveolar-arterial oxygen difference in premature infants breathing 100 per cent oxygen. *Journal of Pediatrics*, **71**, 814–824.

THIBEAULT, D. W., POBLETE, E. & AULD, P. A. M. (1968) Alveolar-arterial O_2 and CO_2 differences and their relation to lung volume in the newborn. *Pediatrics, Springfield*, **41**, 574–587.

THOMAS, D. B. & YOFFEY, J. M. (1962) Human Foetal Haemopoiesis. I. The cellular composition of foetal blood. *British Journal of Haematology*, **8**, 290–295.

THOMSON, A. (1964) Arterial blood sampling in small infants. *Acta Paediatrica Scandinavica*, **53**, 237–240.

THOMSON, A. M., BILLEWICZ, W. Z. & HYTTEN, F. E. (1968) The assessment of fetal growth. *Journal of Obstetrics and Gynaecology of the British Commonwealth*, **75**, 903–916.

TOWBIN, A. (1968) Cerebral intraventricular haemorrhage and subependymal matrix infarction in the fetus and premature newborn. *American Journal of Pathology*, **52**, 121–134.

TSANG, R. C. & OH, W. (1970) Neonatal hypocalcaemia in low birth weight infants. *Pediatrics, Springfield*, **45**, 773–781.

TURNBULL, E. P. N. & WALKER, J. (1955) Haemoglobin and red cells in the human foetus. II. The red cells. *Archives of Disease in Childhood*, **30**, 102–110.

USHER, R. (1959) The respiratory distress syndrome of prematurity. I. Changes in potassium in the serum and the electrocardiogram and effects of therapy. *Pediatrics, Springfield*, **24**, 562–576.

USHER, R. (1967) Comparison of rapid versus gradual correction of acidosis in RDS of prematurity (Abstract). *Pediatric Research*, **1**, 221.

USHER, R. (1968) Controlled series evaluation of oxygen therapy in respiratory distress syndrome. I. Radiological evidence of pulmonary oxygen toxicity (Abstract). *Pediatric Research*, **2**, 429–430.

USHER, R. H. (1970) Liberal versus restricted indications for oxygen in RDS: A controlled trial (Abstract). *Pediatric Research*, **4**, 469.

USHER, R. H. (1971) Clinical implications of perinatal mortality statistics. *Clinical Obstetrics and Gynecology*, **14**, 885–925.

USHER, R. H., ALLEN, A. C. & McCLEAN, F. H. (1971) Risk of respiratory distress syndrome related to gestational age, route of delivery and maternal diabetes. *American Journal of Obstetrics and Gynecology*, **111**, 826–832.

USHER, R., SAIGAL, S., O'NEILL, A., CHUA, L. & SURAINDER, Y. (1971) Red cell volume in respiratory distress syndrome (Abstract). *Pediatric Research*, **5**, 415.

VALMAN, H. B., BROWN, R. J. K., PALMER, T., OBERHOLZER, V. G. & LEVIN, B. (1971) Protein intake and plasma amino-acids in infants of low birth weight. *British Medical Journal*, **4**, 789–791.

VARGA, F. (1959) The respective effects of starvation and changed body composition on energy metabolism in malnourished infants. *Pediatrics, Springfield*, **23**, 1085–1090.

VENGUSAMY, S., PILDES, R. S., RAFFENSPERGER, J. F., LEVINE, H. D. & CORNBLATH, M. (1969) A controlled trial of feeding gastrostomy in low birth weight infants. *Pediatrics, Springfield*, **43**, 815–820.

VEST, M., STREBEL, L. & HAUENSTEIN, D. (1965) The extent of 'shunt' bilirubin and erythrocyte survival in the newborn infant measured by the administration of (^{15}N) glycine. *Biochemical Journal*, **95**, 11c–12c.

WALKER, C. H. M. & HANWELL, A. E. (1968) Impedance respiratory monitoring in the newborn infant. *Bio-medical Engineering*, **3**, 454–459.

WALKER, D., WALKER, A. & WOOD, C. (1969) Temperature of the human fetus. *Journal of Obstetrics and Gynaecology of the British Commonwealth*, **76**, 503–511.

WALLGREN, G., HANSON, J. S. & LIND, J. (1967) Quantitative studies of the human neonatal circulation. III. Observations on the newborn infant's central circulatory response to moderate hypovolaemia. *Acta Paediatrica Scandinavica*, Suppl. **179**, 43–54.

WALLGREN, G. & LIND, J. (1967) Quantitative studies of the human neonatal circulation. IV. Observations on the newborn infant's peripheral circulation and plasma expansion during moderate hypovolaemia. *Acta Paediatrica Scandinavica*, Suppl. **179**, 55–68.

WEIL, W. B. & MILLER, I. (1971) The role of whole carcass analysis in understanding body composition. *Pediatrics, Springfield*, **47**, 275–288.

WEISS, C. F., GLAZKO, A. J. & WESTON, J. K. (1960) Chloramphenicol in the newborn infant. A physiological explanation of its toxicity when given in excessive doses. *New England Journal of Medicine*, **262**, 787–794.

WELCH, B. E., MORGAN, T. E. & CLAMANN, H. G. (1963) Time-concentration effects in relation to oxygen toxicity in man. *Federation Proceedings of the Federation of American Societies for Experimental Biology*, **22**, 1053–1056.

WENNBERG, R. P., SCHWARTZ, R. & SWEET, A. Y. (1966) Early versus delayed feeding of low birth weight infants: effects on physiologic jaundice. *Journal of Pediatrics*, **68**, 860–866.

WEST, J. R., SMITH, H. W. & CHASIS, H. (1948) Glomerular filtration rate, effective renal blood flow and maximum tubular excretory capacity in infancy. *Journal of Pediatrics*, **32**, 10–18.

WIDDOWSON, E. M., CRABB, D. E. & MILNER, R. D. G. (1972) Cellular development of some human organs before birth. *Archives of Disease in Childhood*, **47**, 652–655.

WIDDOWSON, E. M. & SPRAY, C. M. (1951) Chemical development *in utero*. *Archives of Disease in Childhood*, **26**, 205–214.

WILLIAMS, C. P. S. & OLIVER, T. K. (1969) Nursery routines and staphylococcal colonization of the newborn. *Pediatrics, Springfield*, **44**, 640–646.

WIMBERG, J. & WESSNER, G. (1971) Does breast milk protect against septicaemia in the newborn? *Lancet*, **1**, 1091–1094.

WINICK, M. (1970) Cellular growth in intrauterine malnutrition. *Pediatric Clinics of North America*, **17**, 69–78.

WOOD, C. (1970) Weightlessness: its implications for the human fetus. *Journal of Obstetrics and Gynaecology of the British Commonwealth*, **77**, 333–336.

WOOD, C. B. S. (1972) The development of immunity in fetal life and childhood. *Journal of the Royal College of Physicians of London*, **6**, 246–258.

WRETLIND, A. (1972) (Editor) Complete intravenous nutrition. *Nutrition and metabolism*. Supplement 1, **14**.

WU, B., KIKKAWA, Y., ORZALESI, M. M., MOTOYAMA, E. K., KAIBORA, M., ZIGAS, C. J. & COOK, C. D. (1971) Accelerated maturation of fetal rabbit lungs by thyroxine. *Physiologist*, **14**, 253.

WU, P. Y. K., TEILMAN, P., GABLER, M., VAUGHAN, M. & METCALF, J. (1967) 'Early' versus 'late' feeding of low birth weight neonates: effect on serum bilirubin, blood sugar and responses to glucagon and epinephrine tolerance tests. *Pediatrics, Springfield*, **39**, 733–739.

YAO, A. C., HIRVENSALO, M. & LIND, J. (1968) Placental transfusion rate and uterine contraction. *Lancet*, **1**, 380–383.

YAO, A. C. & LIND, J. (1969) Effect of gravity on placental transfusion. *Lancet*, **2**, 505–508.

YAO, A. C., LIND, J. & VUORENKOSKI, V. (1971) Expiratory grunting in the late clamped normal neonate. *Pediatrics, Springfield*, **48**, 865–870.

YEUNG, C. Y. & YU, V. Y. H. (1971) Phenobarbitone enhancement of brom-sulphalein clearance in neonatal hyperbilirubinemia. *Pediatrics, Springfield*, **48**, 556–561.

ZEISBERGER, E., BRÜCK, K., WÜNNENBERG, W. & WIETASCH, C. (1967) Das Ausmass der zitterfreien Thermogenese des Meerschweinchens in Abhängigkeit vom Lebansalter. *Pflügers Archiv für gesammte Physiologie*, **296**, 276–288.

ZUELZER, W. W. (1971) The pediatrician and the species: some implications of our achievements. *Pediatrics, Springfield*, **47**, 339–351.

9 The Road Ahead

R. A. McCANCE

All mammals, including man, undergo considerable development inside the uterus. This is a great advantage to them, for it provides their young with almost complete protection until they are born, by which time some of them can almost fend for themselves. Protection of a similar but less satisfactory kind has been achieved in the course of evolution by a number of different phyla, in a number of different ways. Each of these has presented its own problem of how the developing organism shall be nourished, and of how provision can be made for its excretions before it is capable of taking over these functions, both of which are essential for its well being. The production of an egg with a built-in food supply and some protection against the environment is probably the best known way this has been done, but there are others. Some fish and reptiles, the British adder for example, retain the fertilized egg in the oviduct till the young are hatched and well enough developed to fend reasonably well for themselves, and the marsupials have developed a system of premature birth and the provision of an external pouch on the mother's abdomen in which the young can be fed and protected. Only the mammals, however, have succeeded in bringing the development of their offspring before birth to the pitch of perfection we know it today. The young of some of them, born in an immature state, depend for their survival on the protection of a nest and the warmth provided by their mother. Both of these can be regarded as extensions of the protection afforded by the mother's uterus.

Extensive development *in utero* has entailed the evolution of a number of special features which are very difficult to emulate *in vitro*, and until recently it has appeared quite impossible to do so. Within the last 40 years, however, two things have happened. Firstly, painstaking studies of physiology and of reproduction, unrestricted by religious or emo-

357

tional considerations, have gradually revealed more and more of the processes concerned: and secondly, the enormous advances in technological skills have made it possible to imitate—in a clumsy way only at present—some of the functions of the essential organs of the body. The excretory functions of the kidney were one of the earliest to be successfully replaced by an external substitute, but now we have a number of others. Prolonged operations are carried out on patients' hearts, for example, while the circulation of the blood to all the organs in the body is maintained by an external and entirely artificial system.

Most of the articles in this book deal with the progress that has been made so far in maintaining the development of a non-viable mammalian fetus in a man-made environment. People will turn to this book for a variety of reasons. Scientists working in the field will go to it for summaries of the available information, the views of other workers, references to papers they have missed and so on. Lecturers will use it to collect the facts they require for their presentations to students. Such readers will require no direction. Those who are not familiar with the subject, however, would do well perhaps to read the book backwards so to speak and to begin with the article by Edmund Hey. This sets out clearly and at some length the results of the most careful and conservative neonatal practice at the present time. The problems presented by immaturity are discussed in turn, breathing, low environmental temperatures, nutrition and the rest of them, and the layman will soon find out what a wealth of technology and human service are being devoted to the care of these unfortunate infants. They constitute a very small percentage of the total number of births in this country at present, for only some 1% of the babies born alive weigh less than 1·5 kg and only 0·35% less than 1 kg. Many of these babies die, however, in spite of all the care lavished upon them at the present time, and those that survive are not infrequently mentally or physically handicapped. Unfortunately their numbers are not likely to decrease and, unless the abortion rate can be reduced, they may well increase; for after one or two abortions it seems to become increasingly difficult for a woman to carry her baby to term. The cost of these children to the health service and to the country is great, and the prevention of prematurity and an improved outlook for its results are clearly attainments very much to be desired.

Not so long ago the same doctor was summoned to take care of every generation in the family if they fell ill. With the rise of specialization, however, the pregnant woman and her offspring became the charge of the obstetrician, and the paediatrician was first summoned when the child seemed to require medical attention. In all progressive establishments the paediatricians now, quite properly, take charge of the babies from birth. The care of the developing fetus, however, should have begun long before this, and many consider that the paediatricians who will have to look after the babies, premature or otherwise, after birth should take charge of them from conception and be responsible also for the genetic counselling of their parents. Consultations with the obstetricians will be necessary throughout gestation and all the critical decisions that concern both mother and child will have to be made jointly. It is not quite clear at present how easily this will be brought about, but some reorganization of responsibilities will surely come.

The chapter by Colin Walker and Danesh is very different from that of Hey. The object of both parties has been the maintenance of the immature human infant outside the uterus, but whereas Hey's approach has been essentially conventional and conservative, that of the two Dundee physicians may well be the first real assault of progressive thought, backed by modern technology, on traditional obstetrics and neonatal paediatrics. It is only fair to say at this point that Chamberlain (1968) had been at work in the same field before them, but with a much less sophisticated method. Walker and Danesh have used products of human conception weighing 820 to 3000 g, the last being an anencephalic monster. They have used puppies as their practice material which were probably not the best animals to choose. Lambs or piglets might have been better had the authors been able to get them. Nevertheless, they seem to have made some remarkable improvements in the techniques required for this kind of work. Among them is an ingenious system of alternating membranous plates which cover the functions of both a gas exchanger and a kidney. These are built up on the lines of a more elementary set described by Lawn, McCance & Thorn in 1967 to act only as a gas exchanger. With this equipment the authors seem to have solved the biochemical problems satisfactorily, and to be able to maintain internal stability. They discuss some of the difficulties they have still to surmount. One of these is the slow rate of the blood flow through the fetus, but in so far as they have hitherto only been catheterizing one of the two umbilical arteries they may find that cannulating the second will be a help. Cannulating a large artery would seem to be the only alternative. Sequestration of the perfusion fluid and fetal oedema has been a trouble, but this is probably only a specialized aspect of the problem that faces all those involved in long heart–lung bypass operations. It may, therefore, be solved for them by others. The haematological problems may be more difficult to solve—as may the regulation of the clotting properties of the blood.

We know much more about the metabolism and physiology of the fetal lamb, on which much work has been done, than we do about those of non-viable human fetuses. Unfortunately the fetal lamb differs considerably from its human counterpart in its placentation and metabolism, and in its inability to survive an early premature delivery. Nevertheless, it has many advantages as an experimental model, and in its own right. History has been made upon it since the last century; Barcroft, Huggett, Barron and Dawes have all made great contributions to knowledge by its use, and many papers about the fetal lamb were read at the meeting in Cambridge in 1972 to commemorate the centenary of Sir Joseph Barcroft's birth. In the present book the chapter by D. A. Nixon and the one by Warren Zapol and Kolobow are both devoted to it.

At first the sheep fetuses were usually delivered into a saline bath and studied while still attached to the mother, but with the improved techniques and skills of recent years quite remarkable experiments have been performed involving extensive cannulations of the blood vessels and fluid chambers of the fetus and mother for long periods of time, with the fetus still *in situ* and the uterus and abdomen closed. ECG and EEG recordings have been made, and there is no very obvious limit to the lengths to which these experiments may go or the knowledge that may accrue from them. Organs may be removed, nerves transected and subsequent development studied. The first appearance of enzymes and proteins may be determined and in some cases their later replacement by others. In some ways the sheep is not such a good animal as the rat, pig or guinea-pig for some of these experiments, because an unoperated control cannot always be provided from the same litter, and much of this work was pioneered on the rat, but the lamb has compensatory advantages.

By suitable cannulations of the umbilical and other vessels the functions of the fetus and of the placenta have been studied separately and in combination. If the fetus is to be successfully cannulated in this way an external oxygenator at least must be included in the circuit and usually a pump, although the fetuses of some species have been maintained for some hours by the action of the fetal heart alone. With a refined version of such an apparatus lamb fetuses near term have been kept alive for up to 48 h or so, subsequently delivered and reared to become normal animals.

The possibilities opened up by all this are great, but the difficulties are equally so, and those interested should look at the cautious words in the concluding sections of both these chapters. There are wise words in both of them.

The remaining chapters in the book deal with earlier stages in development. Macnaughton has used mostly human fetuses at or about mid-gestation. This is unquestionably a very important period of human growth and one about which we should like to know more. It merges imperceptibly, moreover, with the stage dealt with by Hey and by Colin Walker and Danesh, at which the fetus may survive outside the uterus and become a functional member of society.

The first problem is, as always, that of keeping the fetus alive and healthy, and the second—in the event of its being certainly non-viable—is to decide which physiological or endocrine function one wishes to study. Following the Swedish lead, most of the work at this stage of development has so far been devoted to the study of steroid metabolism. The apparatus used has been relatively simple, and in some ways not unlike that used by Westin, Nyberg & Enhörning (1958) or Lawn & McCance (1964). Macnaughton's group have also catheterized only one of the two umbilical arteries and flow rate is not yet what it should be, particularly in the smaller fetuses, nor have the perfusion fluids always had the ideal composition. The temperatures used, moreover, have sometimes been well below the physiological ones. Macnaughton makes an interesting point for future investigators, namely that several workers, themselves among them, have found that the heart rate in these small fetuses is no guide to the rate at which the perfusion fluid may be flowing. At this stage of fetal development the technical perfection required for a long perfusion is still far ahead, but the contributions to knowledge already made have been immense.

New and Sharman have both been dealing with much earlier phases of development, Denis New in fetuses belonging to several species, Geoffrey Sharman only in marsupials. This is the period when organs are beginning to take shape and the provision of nutrients and the energy to make use of them in their artificial environment become critical. This is the stage too when drugs and toxins are most likely to do irreparable damage to the growing organism, and to be able to study this *in vitro* would be an immense step forward. Here again technique is everything. Growth, for example, is measured microchemically rather than with a measuring tape, and New has described several ways of cultivating such tiny embryos. Working and thinking as he has been, moreover, with several species, he has been able to bring out some examples of baffling differences between them. Hamsters, for example, have been explanted much more successfully on the eighth rather than on the seventh or ninth days of gestation. Rat and mouse embryos can be grown successfully at present after the time of implantation, but till that time mice present far fewer problems than rats. Rats do equally well, however, after the seventh or eighth day of gestation and have been more used. The satisfactory growth of such embryos has only been achieved by a careful study of their requirements in culture, but there is no reason to suppose that progress in this respect is at an end.

Some of the work described by New on the oxygen consumption of developing rat fetuses is interesting and original, and its explanation in terms of growth rates, metabolic pathways and developing enzyme systems is stimulating. Let us hope that this work can soon be extended to the human species, and perhaps some link forged between it and the actions of certain drugs.

So much for the past, for, up to date as the work described in this book will be to many of its readers, it is already behind the investigators and will be further behind by the time the book is published. Where then do we go from here ? What has the road ahead in store for us ? One thing is certain. Discoveries will continue to be made and knowledge to expand, and probably at an increasing rate, and nothing that any religious sect or other conservative pressure group may say or do can prevent this from happening.

Some of the discoveries that are likely to be made in the near future, both in the fields covered in this book and, especially, in those that concern the transplantation of fertilized mammalian eggs, are certain to attract the attention of the press. As in the past (Hudson, 1972), they will be front-page news for the sensational publications and will arouse comment, most of it uninformed. Where do the investigators and the public stand on all this and where do their feelings about medical ethics and the law come in ? The first matter to bear in mind is that medical ethics are only expressions of personal feelings, and ephemeral ones at that (McCance, 1950). Opinions about what are and are not justifiable things to do vary from one person to another at any one time, and still more from one period of time to another. Some of the experiments, carried out by individuals in all good faith and with the full connivance of all concerned in England and other countries during and after the war, might not be carried out today and would certainly be difficult to get published.

The consensus of general or 'informed' opinion is also liable to change with time, because of alterations in social custom, religious doctrine, economic pressure or for many other causes. There have been great changes of this sort in attitudes towards abortions, contraception, illegitimacy and the execution of criminals, for example, in the very recent past, and they will undoubtedly continue to occur. No ethic, medical or otherwise, can remain widely acceptable for a long period, unless it is couched in general terms and avoids being too dogmatic about detail. This has been the strength of the ethical traditions of medical training and practice, for they have provided just those broad general principles that have allowed the man on the spot his freedom for decision and action.

The law can change too in response to custom and the feelings of society but its effects can be curiously permanent and having it changed is much more difficult. A Mental Deficiency Act was passed in 1913 for instance, based upon the then current ideas of right and wrong, by which an unmarried mother could be admitted to an institution as a morally defective person. In May, 1972, it was discovered that several women had been committed to such an institution in 1920 and were still there! A similar position seems to have been reached over the care of elderly and mentally subnormal patients.

The sanctity of human life had always been regarded in law as commencing at the time of conception. The production of an abortion, however, was permitted by law in the case of a non-viable fetus for rigidly defined therapeutic reasons and originally only for preserving the mother's life. To procure an abortion for any other reason was a criminal act. By the Abortion Act of 1967, however, the reasons for permitting a therapeutic abortion were considerably extended, and have been accorded a much wider interpretation than those in the previous law. Many abortions are now carried out in hospitals in this country, as they have been in Sweden for a considerably longer time, and a prosecution in Britain has not yet been brought against any of those involved.

Since the time of Hippocrates the medical profession has always had two main objectives, which are indeed inseparable; the study of disease and the preservation of human life with all that this entails. It goes against the whole training and nature of a doctor, and above all an obstetrician, to be asked to terminate a life which he or she believes may be valuable and productive. Many doctors, for instance, in a slightly different setting, have refused to participate in the trial of a drug which they believe does good—although others may doubt it—if it involves withholding this drug and giving some valueless material to half the number of their patients who are similarly affected. The wider scope of the Abortion Act, therefore, caused many of the profession much distress, which was intensified by the way it was interpreted. The College of Obstetricians and Gynaecologists set up a working party in 1969 to study the implications of an 'unplanned pregnancy' and how it could or should be viewed by the profession. With this the present book is not concerned, but the College was subsequently invited by the Secretary of State for the Social Services and the Secretaries of State for Scotland and Wales to consider the ethical, medical and legal implications of using fetuses and fetal material for research. With this Report (1972) the writers of a number of the chapters in this book are very much concerned. The committee was a distinguished and representative one except that, perhaps unfortunately, there was no one on it actively interested in research on the human fetus.

It may be stated at once that the use of fetal material for research and hence for the future welfare of mankind has already been shown to be most valuable, but it differs in no way in this respect from research into the well being and disease of the newly born, the adolescent or the adult, and exactly the same general principles apply to its conduct.

McCance (1951) took the line that anything done to a patient, not generally accepted as being for his direct therapeutic benefit or as contributing to the diagnosis of his disease, constituted an experiment. The experiments visualized might be one of omission and consist of withholding treatment from a 'control', or it might be one of commission and consist of making some test on a patient for which there was no obvious need. One has only to read the chapter by Hey to realize that hundreds of such experiments, most of them to compare methods of treatment, must have been made on very premature infants before the present techniques for dealing with them were evolved. It is probable that before long investigators will wish to apply the techniques outlined by Walker & Danesh in a similar way, and how is one to class the work already done by these and other workers ? Some of it has concerned viable fetuses and infants born alive, but certainly doomed to die within a day or two. The pioneer work on neonatal renal function was carried out on such infants over 30 years ago. It did them no harm and has been followed, confirmed and extended by many similar experiments on healthy babies.

The Committee responsible for the report on the use of fetal material for research has recommended that in future a fetus of over 20 weeks' gestation should be considered viable. In the opinion of the writer this decision was not in the best interests of society. Very, very few fetuses of less than 28 weeks are viable, or can be made so, at the present time, and this decision will create a shadowy period of 8 weeks during which many abortions will necessarily be performed on fetuses that are certainly not viable, yet in which their use for research of any kind may become illegal. Yet this is the very material on which most valuable work can and should be done. Only the paediatrician on the spot can decide whether a fetus is likely to be viable or not, and only the investigator whether an experiment would be justifiable or not. No ethical committee can decide. If, moreover, a fetus is to be cannulated successfully to study the therapeutic possibilities of doing so, it may be highly desirable to act immediately it is born, and there should be no ethical objection to this being done in the delivery room or the operating theatre, although the Report (1972) was strongly against this. The obstetrical nursing staff need not participate if it is distasteful to them. No one need take part in experimental work of any kind against their wishes, any more than they need to take part in the slaughter and subsequent butchery of the ox from which their Sunday sirloin will be cut and laid out for their inspection on a marble slab.

The tone of the report suggests that the public must be protected from persons who are bent on subjecting fetuses alive or dead to experiments that are an outrage to decent feelings. This is just not so. Investigators with rare exceptions are honourable men and women, and the public are already fully protected by law against anyone who violates the rights of anyone else, old or young. The principles involved in research on human subjects have been set out by many authors. The importance of obtaining the permission and co-operation of the persons responsible for their welfare has been laid down many times (see McCance, 1959).

What then has gone wrong ? Nothing so far as the law and the investigators are concerned. Something certainly so far as their image in the eyes of the public is concerned. At one time it probably never occurred to a patient that a doctor might be taking advantage of him in any way, but with the increase in general education, and above all with the search for sensationalism and its representation through the mass media, the public have become too wise, in the sense that a little knowledge can be a dangerous thing. The great advantages and help the man in the street has derived from medical research have not been represented to him enough in an unsensational way, and further, he has not been told what he can do to help. This is a difficult thing to put across, particularly for the honest investigator, who usually prefers to keep out of the limelight and to suspect those who court it. Nevertheless, investigators

365

have allowed themselves to be represented incorrectly and forced into positions of defence by not making use of the modern methods of communication, which their opponents and agitators do not hesitate to use.

Tolerance and co-operation seem to be the best hope for the future of clinical research. The specialists and the whole-time investigators, who now stand behind the family doctors, are essential for the progress of medicine from which all will ultimately benefit, but, in order to make this progress, doctors, patients and the public must be prepared to work together in a field that is only just beginning to be harvested. Patients must be ready to do what they can to help, and they must never be allowed to feel that their illness is being exploited to someone else's advantage, or that their chosen physician is not being straight with them. Investigators must be honest with their subjects; take them into their confidence so far as they can and scrupulously respect their fears and weaknesses. If patients know that this will be done—and the public know—trust and respect will be enhanced and all will be well. Discord will get us nowhere, so let us make progress together.

References

CHAMBERLAIN, G. (1968) An artificial placenta. *American Journal of Obstetrics and Gynecology*, **100**, 615–626.

HUDSON, D. (1972) Reckless scientists experimenting with manufacturing babies in test tubes will destroy the human race. *National Enquirer*, **26**, 22.

LAWN, L. & McCANCE, R. A. (1964) Artificial placentae. *Acta Paediatrica Stockholm*, **53**, 317–325.

LAWN, L., McCANCE, R. A. & THORN, A. E. (1967) Artificial placentae: comparative results with two gas exchangers. *Quarterly Journal of Experimental Physiology*, **52**, 157–167.

McCANCE, R. A. (1951) The practice of experimental medicine. *Proceedings of the Royal Society of Medicine*, **44**, 189–194.

McCANCE, R. A. (1959) Reflections of a medical investigator. *Scripta Academica Groningana*. Gröningen: J. B. Wolters.

REPORT OF THE ADVISORY GROUP (1972) *The Use of Fetuses and Fetal Material for Research*. London: H.M.S.O.

WESTIN, B., NYBERG, R. & ENHÖRNING, G. (1958) A technique for the perfusion of the previable human foetus. *Acta Paediatrica, Stockholm*, **47**, 339–349.

Author Index

376

Subject Index